Primer on Climate Change and Sustainable Development
Facts, Policy Analysis, and Applications

Climate change and variability have become the primary environmental concern of the twenty-first century; the potential impacts and mitigation of climate change need to be analysed within the context of sustainable development. How does climate change affect sustainable development prospects? How can climate change response measures best be incorporated into broader development strategies?

This *Primer on Climate Change and Sustainable Development* gives an up-to-date, comprehensive and accessible overview of the links between climate change and sustainable development. Building on the main findings of the last series of Intergovernmental Panel on Climate Change (IPCC) assessment reports, in which both authors were involved, the book summarizes the latest research linking the two. Our current knowledge of the basic science of climate change is described, before moving on to future scenarios of development within the context of climate change. The authors identify opportunities for synergies and resolving potential trade-offs. Discussing theory, policy analysis and applications, they analyse effective implementation of climate policy at scales ranging from the global to the local. The book contains a foreword from the chair of the IPCC.

The book will be compulsory reading for those policy-makers, scientists, students and engaged citizens around the world who have an interest in any aspect of climate change and sustainable development.

Professor MOHAN MUNASINGHE has postgraduate degrees in engineering, physics, and development economics, from Cambridge University, the Massachusetts Institute of Technology, McGill University, and Concordia University. Presently, he is chairman of the Munasinghe Institute for Development (MIND), Colombo; vice-chair of the IPCC, Geneva; visiting professor at Yale University, USA; and honorary chief energy advisor to the Government of Sri Lanka, Colombo. From 1974 to 2002, he worked for the World Bank, Washington DC, in various positions, including director and senior advisor.

From 1982 to 1987, he was the senior energy advisor to the president of Sri Lanka. From 1990 to 1992, he served as advisor to the United States President's Council on Environmental Quality. He has implemented international development projects for three decades, and contributed to IPCC work for 15 years. He has won a number of international awards and medals for his research, authored over eighty books and several hundred technical papers, and serves on the editorial boards of a dozen international journals.

Doctor ROB SWART was trained as an environmental engineer at Delft Technological University and received his Ph.D. on the risks of climate change from Amsterdam Free University. He has held various positions at The Netherlands National Institute for Public Health and Environment (RIVM) since 1980, and also spent time working for the World Health Organization and the US Environmental Protection Agency. He has also worked on projects in the area of global change and sustainability for the Stockholm Environment Institute, the Organization of Economic Development (OECD) and the United Nations Environment Programme (UNEP). He was head of the Technical Support Unit of Working Group III of the International IPCC, and co-editor of the IPCC's *Climate change 2001: mitigation. Contribution of Working Group III to the Third Assessment Report*, Cambridge University Press, 2001) and the IPCC's *Emissions scenarios: A Special Report of the IPCC Working Group III* (Cambridge University Press, 2000). Currently, he is manager of the European Topic Centre on Air and Climate Change of the European Environment Agency.

Primer on Climate Change and Sustainable Development

Facts, Policy Analysis, and Applications

MOHAN MUNASINGHE
ROB SWART

CAMBRIDGE UNIVERSITY PRESS

CAMBRIDGE UNIVERSITY PRESS
Cambridge, New York, Melbourne, Madrid, Cape Town, Singapore, São Paulo

Cambridge University Press
The Edinburgh Building, Cambridge CB2 2RU, UK

www.cambridge.org
Information on this title: www.cambridge.org/9780521810663

First published 2005

Printed in the United Kingdom at the University Press, Cambridge

A catalogue record for this book is available from the British Library

ISBN-13 978-0-521-81066-3 hardback
ISBN-10 0-521-81066-3 hardback

ISBN-13 978-0-521-00888-3 paperback
ISBN-10 0-521-00888-3 paperback

Contents

Foreword

It is becoming increasingly clear that the problem of climate change cannot be viewed in a narrow and limited context. Scientifically, it is now established that, irrespective of what the world does to reduce emissions of greenhouse gases in the immediate future, the concentration of these gases (and particularly of carbon dioxide) would remain high for a long period of time, thereby making future climate change inevitable. Hence, whatever the human race does to reduce emissions of greenhouse gases, the world is committed to continuing climate change for several decades, and possibly for centuries.

The work of the Intergovernmental Panel on Climate Change (IPCC) has highlighted both the scientific evidence for climate change, and the twin challenges it poses for humanity. Firstly, we must display our responsibility to all living beings and the delicate ecosystems that nature has kept in balance for aeons. Secondly, we need to ensure that future generations are able to survive without undue hardship and intolerable risk. Our development path has to ensure minimal interference with the world's climate system and adequate adaptation to the change that is inevitable.

This book is co-authored by Professor Mohan Munasinghe, vice-chair of the IPCC, who is recognized as a leading international expert in assessing the nexus between climate change and sustainable development, and Dr Rob Swart, former head of the IPCC Working Group 3 Technical Support Unit, and a very knowledgeable and accomplished researcher. The book fills an important vacuum on a subject that the world cannot ignore any longer. It is essential that we make development truly sustainable such that climate change remains within acceptable limits.

The volume is written in a simple and lucid style, providing an in-depth analysis of the impacts of climate change on the poor, including the worsening of existing income inequalities. It also applies the powerful new sustainomics framework, unveiled by Professor Munasinghe at the 1992 Earth Summit in Rio, to analyse and explain the imperative of making existing development patterns more sustainable.

Important elements of such a strategy include the mitigation of greenhouse gas emissions, and adaptation to minimize future climate change impacts. All in all, this is a book that, if read carefully, truly can make a difference. For this, the authors deserve acclaim. Furthermore, the human race deserves a second chance to set a course that would not pose a danger to all forms of life on the planet. This book will help a wide range of stakeholders, including decision-makers, policy analysts, researchers, students, practitioners and the informed public, to seize the moment for that second chance.

RAJENDRA PACHAURI
Chairman, Intergovernmental Panel on Climate Change, Geneva
and director-general, The Energy and Resources Institute, Delhi

January 2004

Preface

The genesis of this book may be traced to the work of the Intergovernmental Panel on Climate Change (IPCC), and several decades of our own independent experiences in the climate and development fields. We have both been involved in the work of the IPCC since the beginning, and continue to have confidence in the comprehensiveness, high quality and neutrality of its assessments.

However, we have found also that the sheer volume and number of the reports, as well as the overly technical language, make them less attractive to non-scientists, including key decision-makers, students and other interested audiences around the world, especially in the developing countries. Thus, one important motivation behind the writing of this book is our belief that a more concise, less technical, but nevertheless comprehensive and rigorous account of the findings of the IPCC, as well as more recent developments, will be attractive to an even wider audience.

Another objective of this book is to strengthen the growing conviction that climate change and broader sustainable development issues cannot be treated in isolation. Indeed, policies to address both topics will be mutually beneficial, to the extent that they are formulated and implemented within a common and consistent framework, along the lines that we present in this volume.

We hope that our efforts will help policy-makers, researchers, students and the concerned public to become more aware of the salient issues arising from the nexus of climate and development. While we draw heavily on IPCC work, the responsibility for the selection of issues from its reports and their interpretation rests solely with us. Our ideas do not necessarily represent the views of the IPCC or the institutions to which we are affiliated.

This book could not have been written without the basic groundwork laid by leading colleagues and researchers from around the world. We are grateful particularly for the commitment shown by thousands of experts who contributed voluntarily to the IPCC assessment reports, as well as various national and international research programmes in the areas of global change and sustainable development.

The list of specific persons who have provided us both with ideas and encouragement is too long to mention here. Nevertheless, we would like to express our gratitude to members of the IPCC bureau who have guided the development of the IPCC reports over recent years, especially those individuals who inspired and supported our goals of bringing climate change and sustainability together. Thanks are also owed to Yvani Deraniyagala and Nishanthi de Silva of the Munasinghe Institute for Development, who helped to prepare the draft manuscript. Support received from the government of the Netherlands, RIVM, Netherlands and MIND, Sri Lanka, is gratefully acknowledged.

1

Climate change: scientific background and introduction

1.1   Objectives and background

Is the climate changing? If so, how does it affect sustainable development opportunities? What are the options to respond to the changes: adapting, mitigating, both? What do these response options cost? How can we maximize synergies between the climate change response and broader sustainable development strategies, and minimize trade-offs between the two? These are the central questions addressed in this book.

Since global climate change was put on the international political agenda in 1992, developments in both climate science and climate policies have been swift, taking into account the complexity of the issues at stake. On the political side, the United Nations Framework Convention on Climate Change (UNFCCC) was agreed upon in 1992, and the subsequent Kyoto Protocol with legally binding commitments was signed in 1997. Progress was stalled after the repudiation of the Protocol by the US government in 2001. In a world divided by a wide variety of stakeholders, the UNFCCC continues, however, to be the main negotiations platform and the only available international regime that has to meet the challenge of achieving global co-ordination of national climate change responses in the direction of a common goal: avoiding dangerous interference with the climate system (see also Box 1.1). During negotiations in The Hague in 2000, in Bonn in the summer of 2001, and finally in Marrakech in autumn 2001, barriers were overcome and a very detailed cookbook was agreed describing the implementation modalities of the Protocol: the Marrakech Accords. In 2002, it is hoped that sufficient countries ratify the Protocol to make it enter into force even without the participation of the USA.[1] The political deliberations were supported by scientific assessments,

---

[1] In order to enter into force, fifty-five countries should have ratified the Protocol, incorporating Annex-I countries accounting for at least 55 per cent of the total Annex-I emissions in 1990.

Box 1.1   The United Nations Framework Convention on Climate Change and the Kyoto Protocol

The UNFCCC was signed at the United Nations Conference on Environment and Development (UNCED) in Rio de Janeiro in 1992. Its ultimate objective (Article 2) is 'stabilisation of greenhouse gas concentrations in the atmosphere at a level that would prevent dangerous anthropogenic interference with the climate system. Such a level should be achieved within a timeframe sufficient to allow ecosystems to adapt naturally to climate change, to ensure that food production is not threatened, and to enable economic development to proceed in a sustainable manner.' The UNFCCC also explicitly acknowledges a number of principles (Article 3), such as the precautionary principle, protection of the climate system on the basis of equity, the need for developed countries to take the lead in combating climate change and its adverse effects, full consideration of the specific needs and special circumstances of developing countries, and the need for pursuance of sustainable development. The UNFCCC also states that 'where there are threats of serious or irreversible damage, lack of full scientific certainty should not be used as a reason for postponing such measures, taking into account that policies and measures to deal with climate change should be cost-effective so as to ensure global benefits at the lowest possible cost'. The UNFCCC came into force in 1994, and as of May 2004, 189 parties had ratified it.

After 5 years of negotiations, the Kyoto Protocol was agreed in 1997. This Protocol has legally binding obligations, e.g. 'The Parties included in Annex I shall, individually or jointly, ensure that their aggregate anthropogenic carbon dioxide equivalent emissions of the greenhouse gases listed in Annex A do not exceed their assigned amounts, calculated pursuant to their quantified emission limitation and reduction commitments inscribed in Annex B and in accordance with the provisions of this Article, with a view to reducing their overall emissions of such gases by at least 5 per cent below 1990 levels in the commitment period 2008 to 2012.' The Protocol also includes three international mechanisms to facilitate its implementation: International Emissions Trading, Joint Implementation and the Clean Development Mechanism. Notwithstanding the fact that these mechanisms specifically had been developed to satisfy concerns from the USA, it announced in early 2001 that they would not ratify the Protocol. Unexpectedly, the other countries managed to reach agreement on the implementation details of the Protocol in November 2001: the Marrakech Accords. Many view these agreements as watered down versions of the original Kyoto Protocol commitments. However, by relaxing the original targets and offering countries various escape routes, the Marrakech Accords increase the possibility that the Protocol will be ratified by sufficient countries to keep it alive. This would avoid the renegotiation of a different international climate change response framework, which would be likely to take many more years, if not decades. According to the UNFCCC Secrtariat, 126 countries had ratified the Kyoto Protocol by mid 2004, representing over 44 per cent of total global emissions in 1990. With the subsequent ratification in October 2004, by Russia (representing 17 per cent of emissions), the Kyoto Protocol takes effect in February 2005, despite absence of the USA. This will also give impetus to discussions about the next commitment period for further emissions reductions.

notably those of the Intergovernmental Panel on Climate Change (IPCC). The IPCC was established in 1988 by the World Meteorological Organization and the United Nations Environment Programme (UNEP). It has developed three major comprehensive assessment reports in 1990, in 1995 and in 2001, in addition to a series of technical papers and special reports addressing specific issues (see also Box 1.2). These interdisciplinary reports reflect the rapid progress made in the scientific understanding of the climate system, the impacts of, and vulnerability to, climatic change, and the options to respond to these changes, both through adaptation and mitigation.

A key reason why it is so difficult to achieve a co-ordinated international response to the climate change threat is that countries have more urgent priorities, the foremost being economic development. But also in the area of environmental problems, climate change is not the only, or for many countries the most important environmental problem, nor is it detached from others. The UNFCCC was just one of the multilateral environmental agreements agreed upon during the 1992 UNCED, together with the Convention to Combat Desertification and the Convention on Biological Diversity. At UNCED, the challenge to reconcile the pursuance of an equitable economic development with the preservation of the natural resources of the Earth was addressed in the action programme Agenda 21, which laid out a blueprint for a just and sustainable world in an integrated, holistic manner. Agenda 21 covered the three global environmental problems mentioned above, and others, at global, regional and local levels.

After 1992 however, progress in implementing this agenda was slow, and the debate on the various global environmental problems largely was organized independently. In many areas (e.g. the availability of sufficient safe fresh water resources and fertile land for food production) the situation in many regions in the world actually has deteriorated, not withstanding overall positive

Box 1.2   IPCC organization and mandate

The IPCC was established in 1988 by the World Meteorological Organization and UNEP to assess periodically the scientific, technical, and economic knowledge pertinent to the problem of global climate change. The IPCC has issued comprehensive Assessment Reports in 1990, 1996, and 2001. It has published a series of technical papers, and special reports on specific issues, since 1999, e.g. regional impacts, aviation and the global atmosphere, emissions scenarios, methodological and technological issues in technology transfer, and land use, land-use change and forestry. Currently, there are three working groups: (a) on the science of the climate system; (b) on impacts, vulnerability and adaptation, and (c) on mitigation, respectively. Each working group has two co-chairs, one from a developed country and one from a developing country. The substance of all IPCC reports is the full responsibility of interdisciplinary writing teams of experts from all parts of the world. The rules of the IPCC ensure a rigorous scientific peer-review process, including an extensive scientific review of drafts of the report by independent experts. But also, the intergovernmental nature of the IPCC is important at three stages of report development: (a) governments approve the terms of reference or main outline of the reports; (b) they participate in the review of the second draft of the report (in addition to scientific expert reviewers), and (c) finally approve the so-called *Summary for Policymakers* line by line. Any changes in the *Summary for Policymakers* at the approval stage should be completely consistent with the underlying document, which is confirmed by the authors who are present at this stage. In this way, governments acquire ownership of the assessment reports, while the scientific integrity is maintained fully. The IPCC reports are important background documents in support of the negotiations in the context of the UNFCCC, notably for the Subsidiary Body on Scientific and Technological Advice. They may, directly or indirectly, have influenced important advances in the negotiations. The 1990 First Assessment Report (IPCC 1990) preceded the agreement of the UNFCCC, the 1996 Second Assessment Report (IPCC 1996a, b) the agreement of the Kyoto Protocol, and the Third Assessment Report (IPCC 2001a, b, c) the Marrakech Accords.

economic development in most regions. Thus, the implementation of the actions proposed in Agenda 21 is as important and urgent now as it was in 1992, if not even more so. However, the debate about the implementation of Agenda 21 basically left the climate change issue to the UNFCCC, while the UNFCCC negotiations –

notwithstanding the mentioning of sustainable development in the Convention and the associated Kyoto Protocol – did not make significant advances in linking the two issues. Some parties in the UNFCCC see the linking even as a threat, drawing the attention away from their main negotiation issue: climate change. Also, the discourse in science about sustainable development and climate change has progressed largely independently. One reason is that the framing of climate change in the late 1980s by natural scientists with their climate models divorced the issue from its social context and normative aspects have long been ignored (Cohen *et al.* 1998). Although, in the 1990s, social elements increased in climate change research and assessments, these were related mainly to quantitative economic analyses rather than including other social sciences and humanities. Only recently, the importance of looking closer at the linkages with social issues has been acknowledged by the IPCC in its Third Assessment Report (IPCC 2001a, b, c).

The IPCC notes in the Report that both in terms of natural processes and in terms of policy responses, climate change is linked closely to other environmental and socioeconomic problems. Climate change response strategies can be made more effective, if they are integrated with broader sustainable development efforts (Munasinghe 2000; Munasinghe & Swart 2000). This notion is at the core of this book.

The objective of the book is to provide a comprehensive and up-to-date overview of the options to adapt to and mitigate climate change and their economic, social and environmental implications. Also, in Chapter 1, a concise update of the science of climate change, and its possible impacts, is provided. The starting-point is that climate change response should be guided by broader objectives of development, equity and sustainability. As authors, we have both been involved closely in the IPCC process since its inception in 1988 in various roles and therefore the book to a large extent builds upon the authoritative IPCC reports, but is not constrained by those assessments. We recognize that climate change is a very serious problem threatening natural and human systems, but also that humankind has the ingenuity to develop and apply technologies to address it effectively and the ability to adjust lifestyle patterns associated with those technologies in order to remain within the carrying capacity of our globe. It is a matter of choice.

1.2   Chapter outline

The issues discussed in this book are focused on the linkages between climate change and sustainable development. While the volume covers the science of climate change, its potential impacts, and climate change response options comprehensively, the emphasis is on the latter – in particular adaptation and mitigation options, which are related to sustainable development. The framework

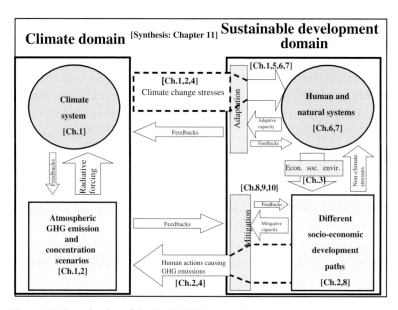

Figure 1.1. Organization of the book. GHG = greenhouse gas.

according to which the book is organized is depicted in Figure 1.1. The state of knowledge of the science of climate change and its potential impacts is summarized in the next section. Climate change is with us today, but its main impacts are expected to occur in the future. Not only the climate system, but also socioeconomic systems, are characterized by important inertia (see Box 1.5). Therefore long-term scenarios are indispensable for the analysis and assessment of climate change, its impacts, and possible response options. Chapter 2 reviews a number of long-term scenarios for the development of the global economy, population, technology, and the use of key resources, e.g. energy and land. It focuses on the greenhouse gas emissions scenarios developed by the IPCC in the *Special Report on Emissions Scenarios (SRES)* and broader scenarios, notably those developed by the Global Scenario Group. Both these sets of scenarios combine narratives about alternative development pathways with quantitative information about key indicators of development and forces driving climate change. The chapter also provides an overview of the future climate changes and impacts that can be associated with these emissions scenarios. Chapters 3 and 4 have sustainable development as the main focus. Chapter 3 lays out the sustainomics transdisciplinary framework for making development more sustainable, including some key methodological elements of sustainable development, different conceptual definitions and approaches to sustainable development, and its linkages with climate change. It also addresses different decision-analytical tools used in the area of climate change analysis, and discusses their relevance for sustainable development issues. Chapter 4 draws the findings from the results of studies applying the methodologies discussed in Chapter 3 to make development more sustainable while the climate changes. What are the opportunities for synergy, and what are the pitfalls of trade-offs between climate change response and sustainable development policies?

Chapters 5 to 10 focus on climate change adaptation and mitigation, paying as much attention as possible to their linkages with broader issues of sustainable development. Chapters 5, 6, and 7 discuss adaptation, while Chapters 8, 9, and 10 cover mitigation. In Chapter 5, we explain how development choices influence vulnerability to climate change and can enhance adaptive capacity. In particular, it reviews what the future costs and benefits of climate impacts and adaptation might be. While in Chapter 5 we discuss climate change adaptation in a generic sense, in Chapter 6 we evaluate options to adapt to climate change in the areas of: (a) hydrology and water resources; (b) natural and managed ecosystems; (c) coastal zones and marine ecosystems; (d) energy, industry and settlements; (e) financial resources and services, and (f) human health. Adaptation options differ not only across various economic sectors but also across regions. Therefore, in Chapter 7, an overview is given of adaptation options in the various regions of the world, both developed and developing.

After these three Chapters on adaptation, Chapters 8 to 10 deal with options to mitigate climate change. Chapter 8 starts with an overview of basic concepts, methods and approaches. It also discusses possible long-term goals that could be associated with the ultimate objective of the UNFCCC. In Chapter 9, the core of the information about mitigation options for short, medium and long term is presented, e.g. (a) which technologies and practices to abate greenhouse gas emissions or enhance sinks are known; (b) which social, economic or institutional barriers prevent these options from being implemented, and (c) which policies, measures and instruments are available to overcome these barriers? Methodologies for quantifying the costs of mitigation at different levels are described in Chapter 10, and results are discussed. This review includes important ways of reducing costs, including removal of market barriers, taking into account ancillary benefits, and revenue recycling in order to achieve an economic and environmental 'double dividend'. In Chapter 11, we conclude with a synthesis of the information regarding climate change response strategies in the context of sustainable development: how adaptive and mitigative capacity can be enhanced while at the same time accelerating efforts towards development, sustainability and equity. This final chapter would not be complete without our views on the future directions in research and climate policy assessment.

1.3 Historical record, recent observations, and climate system outlook

1.3.1 Introduction: new and stronger evidence

Observed climatic changes

In 1995, the IPCC concluded that 'the balance of evidence suggests a discernable human influence on climate' (Houghton *et al.* 1996). In 2001, this finding was strengthened, referring to 'new and stronger evidence that most of the warming observed over the last 50 years is attributable to human activities' (Houghton *et al.* 2001). Because of climate variability, it is difficult to detect changes in climate, and if such changes can be detected, to attribute them to either natural or anthropogenic influences. This area of research has received much attention and, on the basis of detailed analysis of available datasets, rigorous evaluation of their quality, and comparisons amongst data from different sources, researchers have been able better to understand climatic change. Figure 1.2 shows the variations in the surface temperature of the Earth for the past 140 and 1000 years. In the twentieth century, global average surface temperature has increased by about 0.6 °C. This number is 0.15 °C higher than assessed in Houghton *et al.* (1996), due primarily to the relatively warm years since 1995, and improved data-processing. The global mean surface temperature in 2002 was 0.48 °C above the 1961–90 annual average. This places 2002 as the second warmest year since 1861; 1998 was the warmest year on record, and the 1990s the warmest decade. The warming was not equal over time and space. For example, in the latter half of the twentieth century, minimum night time air temperatures over land increased twice as fast as maximum daytime temperature changes (0.2 versus 0.1 °C/decade), and the increase in sea surface temperature was about half as much as the increase in mean land surface temperature. In contrast with most regions in the Northern Hemisphere, parts of the Southern Hemisphere oceans and Antarctica have not warmed recently. Panel (a) of Figure 1.3 shows schematically how temperature changes are different for land and ocean areas.

But climate change is not limited to changes in temperature. Although less certain, it is also considered to be very likely that precipitation has increased by 0.5–1 per cent per decade in the twentieth century over most of the mid and high latitudes of the Northern Hemisphere, and by 0.2–0.3 per cent in tropical areas, while it has decreased by 0.3 per cent per decade in the subtropical regions of the Northern Hemisphere (see Figure 1.3, panel (b)). These changes in precipitation are particularly important for impacts because of their importance for ecosystems and agriculture (see Section 1.4). Changes in climate variability, extreme weather and other climate events also have been observed, e.g. increased occurrence of heavy precipitation events in the Northern Hemisphere, an increase in cloud cover over mid and high latitudes, a reduction in the frequency of very low temperatures, and

Variations of the surface temperature of the Earth for:

(a) The past 140 years

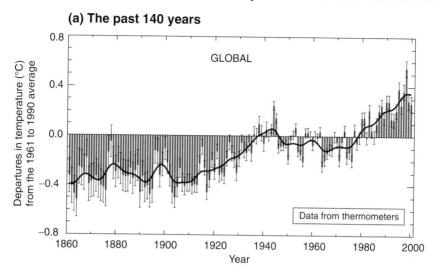

(b) The past 1000 years

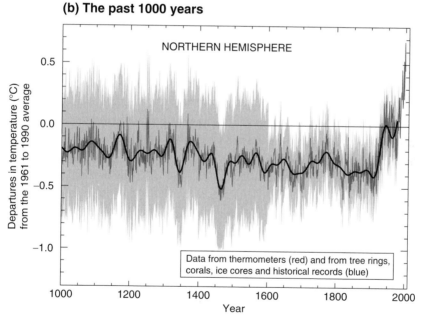

Figure 1.2. Variations of the surface temperature of the Earth over the last 140 years and the last millennium. The black line in (a) represents a decade by decade curve, in (b) the 50 years average variations. The thin whisker bars in (a) and the grey shaded area in (b) represent the 95% confidence range. For the difference between the direct data sources (thermometer, red) and indirect data sources (tree rings, corals, ice cores, historical records, blue), we refer to the original publication or website http://www.grida.no/climate/ipcc_tar/wg1/figspm-1.htm.
Source: Houghton *et al.* (2001).

Box 1.3   Key conclusions of the IPCC's Third Assessment Report, Working Group I: The scientific basis

Key conclusions

(1) An increasing body of observations gives a collective picture of a warming world and other changes in the climate system as follows.

   (a) *The global average surface temperature has increased over the twentieth century by about 0.6 °C.*

   (b) *Temperatures have risen during the past four decades in the lowest 8 km of the atmosphere.*

   (c) *Snow cover and ice extent have decreased.*

   (d) *Global average sea-level has risen and ocean heat content has increased.*

   (e) *Changes also have occurred in other important aspects of climate, e.g. precipitation, cloud cover, the frequency of extreme low temperatures, warm episodes of the El Niño–Southern Oscillation phenomenon, global land areas experiencing severe drought or severe wetness and frequency and intensity of droughts.*[1]

   (f) *Some important aspects of climate appear not to have changed, e.g. temperatures in some areas of the world, Antarctic sea ice extent, changes globally in tropical and extra-tropical storm intensity and the frequency of tornadoes, thunder days, or hail event.*[1]

(2) Emissions of greenhouse gases and aerosols due to human activity continue to alter the atmosphere in ways that are expected to affect the climate as follows.

   (a) *Concentrations of atmospheric greenhouse gases and their radiative forcing have continued to increase as a result of human activities.*

   (b) *Anthropogenic aerosols are short-lived and produce mostly negative radiative forcing. Natural factors have made small contributions to radiative forcing over the past century.*

(3) Confidence in the ability of models to project future climate has increased.

(4) There is new and stronger evidence that most of the warming observed over the last 50 years is attributable to human activity.

(5) Human influences will continue to change atmospheric composition throughout the twenty-first century.

(6) Global average temperature and sea-level are projected to rise under all IPCC *SRES* scenarios.[2]

(7) Anthropogenic climate change will persist for many centuries.

(8) Further action is required to address remaining gaps in information and understanding.

[1] These conclusions are paraphrased from the text. For exact wording of the – longer – conclusions, see the original source (Houghton *et al.* 2001).

[2] *Special Report on Emissions Scenarios.*

in some regions, such as parts of Asia and Africa, increased frequency and intensity of droughts (Houghton *et al.* 2001). It is no surprise that these climatic changes are accompanied by changes in the extent of snow cover and ice. In addition to the widespread retreat of glaciers in non-polar regions, the winter snow cover of the world has been getting smaller – 10 per cent less since the late 1960s. While no significant change in the Antarctic ice sheets has been identified, sea ice in the North Atlantic has decreased by 10–15 per cent since the 1950s, and during late summer to early autumn sea ice thickness has decreased by about 40 per cent. In 1999, visitors to the North Pole even encountered open water. Whether the causes of this remarkable phenomenon are natural or human-induced, remains to be determined. However, it may be asserted that a situation like this has never occurred before in millions of years.

The enhanced greenhouse effect, or 'global warming'

A main question is whether the observed changes discussed above are of an anthropogenic or natural origin. In order to answer this question, we have to know how natural factors and human actions can affect climate. Let us briefly discuss the so-called 'enhanced greenhouse effect' theory, a well-established theory on how increasing greenhouse gas concentrations affect climate (see Figure 1.4). The climate of the Earth is influenced by the global radiation budget. The Earth absorbs the radiation from the Sun and redistributes the embodied energy through atmospheric and oceanic circulation. The energy is radiated back in the form of long-wave radiation (infrared). For the annual mean and the Earth as a whole, the incoming energy and outgoing energy are more or less balanced. A change in the net energy available to the global atmosphere system is called 'radiative forcing': it can be natural or human-induced, and be positive (warming) or negative (cooling). Greenhouse gases play an important role in this balance: without them, the Earth temperature would be about 34 °C colder than it is today, making life on Earth as we know it today impossible. If the concentration of these greenhouse gases in the atmosphere increases, global average temperature in the lower atmosphere can be expected to rise, a phenomenon that is indeed also being observed. Since less energy is emitted from the lower levels of the atmosphere, temperatures at higher altitudes (e.g. in the stratosphere) can be expected to fall. Also, this has indeed been

(a) Temperature indicators

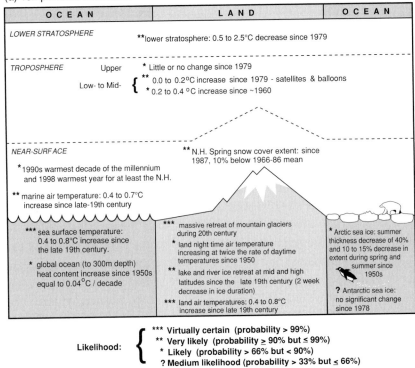

OCEAN	LAND	OCEAN

LOWER STRATOSPHERE **lower stratosphere: 0.5 to 2.5°C decrease since 1979

TROPOSPHERE Upper * Little or no change since 1979

Low- to Mid- { ** 0.0 to 0.2°C increase since 1979 - satellites & balloons

* 0.2 to 0.4 °C increase since ~1960

NEAR-SURFACE ** N.H. Spring snow cover extent: since 1987, 10% below 1966-86 mean

*1990s warmest decade of the millennium and 1998 warmest year for at least the N.H.

** marine air temperature: 0.4 to 0.7°C increase since late-19th century

*** sea surface temperature: 0.4 to 0.8°C increase since the late 19th century.

* global ocean (to 300m depth) heat content increase since 1950s equal to 0.04°C / decade

*** massive retreat of mountain glaciers during 20th century

* land night time air temperature increasing at twice the rate of daytime temperatures since 1950

** lake and river ice retreat at mid and high latitudes since the late 19th century (2 week decrease in ice duration)

*** land air temperatures: 0.4 to 0.8°C increase since late 19th century

* Arctic sea ice: summer thickness decrease of 40% and 10 to 15% decrease in extent during spring and summer since 1950s

? Antarctic sea ice: no significant change since 1978

Likelihood:
{ *** Virtually certain (probability > 99%)
** Very likely (probability ≥ 90% but ≤ 99%)
* Likely (probability > 66% but < 90%)
? Medium likelihood (probability > 33% but ≤ 66%)

(b) Hydrological and storm related indicators

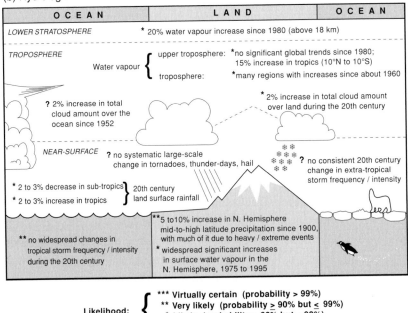

OCEAN	LAND	OCEAN

LOWER STRATOSPHERE * 20% water vapour increase since 1980 (above 18 km)

TROPOSPHERE Water vapour { upper troposphere: *no significant global trends since 1980; 15% increase in tropics (10°N to 10°S)

troposphere: *many regions with increases since about 1960

? 2% increase in total cloud amount over the ocean since 1952

* 2% increase in total cloud amount over land during the 20th century

NEAR-SURFACE ? no systematic large-scale change in tornadoes, thunder-days, hail

? no consistent 20th century change in extra-tropical storm frequency / intensity

* 2 to 3% decrease in sub-tropics
* 2 to 3% increase in tropics
} 20th century land surface rainfall

** no widespread changes in tropical storm frequency / intensity during the 20th century

**5 to10% increase in N. Hemisphere mid-to-high latitude precipitation since 1900, with much of it due to heavy / extreme events
* widespread significant increases in surface water vapour in the N. Hemisphere, 1975 to 1995

Likelihood:
{ *** Virtually certain (probability > 99%)
** Very likely (probability ≥ 90% but ≤ 99%)
* Likely (probability > 66% but < 90%)
? Medium likelihood (probability > 33% but ≤ 66%)

Figure 1.3. (a) Schematic of observed variations of the temperature (b) the hydrological and storm-related indicators. *Source:* Houghton *et al.* (2001).

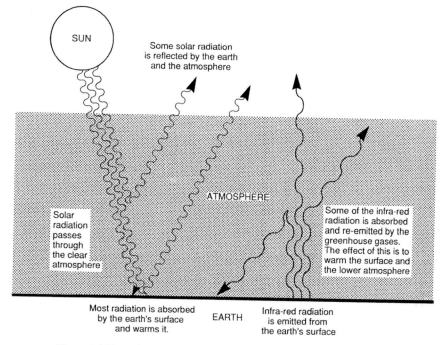

Figure 1.4. The enhanced greenhouse effect: the earth's radiation and energy balance.
Source: Houghton *et al.* (1990) *Climate Change: The IPCC Scientific Assessment.*

observed: average temperatures in the lower stratosphere have recently fallen from about 0.3 to 0.5 °C per decade.

Important compounds influencing the radiative budget positively (i.e. causing warming) are water vapour, carbon dioxide, methane, nitrous oxide, halocarbons (e.g. chlorofluorocarbons, hydrochlorofluorocarbons, hydrofluorocarbons, and perfluorocarbons), sulphurhexafluoride, and tropospheric ozone. Except for some manmade halocarbons, these gases have been in the atmosphere naturally for billions of years. Aerosols, tiny particles or droplets in the atmosphere often caused by burning of biomass or fossil fuels, can have a negative radiative effect, causing cooling. Because of the complexity and variability of the climate system and the unpredictability of human society and its greenhouse gas emissions, it is unknown by how much exactly, where, and how fast increases in greenhouse gas concentrations will change temperature and other climatic aspects in the future (Houghton *et al.* 1996; 2001).

Radiative forcing

The key indicator of how much a particular greenhouse gas contributes to global warming is its radiative forcing in watts per square metre: the change

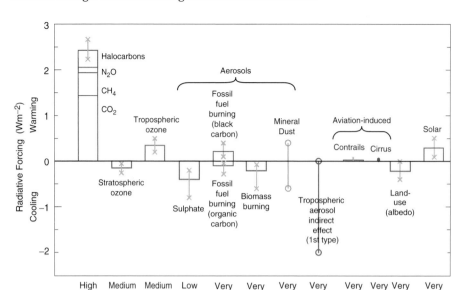

Figure 1.5. Global mean radiative forcing (W/m²) due to a number of agents for the period 1750 to present, including level of scientific understanding. *Source:* Houghton *et al.* (2001).

in the net irradiance at the tropopause due to an internal change, or, as in case of the enhanced greenhouse effect, a change in the external forcing of the climate system (Houghton *et al.* 2001). Different greenhouse gases have very different radiative forcing effects, as shown in Figure 1.5, in which the forcing agents have been organized according to the level of scientific understanding. Figure 1.5 shows that the direct forcing by carbon dioxide, methane, nitrous oxide and the halocarbons is positive and the most certain. Carbon dioxide contributes about 60 per cent to the total radiative forcing of the globally mixed long-lived greenhouse gases, methane about 20 per cent, nitrous oxide about 6 per cent, and halocarbons about 14 per cent. Other anthropogenic positive forcing is caused by tropospheric ozone and black carbon. Variations in solar irradiance are estimated to have had a positive radiative forcing too. The magnitude of this forcing is much more uncertain. But not all forcings are positive. Changes in stratospheric ozone due to ozone depletion and changes in the world's albedo due to land-use changes are estimated to produce a small negative forcing. The most important negative forcing is due to aerosols, both from human sources (like fossil fuel combustion and biomass burning) and natural sources, e.g. volcanic eruptions. Although the scientific understanding of the radiative processes associated with aerosols has improved, the actual effect of aerosols is still very uncertain, since their concentrations vary very much by region and they respond very quickly to changes in emissions. Most aerosols (with the exception of fossil fuel black carbon) have a negative direct and indirect effect, i.e. they have a cooling effect counteracting the warming effect of the greenhouse gases, both directly through their radiative properties and through their indirect effects on cloud cover and cloud characteristics.

From a policy point of view, it is useful to be able to compare the radiative forcings of different greenhouse gases, or to add them to get a sense of the total radiative forcing of various atmospheric substances taken together. To address this need, the concept of global warming potential (GWP) was introduced, i.e. a measure of the relative radiative effect of a given substance compared to carbon dioxide, integrated over a given time horizon. This measure has been used successfully for a decade. Table 1.1 provides the latest IPCC estimates for this indicator. It should be noted that the indicator is not without problems and has been criticized. For example, for many substances, uncertainties are very large. Also, the indicator is not unequivocal, as it is dependent on an arbitrarily selected time horizon. The GWP is calculated as the radiative forcing of an emissions pulse. For greenhouse gases with a short lifetime, the effect dies out more quickly, so they become 'less important' as the selected time horizon is greater. Usually, a century timescale is selected as a compromise between the importance of the short term for policy-making, and in the long term of eventual climate change. The UNFCCC adopted a 100 year timescale for GWP application in the comprehensive approach to green-house gas emissions in the Kyoto Protocol. In reality, radiative forcing will not be caused by an emissions pulse, but by a continuous release of the substance, and the radiative forcing over time would be dependent on past and future emissions (and thus be time- and scenario-dependent). The GWP concept has been criticized for this and other reasons, but there is no consensus on any better alternative metric for comparison of greenhouse gases. Box 1.4 provides a comprehensive assessment of the GWP concept and the proposed alternative measures, including several measures taking into account economic factors related to impacts. While such measures would bring the measure closer to what is important to policymakers, the uncertainties and the need for often subjective assumptions increase. In Table 1.1, for methane and nitrous oxide it has been taken into account that releases of these gases affect their own lifetime in the atmosphere. For example, additional methane in the atmosphere reduces the amount of hydroxyl-radicals responsible for oxidizing methane, therewith reducing its own lifetime. Various substances that have no, or only small, radiative forcing themselves, can have indirect effects on radiative forcing, e.g. the ozone precursors: nitrogen oxides, carbon monoxide and non-methane volatile organic compounds. Thus far, they have not been included in policy packages of greenhouse gases, nor have they been

Box 1.4   Metrics to compare the climatic effects of different
greenhouse gases: GWP and alternatives

Anticipating a need to compare the climatic effects of changing
concentrations of different greenhouse gases, the concept of GWP was
introduced in the 1980s and taken up by the IPCC in its First Assessment
Report (IPCC 1990). The GWP reflects the time-integrated warming effect of a
pulse of a given greenhouse gas in today's atmosphere, relative to carbon
dioxide. In theory, a metric for comparing these effects could be chosen at
any step of the chain from emissions through concentration changes,
radiative forcing, climatic changes (e.g. temperature), climate change
impacts, and eventually to the damage caused. Although the political
relevance would increase as one goes further along the chain, the
uncertainties and required assumptions increase. Interestingly, there has
not been a serious political debate about the most appropriate metric to be
used, and the Kyoto Protocol prescribes the usage of the 100 years GWPs as
reported in the IPCCs Second Assessment Report (IPCC 1996a, b).
Nevertheless, there has been a continuous scientific debate about the
weaknesses and strengths of the GWP and alternative metrics. Only those
elements of this debate pertaining to the GWP concept itself have been
discussed in the IPCC reports, focusing on the natural science aspects
assessed by IPCC Working Group 1.

A first criticism is related to the fact that it is difficult to assign GWPs to
all substances that are radiatively active, especially those that have an
indirect effect on climate, e.g. ozone and aerosol precursors. Second, the
simplicity of the concept and the linkage with temperature suggested by the
name (while GWPs are really about radiative forcing) could be considered as
misleading to policymakers not aware of the scientific complexities
involved. For example, even if the effects further down the cause–effect
chain are ignored, the radiative effects of emissions are dependent on the
evolution of concentrations over time and thus scenario-dependent. The
choice of a time horizon for integration is basically arbitrary, while this
choice has a large influence on the GWP values. Third, it is questioned if the
GWP metric really addresses what is politically relevant, i.e. costs and
damage related to climate change.

The most important type of the last criticism comes from economic
analysts, who argue that what is really important for policy is minimizing
costs of abating a combination of greenhouse gases or minimizing damages.
Fuglestvedt *et al.* 2001 give a valuable overview of various alternative
indicators proposed. There is no consensus on the superiority of any of these

alternatives, since usually they all have different weaknesses and strengths. Bringing in economic aspects introduces dependency of the values of the metric on economic models used, and assumptions in the shape of damage and abatement cost curves, and on the appropriate discount rate. For example, using a cost-effectiveness approach, one needs to assume climate targets. While such targets are politically sensitive, choosing different values would lead to different outcomes. For example, setting a constraint on long-term temperature change would lead to the higher importance of long-lived greenhouse gases, while a constraint on the rate of temperature change would lead to the opposite. For a damage-minimization approach, the shape of the damage curve is crucial: for very convex curves (exponentially increasing impacts for large climatic changes) long-lived substances are of higher importance, because in this case long-term impacts are more important than short-term impacts. While it seems uncertain if alternative metrics to the GWP concept can enter the UNFCCC debate, it seems relevant if the IPCC would broaden its assessment to include economic considerations. Such considerations would provide important additional information to policy-makers for the evaluation of policy options or investment decisions involving several greenhouse gases.

Sources: Fuglestvedt *et al.* (2001); Houghton *et al.* (1990; 1996).

included in Table 1.1.[2] Also negative GWPs of aerosols or aerosol precursors have not been established, since these would be meaningless – they would suggest that the warming of the greenhouse gases could be compensated by aerosol cooling, while the mechanisms and spatial distributions are very different – most greenhouse gases are well mixed in the global atmosphere, while the aerosol effect is mainly local. Many alternative methods to quantify the relative contribution of greenhouse gases have been proposed, but most of them suffer from similar scientific uncertainties and a level of arbitrariness. Until a more acceptable and better indicator is found, the GWP can be expected to be used, but it is important to be aware of its shortcomings.

Greenhouse gas concentrations

In order to evaluate the possible anthropogenic influence on observed climatic changes, it is important to know how concentrations of greenhouse gases have changed over the years. Fortunately – through a global network of measuring stations and the well-mixed nature of most (long-lived) greenhouse gases – this

---

[2] In fact, for carbon dioxide, indirect GWP calculations have been made (IPCC 2001e), with a range of outcomes, leading to a GWP of between 1 and 3 over a 100 year time-frame.

Table 1.1. *Direct global warming potentials (GWPs) relative to carbon dioxide. They form an index for estimating relative global warming contribution due to atmospheric emission of 1 kg of a particular greenhouse gas compared to emission of 1 kg of carbon dioxide. The GWPs for different time horizons show the effects of atmospheric lifetimes of the different gases.* Source: *Houghton* et al. *(2001).*

Gas		Lifetime (yr)	Global warming potential (time horizon yr)		
			20	100	500
Carbon dioxide	CO_2		1	1	1
Methane[a]	CH_4	12.0[b]	62	23	7
Nitrous oxide	N_2O	114[b]	275	296	156
Hydrofluorocarbons					
HFC-23	CHF_3	260	9 400	12 000	10 000
HFC-32	CH_2F_2	5.0	1 800	550	170
HFC-41	CH_3F	2.6	330	97	30
HFC-125	CHF_2CF_3	29	5 900	3 400	1 100
HFC-134	CHF_2CHF_2	9.6	3 200	1 100	330
HFC-134a	CH_2FCF_3	13.8	3 300	1 300	400
HFC-143	CHF_2CH_2F	3.4	1 100	330	100
HFC-143a	CF_3CH_3	52	5 500	4 300	1 600
HFC-152	CH_2FCH_2F	0.5	140	43	13
HFC-152a	CH_3CHF_2	1.4	410	120	37
HFC-161	CH_3CH_2F	0.3	40	12	4
HFC-227ea	CF_3CHFCF_3	33	5 600	3 500	1 100
HFC-236cb	$CH_2FCF_2CF_3$	13.2	3 300	1 300	390
HFC-236ea	CHF_2CHFCF_3	10	3 600	1 200	390
HFC-236fa	$CF_3\,CH_2CF_3$	220	7 500	9 400	7 100
HFC-245ca	$CH_2FCF_2CHF_2$	5.9	2 100	640	200
HFC-245fa	$CHF_2CH_2CF_3$	7.2	3 000	950	300
HFC-365mfc	$CF_3CH_2CF_2CH_3$	9.9	2 600	890	280
HFC-43-10mee	$CF_3CHFCHFCF_2CF_3$	15	3 700	1 500	470
Fully fluorinated species					
SF_6		3 200	15 100	22 200	32 400
CF_4		50 000	3 900	5 700	8 900
C_2F_6		10 000	8 000	11 900	18 000
C_3F_8		2 600	5 900	8 600	12 400
C_4F_{10}		2 600	5 900	8 600	12 400
c-C_4F_8		3 200	6 800	10 000	14 500
C_5F_{12}		4 100	6 000	8 900	13 200
C_6F_{14}		3 200	6 100	9 000	13 200

Table 1.1. (*cont.*)

Gas		Lifetime (yr)	Global warming potential (time horizon yr)		
			20	100	500
Ethers and halogenated ethers					
CH_3OCH_3		0.015	1	1	$\ll 1$
HFE-125	CF_3OCHF_2	150	12 900	14 900	90 200
HFE-134	CHF_2OCHF_2	26.2	10 500	6 100	2 000
HFE-143a	CH_3OCF_3	4.4	2 500	750	230
HCFE-235da2	$CF_3CHClOCHF_2$	2.6	1 100	340	110
HFE-245fa2	$CF_3CH_2OCHF_2$	4.4	1 900	570	180
HFE-254cb2	$CHF_2CF_2OCH_3$	0.22	99	30	9
HFE-7100	$C_4F_9OCH_3$	5.0	1 300	390	120
HFE-7200	$C_4F_9OC_2H_5$	0.77	190	55	17
H-Galden 1040x	$CHF_2OCF_2OC_2F_4OCHF_2$	6.3	5 900	1 800	560
HG-10	$CHF_2OCF_2OCHF_2$	12.1	7 500	2 700	850
HG-01	$CHF_2OCF_2CF_2OCHF_2$	6.2	4 700	1 500	450

[a] The methane GWPs include an indirect contribution from stratospheric water and ozone production.

[b] The values for methane and nitrous oxide are adjustment times, which incorporate the indirect effects of emission of each gas on its own lifetime.

information is available. Concentrations of most greenhouse gases have increased rapidly. Carbon dioxide concentrations increased from 270 to about 360 ppmv since pre-industrial times (more than 30 per cent, see Figure 1.6) with a rate of increase faster than at any time before during the past 20 000 years. The current concentration has not occurred since 420 000 years ago. The increase during the last decades has been caused primarily by burning of fossil fuels in our industrial society (about 75 per cent) and by converting forests to other land uses (about 25 per cent). Similarly, concentrations of methane have increased by more than 1000 ppb, from 750 to 1750 ppb, a concentration level again higher than at any time in 420 000 years. Recently, the rate of increase in concentrations of methane has decreased and become more variable, for as yet unknown reasons. There is more uncertainty about methane sources than there is about those of carbon dioxide. It is estimated that about half comes from anthropogenic sources such as cattle, landfills, rice production and production, transport and combustion of fossil fuels. Also, through atmospheric chemical reactions, methane concentrations increase due to increasing emissions of carbon monoxide, competing with methane for hydroxyl-radicals in the oxidation process. A third greenhouse gas, nitrous oxide, has increased by about 17 per cent since pre-industrial times, to about 310 ppb, a

Indicators of the human influence on the atmosphere during the Industrial era

(a) Global atmospheric concentrations of three well mixed greenhouse gases

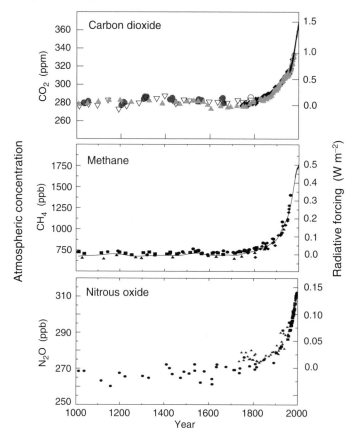

(b) Sulphate aerosols deposited in Greenland ice

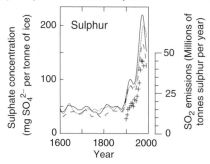

Figure 1.6. Long records of past changes in the atmospheric composition provide the context for the influence of anthropogenic emissions. *Source:* Houghton *et al.* (2001).

level not attained in the last few thousands of years. Approximately one-third of the emissions are estimated to come from anthropogenic sources, notably combustion, some industrial sources and the usage of nitrogen fertilizers. Uncertainties here are larger than for the other gases. Most gases controlled by the Montreal Protocol on the depletion of the ozone layer (e.g. chlorofluorocarbons and their initial replacements, hydrochlorofluorocarbons) are potent greenhouse gases, but their earlier increase in concentrations has either turned into a decrease or slowed down, in response to production controls agreed under the Montreal Protocol. Concentrations of HFCs, which are increasingly produced as alternatives for the Montreal Protocol compounds and which do not deplete the ozone layer but do affect radiative forcing, are increasing in the atmosphere, similarly to other compounds such as perfluorocarbons and sulphurhexafluoride. Finally, one of the most important greenhouse gases is tropospheric ozone, which is not emitted, but formed in the atmosphere in the presence of so-called ozone precursors such as nitrogen oxides, carbon monoxide, and hydrocarbons or 'volatile organic compounds'. Ozone is much more short-lived than most other greenhouse gases, and as a consequence more variable by region, dependent upon the atmospheric conditions and emissions of precursors. It is estimated that, overall, the amount of tropospheric ozone has increased by more than one-third since pre-industrial times.

The carbon cycle

Greenhouse gas concentrations are increasing generally because their sources exceed their sinks. For most gases the most important sink is chemical oxidation in the atmosphere under the influence of solar radiation. Some gases are relatively short-lived, like methane (a lifetime of about 12 years), others are more long-lived, like nitrous oxide (lifetime more than 100 years) and CFCs (lifetimes up to hundreds of years), while some fully fluorinated compounds like sulphur hexafluoride, chloro hexafluoride, and other fluorocarbons have lifetimes of a thousand to tens of thousands of years. The most important greenhouse gas, carbon dioxide, is not broken down in the atmosphere but is exchanged continuously between the terrestrial biosphere, the oceans, the atmosphere, and – very much slower – with sedimentary rocks. Land, atmosphere and ocean have large pools of stored carbon, the largest being in the oceans. From the perspective of the enhanced green-house effect, the exchange of carbon between the various pools is of key importance (see Figure 1.7). In the last 150 years, mankind has emitted about 405±60 gigatonnes (Gt) of carbon into the atmosphere, with about two-thirds from fossil fuel combustion and one-third from land-use changes (Watson *et al.* 2000). Before the middle of the last century, land-use related carbon emissions are estimated to have originated primarily from the middle and high latitudes (about 40 per cent of the cumulative land-use related carbon emissions), thereafter primarily

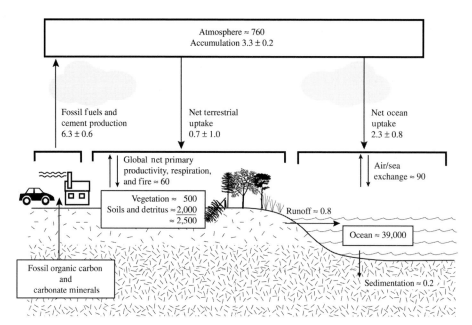

Figure 1.7. The global carbon cycle, showing the carbon stocks in reservoirs (in Gt Carbon) and carbon flows (in Gt/yr) relevant to the anthropogenic perturbation as annual averages over the decade 1989–1998. *Source:* Watson *et al.* (2000).

from the tropics (about 60 per cent), in line with land-use clearing patterns. This has led to an increase in atmospheric carbon dioxide of about 28 per cent, from 285 ppmv around 1850 to approximately 368 ppmv by 2000, with an accurately measured rate of increase of about 3.2 ± 0.1 Gt carbon per year (0.4 per cent per year). Current emissions are about 6.3 ± 0.6 Gt carbon per year from fossil fuel combustion and cement production, and 1.6 ± 0.8 Gt carbon per year from land-use changes (see Table 1.2).

Natural cycles lead currently to the absorption of about 60 per cent (4.7/7.9) of the total emitted carbon into the upper layers of the oceans and the vegetation of terrestrial ecosystems. The uptake by the oceans is modelled to have decreased from about 1.9 ± 0.6 Gt carbon per year in the 1980s to 1.7 ± 0.5 Gt carbon per year in the 1990s. As a consequence, the global uptake by terrestrial vegetation is estimated to exceed the manmade land-use related emissions, net uptake having increased from about 0.2 Gt carbon per year in the 1980s, to 0.7 Gt carbon per year in the 1990s. However, there is great uncertainty surrounding these numbers ($\pm$ 0.7–1 Gt carbon per year). This implies that the total 'residual' uptake of the terrestrial ecosystems may have increased from about 1.9 to 2.3 Gt carbon per year. The reasons for this increase are uncertain, the uptake being dependent on a multitude of natural and human factors. Natural factors include variable climatic conditions. Human-induced factors include forest and other land management, carbon fertilization of vegetation, and plant growth stimulated by deposition of

Table 1.2. *Average annual budget of carbon dioxide perturbations from 1980 to 1989 and 1989 to 1998. Flows and reservoir changes of carbon are expressed in gigatonnes of carbon per year; error limits correspond to an estimated 90 per cent confidence interval.* Source: *Watson* et al. *(2000).*

	1980–89	1989–98
1. Emissions from fossil fuel combustion and cement production	5.5 ± 0.5	6.3 ± 0.6[a]
(a) from Annex I countries[d]	3.9 ± 0.4[a]	3.8 ± 0.4[a]
(i) from countries excluding those with economies in transition	2.6 ± 0.3	2.8 ± 0.3
(ii) from countries with economies in transition[d]	1.3 ± 0.3[a]	1.0 ± 0.3[a]
(b) from rest of world[d]	1.6 ± 0.3[a]	2.5 ± 0.4[a]
2. Storage in the atmosphere	3.3 ± 0.2	3.3 ± 0.2[b]
3. Ocean uptake	2.0 ± 0.8	2.3 ± 0.8[c]
4. Net terrestrial uptake $= (1) - [(2) + (3)]$	0.2 ± 1.0	0.7 ± 1.0
5. Emissions from land-use change	1.7 ± 0.8[e]	1.6 ± 0.8[f]
6. Residual terrestrial uptake $= (4) + (5)$	1.9 ± 1.3	2.3 ± 1.3

[a] Based on emission estimates through 1996 by Marland *et al.* (1999) and estimates derived from energy statistics for 1997 and 1998 (British Petroleum Company 1999).

[b] Based on atmospheric carbon dioxide concentrations measured at Mauna Loa, Barrow, and South Pole (Keeling & Whorf 1999).

[c] Based on ocean carbon cycle model (Jain *et al.* 1995) used in the IPCC Second Assessment Report (IPCC 1996; Harvey *et al.* 1997) consistent with an uptake of 2.0 Gt carbon per year in the 1980s.

[d] Annex 1 countries and countries with economies in transition (a subset of Annex 1 countries) defined in the FCCC. Emissions include emission estimates from geographic regions preceding this designation and include emissions from bunker fuels from each region.

[e] Based on land-use change emissions estimated by Houghton (1999) and modified by Houghton *et al.* (1999, 2000), which include the net emissions from wood harvesting and agricultural soils.

[f] Based on estimated annual average emissions for 1989–95 (Houghton *et al.* 1999, 2000).

nutrients like phosporus and nitrogen. These numbers are small compared to the total exchange of carbon between the atmosphere and terrestrial biosphere. Global gross primary productivity (the amount of carbon fixed by plants) is estimated to be around 120 Gt carbon per year with about 60 Gt carbon per year recycled through plant respiration, leading to a net primary productivity of about 60 Gt carbon per year 50 Gt carbon per year lost through decomposition of plant materials, and about 9 Gt carbon per year per year through ecosystem disturbance (fires, droughts, pests), leaving less than 1 Gt carbon per year for long-term carbon storage at the global level (see Figure 1.8).

These issues are not only interesting from a scientific perspective, but increasingly also from a political one: carbon sequestration through (re-)afforestation is one of the ways to reduce net carbon emissions and, for many countries such as

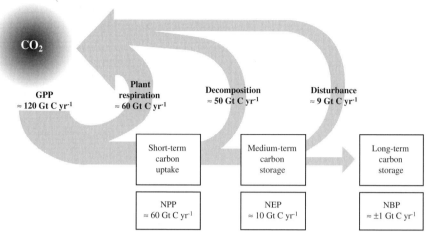

Figure 1.8. Global terrestrial carbon uptake. Plant (autotrophic) respiration releases carbon dioxide into the atmosphere, reducing gross primary productivity (GPP) to net primary productivity (NPP) and resulting in short-term carbon uptake. Decomposition (heterotrophic respiration) of litter and soils in excess of that resulting from disturbance further releases carbon dioxide to the atmosphere, reducing NPP to NEP and resulting in medium-term carbon uptake. Disturbance from both natural and anthropogenic sources (e.g. harvest) leads to further release of carbon dioxide to the atmosphere by additional heterotrophic respiration and combustion – which, in turn, leads to long-term carbon storage. *Source:* Watson *et al.* (2000).

Australia, New Zealand, the USA and Canada, offers important opportunities to compensate for increased emissions from fossil fuel combustion (see Chapter 9). Thus, the longer-term characteristics of the terrestrial carbon sink are of great relevance to policy-makers, but alas, uncertainties abound. According to Watson *et al.* (2000), the current terrestrial carbon sink is likely to be maintained for some decades into the future (under appropriate and sustainable management conditions), but may gradually diminish in the longer term. There are several reasons for this. First, the sequestration capacity of the ecosystems may be saturated. Second, as carbon dioxide concentrations continue to rise, photosynthesis at some point will not increase further, while respiration will increase further with rising temperatures. Finally, uptake of carbon by the oceans is projected to decrease with rising concentrations, further contributing to increases in atmospheric greenhouse gas concentration.

Attribution

Attributing the observed climatic changes to either anthropogenic or natural causes is very interesting from both a scientific and a political perspective. There are various ways to approach this. One is to compare recent temperature changes with the longer-term climate variability. Based on a careful analysis of the

Simulated annual global mean surface temperatures

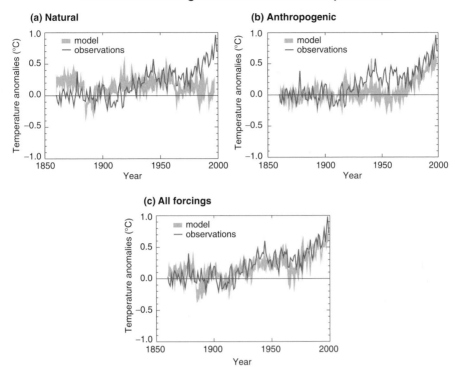

Figure 1.9. Simulating the temperature variations of the Earth ($°$C) and comparing the results to the measured changes can provide insight to the underlying causes of the major changes. *Source:* IPCC (2001).

temperature record and model estimates of climate variability, it was found that it is very unlikely that the warming observed over the last century would be caused by internal variability alone. The warming also seems unusual, taking into account the record for the last 1000 years. A clear view on this issue is given in Figure 1.9. If only natural factors (e.g. solar irradiation and volcanic aerosols) are taken into account, model results and observed temperatures deviate significantly in the last half century as shown in (a). If only what is known about anthropogenic factors is simulated (b), results are consistent during most of the twentieth century, except for the period 1940–60. If both natural and anthropogenic factors are considered (c), the fit becomes much better, making it possible that not only natural, but also anthropogenic, factors play a role in the observed global warming. This kind of evidence cannot give 100 per cent certainty, but was sufficient for the IPCC to strengthen its 1995 conclusion that 'the balance of evidence suggests a discernible human influence on climate' into 'there is new and stronger evidence that most of the warming observed during the last 50 years is attributable to human activities'. (Houghton *et al.* 2001).

The global climate of the twenty-first century

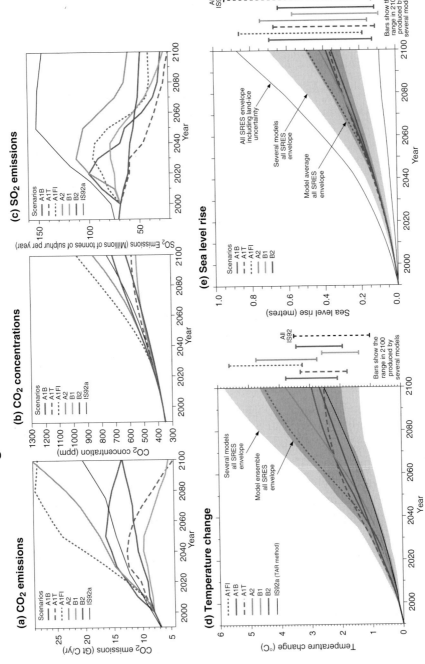

Figure 1.10. The global climate of the twenty-first century will depend on natural changes and the response of the climate system to human activities. The Figure shows the CO_2 emissions, CO_2 concentrations, SO_2 emissions, modelled temperature change and sea-level rise for six SRES scenarios, discussed in Chapter 2. The grey areas represent a fuller set of scenarios as well as model variations. For a coloured version of the graphs see the original publication or website http://www.grida.no/climate/ipcc_tar/wg1/figspm-5.htm. *Source:* Houghton *et al.* (2001).

1.3.2   Future outlook

If anthropogenic activities have contributed to recent climatic changes, as discussed above, it becomes highly relevant how the climate may change further into the future, taking into account that greenhouse gas emissions are still increasing worldwide. Over the last few years, scenarios have been developed for such future changes, based on the socioeconomic and greenhouse gas emission scenarios developed in the *SRES* (Nakicenovic & Swart 2000) and other publications (see Chapter 2 for details). The scenarios in *SRES* are all 'reference' scenarios, i.e. they do not assume any climate policies additional to the ones agreed at the time of their development (thus the Kyoto Protocol targets are not included). These scenarios cover a very wide range of possible future worlds with accordingly a wide range of greenhouse gas emissions, dependent on autonomous developments and deliberate human choices with reference to the rate and type of growth of the economy and population, and technology development. As discussed in more detail in Chapter 2, in all *SRES* scenarios the emission of carbon dioxide remains the most important factor in global warming. The global carbon dioxide emission in these reference scenarios either can return to or be below current levels within a century, or increase by a factor of 5. Because of the dynamics of the climate system and the associated delays (see Box 1.5), all *SRES* scenarios lead to carbon dioxide concentrations that would still be increasing by the end of the century, and temperature and sea-level would continue to increase (see Figure 1.10). For two of the scenarios (B1 and A1T), increases in concentrations will have levelled off by the turn of the century (at about 550 and 600 ppmv, respectively) and stabilization may be achieved relatively easily thereafter. For others, concentrations are still increasing more or less rapidly at the end of the century. The projections show that very different worlds can lead to very different greenhouse gas emissions and associated climatic changes, but also that similar greenhouse gas emissions can be caused by very different worlds. The difference between the scenarios is caused not only by different autonomous socioeconomic factors, but also by human choices, leading to different development pathways. This finding is very important for the core theme of this book – climate change and sustainable development – and is discussed in more detail in Chapter 2.

Compared to earlier assessments of the IPCC and other groups, the range of temperature projections has moved upwards slightly. Averaged surface temperature is now projected to increase globally by 1.4–5.8 °C over the period 1990–2100 (Houghton *et al.* 2001). One reason for the difference with earlier projections is that an updated and improved simple climate model was used, but more importantly, the new *SRES* scenarios differ from the so-called IS92 scenarios in a number of important ways. First, the range of carbon emissions has broadened, especially at the upper end of the range. The main reason for

Box 1.5   Inertia in natural and socioeconomic systems

Maybe the most important reason why citizens and policymakers in the world today may recognize the problem of climate change as real but do not feel any urgency of doing something about it, is that serious adverse effects are not projected to occur until some time well into the future. It appears to be hard to appreciate fully the inertia of the climate system, and many may overestimate the feasibility of future rapid changes in socioeconomic systems (these may be considered necessary to mitigate climate change). Figure 1.11(a) shows some characteristic time-scales of some key processes in climate change and socioeconomic systems, varying from less than a year for plant–physiological processes, to decades for capital stock replacements, to centuries for lifetimes of some key greenhouse gases, to more than 1000 years for sea-level to stop rising. The effect of increasing greenhouse gases in the atmosphere is projected to have consequences that will persist for centuries. The oceans and the ice masses in the world are the main factors determining the inertia of the climate system, because of their large thermal capacity and mass – following a change in radiative forcing, temperature at the surface will only equilibrate in the course a several centuries, according to current climate–ocean models. Figure 1.11(b) shows the results of an imaginary peak and decline of carbon dioxide emissions within a century: the carbon dioxide concentration stabilizes in the course of the following centuries, temperature rise slows down but does not stop completely, and sea-level continues to rise for thousands of years due to thermal expansion and ice melting. If one would like to stabilize carbon dioxide concentrations at 550 ppmv by 2150,[1] then emissions would have to peak between 2020 and 2030 and be brought back below 1990 levels about 100 years before the stabilization date (2030–2100).

    The possibility of stabilizing at this level is influenced not only by natural processes, but determined also by potential rates of socioeconomic change. For example, scenario analysis suggests that the rates of improvements of energy efficiency, and especially fuel, shifts away from carbon-based fuels ('decarbonization') at an aggregate global or regional level, needs both to go beyond historic rates to achieve stabilization at this level and be maintained for a long time. The lower the targeted level, the more system inertia becomes important (e.g. because of the need for premature retirement of capital stock) and the higher the costs would become. Making adjustments in the energy system of the world that are required to stabilize concentrations thus could be compared with changing the direction of a supertanker: one should start taking action to steer away from dangerous objects well ahead! On the other hand, at the level of individual technologies, the rates of penetration in low-emissions scenarios such as B1 and A1T do not go beyond historic penetration rates of energy technologies.

Sources: IPCC (2001); Nakicenovic and Swart (2000).

[1] This is one of the formally accepted objectives of the European Union (EU). A second EU climate objective – limiting global temperature increase to at most $2\,^\circ C$ as compared to pre-industrial levels – would lead to an even stricter target, i.e. in the order of magnitude of 450 ppmv carbon dioxide for intermediate assumptions with respect to the climate variability. For stabilizing carbon dioxide concentrations at 450 ppmv, global emissions would have to peak between 2005 and 2015 (IPCC 2001e).

this is that some *SRES* scenarios – notably the fossil-intensive high-growth scenario A1FI (see Chapter 2) – couple high economic growth with relatively little attention to global environmental problems, giving priority to the development of additional and often unconventional fossil fuel resources. Second, while in 1992 it was projected that sulphur emissions would increase continuously due to the ongoing industrialization process in the currently developing countries, more recent developments in important developing countries such as China and India make such future very unlikely. By taking steps to abate acid deposition in the 1970s, these countries had started already to abate sulphur emissions in order to bring down local pollution, although their per capita income levels were much below those in industrialized countries. The importance of scenario assumptions for sulphate aerosols is great. For example, it is remarkable that for scenarios with relatively high emissions of carbon dioxide, such as A2 and A1FI, temperature projections are not necessarily higher than for other scenarios initially, because these scenarios also assume high sulphur emissions for the coming period associated with rapid fossil-fuel-based industrialization in developing countries. However, in the longer term, the warming effect of carbon dioxide becomes dominant also in these scenarios. Another interesting element is that in some of the scenarios climate forcing due to tropospheric ozone increases becomes as important as the forcing due to methane, while leading to increased local and regional pollution problems.[3] These two examples show the very important linkage between the causes of climate change, and those of local and regional pollution.

[3] A closer and updated study of sulphur emissions since the 1992 IPCC scenarios led to the assumption in the *SRES* scenarios that global sulphur emissions, which can be abated relatively easily with currently available technology, would not rise at much as assessed before. It is not beyond imagining that in the future also the abatement of ozone precursors will become more successful in both industrialized and developing countries (OECD countries are currently implementing abatement policies for the ozone precursors, but are not very successful yet for, for example, traffic-related nitrogen oxides).

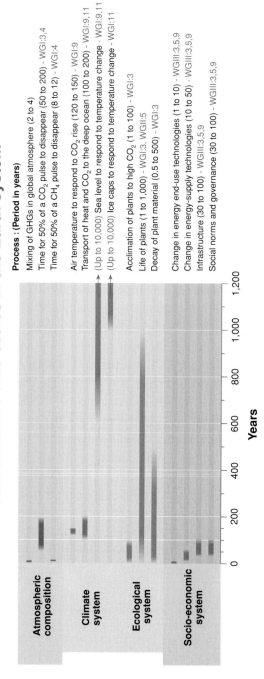

Characteristic time scales in the Earth system

Process : (Period in years)

Mixing of GHGs in global atmosphere (2 to 4)
Time for 50% of a CO_2 pulse to disappear (50 to 200) - WGI:3,4
Time for 50% of a CH_4 pulse to disappear (8 to 12) - WGI:4

Air temperature to respond to CO_2 rise (120 to 150) - WGI:9
Transport of heat and CO_2 to the deep ocean (100 to 200) - WGI:9,11
(Up to 10,000) Sea level to respond to temperature change - WGI:9,11
(Up to 10,000) Ice caps to respond to temperature change - WGI:11

Acclimation of plants to high CO_2 (1 to 100) - WGI:3
Life of plants (1 to 1,000) - WGI:3, WGII:5
Decay of plant material (0.5 to 500) - WGI:3

Change in energy end-use technologies (1 to 10) - WGIII:3,5,9
Change in energy-supply technologies (10 to 50) - WGIII:3,5,9
Infrastructure (30 to 100) - WGIII:3,5,9
Social norms and governance (30 to 100) - WGIII:3,5,9

Atmospheric composition

Climate system

Ecological system

Socio-economic system

Years

0   200   400   600   800   1,000   1,200

Figure 1.11. Inertia: (a) The characteristic time scales of some key processes in the Earth system: atmospheric composition, climate system, ecological system and socioeconomic system. Timescale is defined here as the time needed for at least half of the consequences of change in a driver of the process to have been expressed. (b) After CO_2 emissions are reduced and atmospheric concentrations stabilize, surface air temperature continues to rise by a few tenths of a degree per century or more. Sea level continues to rise for centuries. *Source:* IPCC (2001).

CO₂ concentration, temperature, and sea level continue to rise long after emissions are reduced

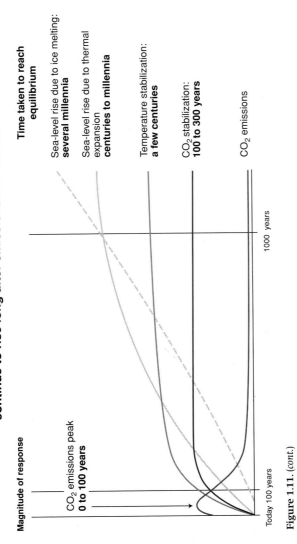

Magnitude of response

CO₂ emissions peak
0 to 100 years

Time taken to reach equilibrium

Sea-level rise due to ice melting:
several millennia

Sea-level rise due to thermal expansion
centuries to millennia

Temperature stabilization:
a few centuries

CO₂ stabilization:
100 to 300 years

CO₂ emissions

Today 100 years

1000 years

Figure 1.11. (cont.)

For all scenarios, globally averaged water vapour, evaporation and precipitation are projected to increase. As to the regional level, projections of precipitation changes for *SRES* A2 and B2 suggest likely precipitation increases in summer and winter over high-latitude regions. For regions such as Australia, Central America and southern Africa the calculations indicate decreases in rainfall in winter. Because of the many uncertainties, it is as yet very difficult to make reliable statements about changes in extreme events. However, it is very likely that more hot days and heat waves will occur over most land areas, and fewer frost days and cold periods. Precipitation extremes are projected to increase more than the mean, as is intensity of precipitation (Houghton *et al.* 2001).

A key uncertainty in these projections is the climate sensitivity, usually defined as the equilibrium change in global mean temperature as a result of a doubling of the carbon dioxide concentration. Already for more than a decade the 'common wisdom' uncertainty range for the climate sensitivity is between 1.5–4.5 °C.[4] Although recent global circulation modelling experiments have suggested numbers toward or beyond the high end of the range, global assessments such as those by the IPCC have not considered these experiments sufficient to change the long-standing range estimate. Evidently, the world will not warm evenly. Regional differences are projected to be large, with the highest increases in the higher latitudes in the Northern Hemisphere (see Figure 1.12 for results for A2 and B2 *SRES* modelling runs). Please note that A2 and B2 do not represent the full range of emissions scenarios. Box 1.6 addresses the potential impacts of changing greenhouse gas concentrations on the El Niño phenomenon.

On the basis of the *SRES* scenarios, global average sea-level is projected to increase by 0.09–0.88 m in the coming century, caused primarily by the thermal expansion of the oceans, and to a lesser extent the melting of glaciers. Models disagree about the contributions from the ice changes on Greenland and Antarctica over this period, the Greenland contribution probably being positive and the Antarctic contribution negative (ice accumulation). Figure 1.10 shows that, up to about 2050, there is very little difference between the scenarios, but after that the scenario projections start to diverge considerably. And even more importantly, even if greenhouse gas concentrations were stabilized, sea-level would continue to rise for centuries, up to 0.5–2 m for stabilising at double pre-industrial carbon dioxide levels, and 1–4 m for 4 times this level. From this ultra-long perspective, one should realize that Greenland and Antarctica contain sufficient ice to have sea-level rise by almost 70 m. So, even smaller effects on their ice masses can have very significant effects. For example, a local warming beyond 3 °C, sustained for thousands of years, could lead to melting of the Greenland ice sheet and would

---

[4] For the ranges in Figure 1.10 a similar range of 1.7–4.2 °C has been used.

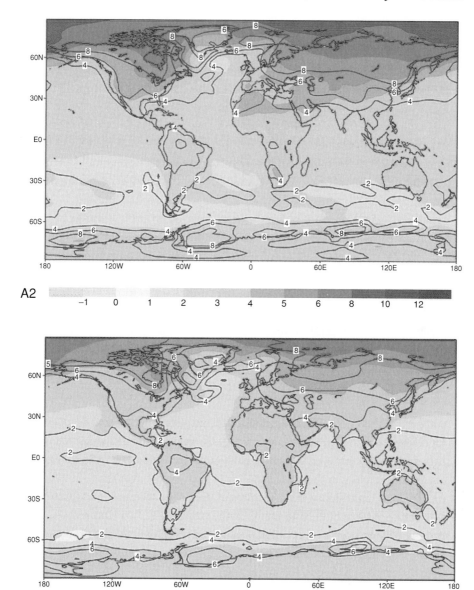

Figure 1.12. Regional differences in temperature change: the annual mean change of temperature and its range for *SRES* scenario A2 (upper panel) and B2 (lower panel). Both *SRES* scenarios show the period 2071 to 2100 relative to the period 1961–1990 and were performed by ocean–atmosphere global circulation models. For a clearer image of the regional temperature differences we refer to the colour figure in the original publication or website http://www.grida.no/climate/ipcc_tar/wg1/figts-20.htm. *Source:* Houghton *et al.* (2001).

> **Box 1.6** How would climate change affect El Niño and associated
> impacts on natural and human systems?
>
> El Niño[1] is a disruption of the ocean–atmosphere system in the tropical
> Pacific having important consequences for weather around the globe. In
> particular years, the waters off the coast of Peru and Ecuador warm, starting
> in late spring or summer and building up towards the end of the year. It is
> associated with a fluctuation of the inter-tropical surface pressure pattern
> and circulation in the Indian and Pacific Oceans: the Southern Oscillation.
> During an El Niño event, the prevailing trade winds weaken and the
> equatorial counter-current strengthens, causing warm surface waters in the
> Indonesean area to flow eastward to overlie the cold waters of the Peru
> current (Houghton *et al.* 2001). The phenomenon disrupts local fisheries, but
> also has consequences such as floods and droughts in many regions around
> the globe that affect and sometimes even disrupt local communities. Many
> natural and managed ecosystems are sensitive to the frequency of El Niño
> events. The upward trend in worldwide numbers of people affected adversely
> by weather disasters has been characterized by peak impacts during El Niño
> events (McCarthy *et al.* 2001).
>
> The El Niño cycle has shown great variability in the past, most of which is
> attributed to natural causes, e.g. variability in the internal climate system
> and external forcings such as orbital variations. Over the last two decades,
> there have been some extreme events, but the reasons for this are not clear.
> According to Houghton *et al.* (2001), current projections show little change,
> or a small increase, in the amplitude for El Niño events over the next 100
> years, but the confidence in projections of the future frequency, amplitude
> and spatial pattern of El Niño events in the tropical Pacific is limited,
> because of the uncertain representation of the complexity of the processes
> involved in models. Different models give different outcomes. However, even
> with little or no change in the amplitude, global warming is likely to lead to
> greater extremes for droughts and precipitation in many regions.
>
> [1] Also called ENSO (El Niño–Southern Oscillation).

raise sea-level by approximately 7 m; for a warming of 5.5 °C, consistent with mid-range stabilization scenarios, about 3 m in 1000 years, and for 8 °C, about 6 m (Houghton *et al.* 2001). Taking into account that local temperatures in Greenland may be 1.2–3.1 times above the global average, in a pessimistic case, even stabilizing at the low level of 450 ppmv could lead to a 3-m sea-level rise because of the ice sheet melting, and stabilizing at 550 ppmv could well be associated with

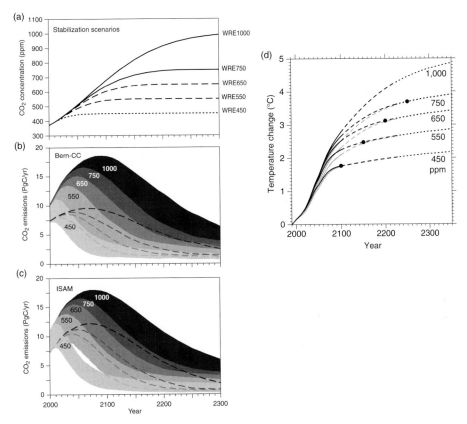

Figure 1.13. Projected relationship between (a): CO_2 concentrations, (b) and (c): CO_2 emissions for two models, and (d) temperature change for a range of stabilization scenarios between 450 and 1000 ppm CO_2. *Source:* Houghton *et al.* (2001).

a sea-level rise of 6 m in the long term (see Figure 1.13). Only for low-stabilization profiles and low climate sensitivity does the risk of melting of the Greenland Ice Sheet appear to be low. Although the workings of the Antarctic ice sheet in terms of ice dynamics are not very well understood, the risk of dramatic melting over the next 1000 years (e.g. because of the disintegration of the West Antarctic icesheet) is considered to be very small.

The objective of the UNFCCC is to stabilize greenhouse gas concentrations. The most important greenhouse gas, carbon dioxide, is also the most researched gas in this respect. Figure 1.12 shows the relationship between time profiles for carbon dioxide emissions, its stabilization at concentration levels between 450 and 1000 ppmv, and the associated temperature change. Delays in the climate system determine the relationships (see also Box 1.5). A general finding is that, in order to stabilize concentrations at 450 ppmv, global emissions have to fall below the current level in the coming decades – for 550 ppmv, this period is approximately a century, and for 1000 ppmv, two centuries. As will be discussed in Chapter 10, the issue of timing of emissions controls from a mitigation perspective is dependent upon the preferred level of stabilization and the time path towards that level. In general, one can say that the eventual concentration–stabilization level depends more on the cumulative carbon emissions than on the time profile of those emissions. Also, the time when concentrations would be stabilized has an impact on emissions profiles: the later concentrations are targeted to be stabilized, the more emissions reduction can be deferred into the future. In general, modellers assume that concentrations would be stabilized within the next few centuries. Stabilization within this century requires earlier emissions reduction. Following stabilization of concentrations, temperature increases will continue for quite some time, albeit they would slow down, the more so as the stabilization level is lower.

1.4   Impacts and vulnerability

1.4.1   Introduction: avoiding dangerous interference with the climate system

As discussed in the previous section, there is increasing evidence that the climate of the world is changing already; it is probable that it will continue to change, and that humans contribute to these changes. What turns this into a problem is that these changes affect the functioning of ecosystems and societies. In UNFCCC parlance: there may be 'dangerous' anthropogenic interference with the climate system. Indeed, the negotiators of the UNFCCC decided as its ultimate objective

> that the Conference of the Parties may adopt is to achieve, in accordance with the relevant provisions of the Convention, 'stabilisation of greenhouse gas concentrations in the atmosphere at a level that would prevent dangerous anthropogenic interference with the climate system. Such a level should be achieved within a timeframe sufficient to allow ecosystems to adapt naturally to climate change, to ensure that food production is not threatened and to enable economic development to proceed in a sustainable manner' (see Box 1.1).

While it is a value judgement which level of impacts is dangerous and should be avoided, scientists have studied the vulnerability of various systems to climate change and the projected impacts on these systems for a variety of scenarios. The results from these studies can help towards an informed judgement of what could be considered 'dangerous'. It is well-established that both natural and human systems are vulnerable to climate change. Because many human systems have a larger capacity to adapt than natural systems, the latter are especially

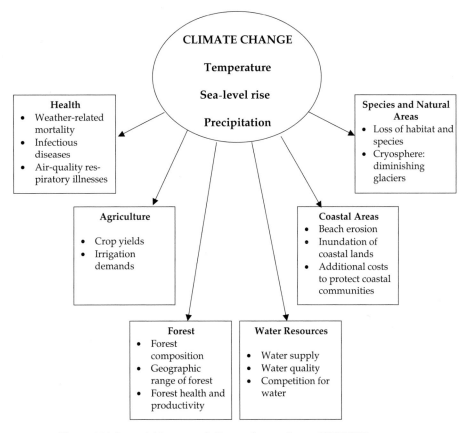

Figure 1.14. Potential impacts of climate change. *Source:* UNEP/GRID

vulnerable. Vulnerability depends on the type of change (e.g. temperature, rainfall, variability, occurrence of extremes), magnitude and rate of the change, exposure, and adaptive capacity. Future climatic changes will affect the level and extent of impacts. The main impact areas are public health, agriculture and food security, forests, hydrology and water resources, coastal areas, biodiversity, human settlements, energy, industry and financial services (see Figure 1.14). Some key findings of IPCC's Third Assessment Report (IPCC 2001a, b, c) with reference to vulnerability and impacts are summarized in Box 1.7.

1.4.2   Observed changes in indicator species

In 1995, the IPCC confirmed that the climate was changing and suggested that there is a 'discernable human influence on climate'. Thus, one may suspect that impacts of these changes, to some unknown extent influenced by human activities, would have to be observable, too. And indeed, important information after 1995 has revealed that there is, indeed, evidence that such impacts are

Box 1.7   Key conclusions of IPCC's Third Assessment Report (IPCC 2001a, b, c), Working Group II: Impact, Vulnerability, and Adaptation

1. Recent regional climate changes, particularly temperature increases, have affected many physical and biological systems already.
2. There are preliminary indications that some human systems have been affected by recent increases in floods and droughts.
3. Natural systems are vulnerable to climate change, and some will be damaged irreversibly.
4. Many human systems are sensitive to climate change, and some are vulnerable.
5. Projected changes in climate extremes could have major consequences.
6. The potential for large-scale and possibly irreversible impacts poses risks that have yet to be quantified reliably.
7. Adaptation is a necessary strategy at all scales to complement climate change mitigation efforts.
8. Those with the least resources have the least capacity to adapt and are the most vulnerable.
9. Adaptation, sustainable development, and enhancement of equity can be mutually reinforcing.
10. The vulnerability of human populations and natural systems to climate change differs substantially across regions and across populations within regions. Regional differences in baseline climate and expected climate change give rise to different exposures to climate stimuli across regions.
11. Further research is required to strengthen future assessments and to reduce uncertainties in order to assure that sufficient information is available for policymaking about responses to possible consequences of climate change, including research in, and by, developing countries.

occurring. Changes in physical, ecological and socioeconomic systems have been identified in many regions. Detecting these changes and associating them with climatic changes poses huge scientific problems, since these systems are usually subject to many stress factors other than climate change, too. Nevertheless, methods have been developed that use vulnerable species and systems, and search

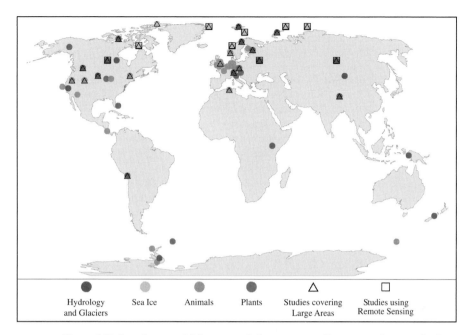

Figure 1.15. Locations at which systematic long-term studies meet stringent criteria documenting recent temperature-related regional climate change impacts on physical and biological systems. For a colour representation of the various types of studies see http://www.grida.no/climate/ipcc_tar/wg2/figspm-1.htm. *Source:* McCarthy *et al.* (2001).

for systematic patterns of change across many studies that are consistent with expectations based on observed or predicted changes in climate (McCarthy *et al.* 2001). Applying such methods, many examples of associations between observed system changes and regional climatic changes have been identified. Examples across the globe for changes consistent with observed temperature changes (see Figure 1.15 for their location) include:

1. Glacier retreat outside the polar regions.
2. Thawing of permafrost.
3. Decrease of the extent and thickness of Arctic sea ice.
4. Later freezing and earlier break-up of river and lake ice.
5. Earlier flowering and longer growing season of plants.
6. Poleward and upward migration of plants, insects, e.g. butterflies, birds and animals.
7. Decline or increase in some animal populations.
8. Earlier arrival and egglaying by birds, and earlier emergence of insects.
9. Increases in coral bleaching events.
10. Higher financial losses due to extreme weather events.

Table 1.3. *Examples of climate change impacts.*

Sensitive system	Examples of impacts
Natural terrestrial and freshwater ecosystems	Loss and fragmentation of habitat and loss of species, especially in vulnerable areas such as mountain and boreal/polar regions
	Change of geographic range and composition of forests and other ecosystems
	Changes in ecosystem (e.g. forest) productivity
	Replacement of wetlands by forests or heathlands
	Disruption of permafrost ecosystems
Managed ecosystems/agriculture and food security	Increasing carbon dioxide increases net primary productivity
	Climate change can lead to a decrease in crop yields; in (sub-)tropics for almost all levels of warming, in temperate zones for warming more than a few degrees
	Increase in crop yields for some crops in some mid-latitude regions for up to a few degrees of warming, as well as increased global timber supply
	Overall, world food supply may not be threatened in the short term in aggregate, but distributional inequities would be exacerbated
	Demand for irrigation water
Human health	Change in distribution of infectious/vector-borne diseases, e.g. malaria, dengue
	Changes in weather-related morbidity and mortality, e.g. heat stress
	Air-pollution-related respiratory illness
	Health problems related to increased food and water scarcity
Water resources	Changes in water supply (surface water runoff and groundwater recharge)
	Associated water scarcity changes, especially adverse in many (but not all) arid and semi-arid regions
	Increases in frequency and intensity of floods
	Changes in water quality related to changes in water temperature
	Distribution of water
Coastal zones, low-lying islands and marine ecosystems	Erosion of beaches
	Damage to coastal mangroves, salt marshes, atolls
	Changes in fish distribution
	Coral reef bleaching
	Inundation of coastal and other low-lying lands
	Costs for coastal protection
Human settlements and economic sectors	Increased energy use due to increased demand for air-conditioning and decreased energy demand due to less heating
	Changes in hydropower energy supply
	Increased risk of population displacement due to floods, sea-level rise, droughts
	Damage to physical infrastructure in areas prone to flooding, permafrost thawing
	Damage to tourism, e.g. wintersports
	Increased insurance losses

Adaptive capacity and vulnerability are very unevenly distributed, with low adaptive capacity and higher vulnerability in the developing world.

1.4.3   Impacts and vulnerability

Over the last two decades, researchers have evaluated potential future impacts on a broad range of sensitive human and natural systems, including natural (terrestrial, freshwater and marine) ecosystems and biodiversity, agriculture and food security, human health, water resources, coastal zones and low-lying islands, human settlements and economic sectors (see Table 1.3). Most of this analysis uses climate scenarios derived from a double carbon dioxide situation, since this is the type of model run most frequently performed with Global Circulation Models. Only more recently, climate scenarios for time profiles determined by scenarios such as the IPCC IS92 (Leggett *et al.* 1992) and the IPCC *SRES* (Nakicenovic & Swart 2000) have become available, and thus impact analysis on the basis of such scenarios is still limited. More limited still is impact work on the basis of mitigation or stabilization scenarios.[5]

1.4.4   Future impacts and reasons for concern

As noted in the introduction to this section, a key question for climate policy is: what represents 'a dangerous level of anthropogenic interference with the climate system'? McCarthy *et al.* (2001) try to provide information pertinent to this question by tying projections of global mean temperature increase for the IPCC scenarios from *SRES* (see also Chapter 2) to five areas of 'reasons for concern' (see Figure 1.16) as follows.

1.   Risks to unique and threatened ecosystems.
2.   Risks from extreme climate events.
3.   Distribution of impacts.
4.   Aggregate (economic) impacts.
5.   Risks from large-scale discontinuities.

These reasons for concern have environmental (1, 2, 3, 5), social (2, 3, 5) and economic (3, 4) aspects and can be associated directly with sustainable development in the context of Article 2 of the UNFCCC.

Risks to unique and threatened systems have been singled out, because not only are these systems often vulnerable especially to climate change because of their relatively small geographic range, but their uniqueness also gives them global significance (Smith, Schellnhuber & Mirza 2001). Unique and threatened systems include physical systems (e.g. tropical and other glaciers), biological systems

---

[5] See Arnell *et al.* (2002) and Parry *et al.* (2001) for a GCM-based analysis for a broad range of impact areas for a reference scenario (IS 92a) and two stabilization scenarios (550 and 750 ppm), and Swart *et al.* (1998) for results using an integrated assessment model – incorporating a simple climate model – for the impacts on natural and managed ecosystems over a range of stabilization scenarios.

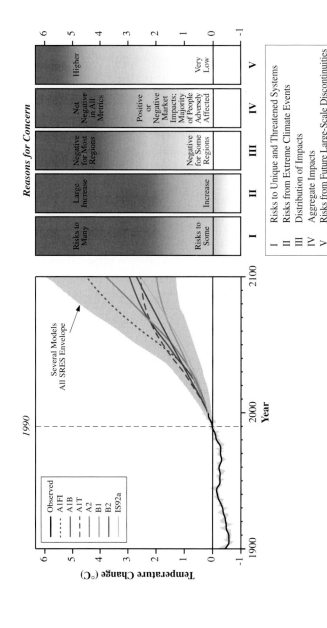

Figure 1.16. Reasons for concern about projected climate change impacts. The risk of adverse impacts from climate change increase with the magnitude of climate change. For a coloured version of this graph we refer to the colour figure in the original publication or website http://www.grida.no/climate/ipcc.tar/wg2/figspm-2.htm. *Source: McCarthy et al.* (2001).

Table 1.4. *Aggregate impacts of climate change on ecosystems.* Source: *McCarthy et al. (2001).*

Impact indicator	Scenario (°C)					
	0.5	1.0	1.5	2.0	2.5	3.0
Temperate cereals, area experiencing						
yield decrease[a]	12	16	18	20	20	22
yield increase[a]	2	3	4	8	12	15
Maize, land area experiencing						
yield decrease[a]	13	18	22	26	29	33
yield increase[a]	2	4	6	9	13	17
Change in natural vegetation[b]	11	19	26	32	37	43
Endangered nature reserves[c]	9	17	24	32	37	42

[a] Yield increase and decrease are percentage area with at least 10 per cent change in potential rain-fed yield. Reference area is current crop area

[b] Change in natural vegetation is percentage of land area that shifts from one vegetation type to another. Reference area is global land area.

[c] Endangered nature reserves are percentage of reserves, where original vegetation disappears, so that conservation objectives cannot be met. Reference is total reserve number.

(e.g. species, ecosystems) and human systems (e.g. small island states, low-lying coastal areas and indigenous peoples). Threats to ecosystems can manifest themselves through species loss, ecosystem shifts, or fragmentation. Particularly sensitive are alpine systems, biodiversity 'hotspots', ecotones (transition areas between different environments), coral reefs and mangrove areas. Many of these systems are already now under heavy stress because of both climate change and reasons other than climate change. This is reflected by the coloured bars in Figure 1.16. There have been very few analyses on how ecosystems may change over time for different scenarios. Table 1.4 shows results of the IMAGE model for risks to natural vegetation and nature reserves as a function of temperature change: the larger the change, the greater the impacts, but also impacts for small changes are significant already (Swart *et al.* 1998, quoted in Smith, Schellnhuber & Mirza 2001). Arnell *et al.* (2002) show that vegetation die-back decreases significantly from an IS92a-like reference scenario to scenarios in which carbon dioxide concentrations stabilize at 550 and 450 ppmv. Not only does the Figure 1.16 show how risks are different across reference scenarios, it also shows that mitigation pays off by reducing ecosystem risks.

Risks from extreme climate events and large scale discontinuities have been identified as important reasons for concern. Many scientific discussions relate to a gradual change in climate. Damage caused by major weather events such as storms,

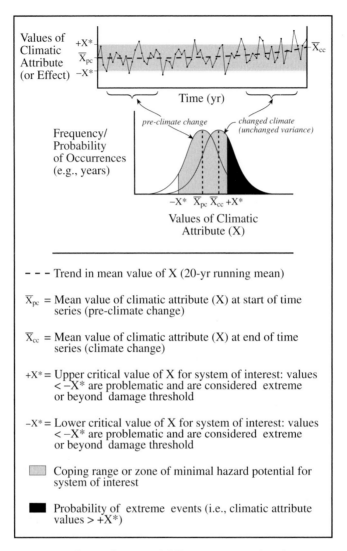

Figure 1.17. Climate change, variability, extremes, and coping range.
Source: McCarthy *et al.* (2001).

droughts and floods demonstrate that climate extremes and climate variability may be even more important. Many societal sectors and communities as well as natural systems are reasonably capable to cope with gradual changes, but can be very vulnerable to extreme events. However, one of the greatest scientific uncertainties remains: how global climate change may manifest itself through changes in extreme weather events and climate variability. Intensity and frequency of some of these events are projected to increase. Assuming that the likelihood of occurrence of an extreme event has a normal distribution, Figure 1.17 shows how an increase in the mean value of a climate indicator such as temperature will also cause an increase in the possibility of extreme events. So, it is likely that climate change will lead to more frequent and severe extreme events in the future. Examples of such possible events are: (a) higher maximum temperatures; (b) more hot days and heat waves; (c) higher minimum temperatures; (d) fewer cold and frost days; (e) fewer cold spells; (f) more intense precipitation events; (g) drought periods; (h) increased tropical cyclone wind and precipitation intensities; (i) intensified droughts and floods associated with El Niño events; (j) increased Asian monsoon precipitation variability, and (k) increased intensity of mid-latitude storms.[6] Please note that some of these examples (e.g. (e) above) can be considered positive.

The risk of non-linear changes and instabilities in components of the climate system is particularly difficult to evaluate. Examples include the weakening of the thermohaline circulation in the northern Atlantic, which would after fundamentally the climate in North America and Europe through changes in the Gulf Stream; disintegration of the West Antarctic ice-sheets; melting of the Greenland ice-sheet; climate-induced release of large volumes of greenhouse gas emissions from terrestrial ecosystems, and saturation of the terrestrial carbon sink. Some of these events (e.g. the thermohaline circulation change) may be irreversible. While many changes associated with climate generally are gradual, some of the fundamental ones may occur at timescales of decades. While these risks are considered to be small-probability large-impacts issues, this perspective seems to be valid mainly for a time horizon of a century. However, this changes if we would look beyond this century (see Section 1.3.2).

Aggregate (economic) impacts and the distribution of impacts influence the opportunities and risks for a sustainable and equitable development. Since the Second Assessment Report (IPCC 1996a, b), monetary estimates of global aggregate impacts have become more modest (from −1.5 to −2 per cent of world GDP to less than −1.5 per cent for a 2.50 °C average global warming (see Table 1.5). There are many uncertainties and controversies regarding the numbers in the table. First, there is disagreement as to what extent valuation of all impacts using a common numeraire is at all possible, notably the valuation of non-market impacts, e.g. loss of biodiversity or risks to human health. Second, there are large knowledge gaps about the impacts themselves, as they may result from geographically uncertain climatic changes. Third, subjective assumptions by the analysts about the shape of the damage function, the discount rate, the time horizon and the choice of welfare criteria determine the outcome over time. Fourth, it is still very difficult to factor-in adaptation into the analyses.

---

[6] The latter is particularly uncertain, with models disagreeing.

Table 1.5. *Indicative world impacts, by region (percentage of current GDP). Estimates are incomplete, and confidence levels in individual numbers is very low.* Source: *McCarthy* et al. *(2001).*

	IPCC (2001)	Mendelsohn *et al.* (2000)		Nordhaus and Boyer (2000)	Tol (1999)
	2.5 °C warming	1.5 °C warming	2.5 °C warming	2.5 °C warming	1 °C warming[a]
North America					3.4 (1.2)
United States			0.3	−0.5	
OECD Europe					3.7 (2.2)
EU				−2.8	
OECD Pacific					1.0 (1.1)
Japan			−0.1	−0.5	
Eastern Europe/FSU					2.0 (3.8)
Eastern Europe				−0.7	
Russia			11.1	0.7	
Middle East				−2.0[b]	1.1 (2.2)
Latin America					−0.1 (0.6)
Brazil			−1.4		
South, Southeast Asia					−1.7 (1.1)
India			−2.0	−4.9	
China			1.8	−0.2	2.1 (5.0)[c]
Africa				−3.9	−4.1 (2.2)
Developed countries	−1.0 to −1.5	0.12	0.03		
Developing countries	−2.0 to −9.0	0.05	−0.17		
World					
Output weighted	−1.5 to −2.0	0.09	0.1	−1.5	2.3 (1.0)
Population weighted				−1.9	
At world average prices					−2.7 (0.8)
Equity weighted					0.2 (1.3)

[a] Figures in parentheses denote standard deviations.

[b] High-income countries in OPEC.

[c] China, Laos, North Korea, Vietnam.

Nevertheless, there is broad agreement about three findings. First, impacts are distributed unevenly, with developing countries being more vulnerable than developed countries. The main reasons for this are that developing countries not only have higher impacts, they also have lower capacity to adapt. Higher impacts result from the greater extent of flood- and drought-prone areas, the larger share of climate-sensitive sectors such as agriculture, a more sensitive population due to poorer health and nutrition, and a more vulnerable shore area. They have lower capacity to adapt because they lack technological, institutional, financial and knowledge capacity. Arnell *et al.* (2002) show how climate impacts affect specific regions such as water-scarce areas, low-lying coasts and islands, and tropical regions (e.g. risk of malaria), more than others. Second, while at low levels of temperature change (less than 2–3 °C) impacts may be mixed (some positive, some negative, dependent on regions and sector), for larger changes (more than 2–3 °C) virtually all regions show adverse impacts. Third, for different magnitudes of climatic changes, distribution of impacts can be different.

Looking at the full picture, Figure 1.16 shows that the risks for unique and threatened ecosystems and from extreme climate events are the greatest of the five reasons for concern: already for currently observed temperature change risks are positive, while they increase rapidly with increasing greenhouse gas emissions and associated increasing temperatures (first and second columns). The last column shows that the risks for such large-scale discontinuities as the slowing of the thermohaline circulation or the decomposition of the west Antarctic ice-sheet (with possible devastating results) is low for low climatic changes, but becomes real for higher changes, which are still in the realm of expectation in the twenty-first century for the high emissions scenarios, becoming larger subsequently. Figure 1.16 raises an important ethical issue for policymakers: if, from a precautionary perspective, for example, 'dangerous' would be defined as avoiding risks to threatened and vulnerable ecosystems or avoiding climatic changes that would exacerbate current inequities, even the lowest of the scenarios could be considered to be dangerous and very low emissions would be required.[7] Looking at the question in that way, the vulnerability of the most sensitive ecosystems (e.g. in alpine or polar regions) or the most vulnerable communities (e.g. in arid or semi-arid developing regions, low-lying coastal areas or island states) would set the standards for a global mitigative response.

It would be extremely useful to have information on the extent to which risks can be reduced by mitigating climate change at different levels, e.g. by stabilizing

[7] This kind of analysis is explored in the so-called 'Tolerable Windows' or 'Safe Landing' approaches, which derive 'safe emissions corridors' from long-term climate risks (see Section 8.1).

Additional millions of people at risk globally due to climate change

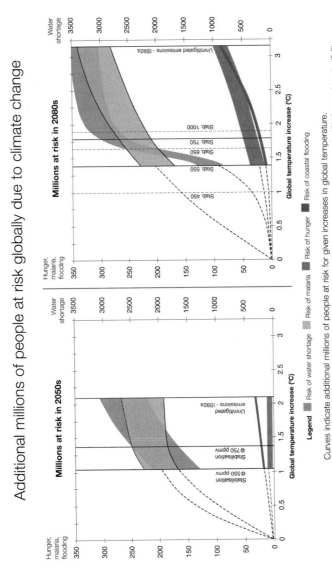

Curves indicate additional millions of people at risk for given increases in global temperature.
The width of the curves indicate one standard deviation around the mean, based on results from four HadCM2 experiments (2,3).
Solid lines indicate model-based estimates. Dotted lines are inferred (6) and intended as schematic.
Global temperature is relative to 1961–90 average. Vertical lines indicate the temperature increase for emissions scenarios with CO_2 concentrations that ultimately stabilize at specified levels.

Figure 1.18. People at risk: levels of impacts as a function of carbon dioxide emissions profiles. *Source: Parry et al.* (2001).

greenhouse gases at different concentrations. Unfortunately, there are, as yet, few studies that analyse systematically damage avoided by stabilizing greenhouse gas concentrations. Figure 1.18 shows a good example of preliminary work in this area. It gives an impression of such impacts in key areas as water shortage, public health (malaria risk), risk of flooding and incidence of hunger as a function of stabilisation levels (Parry *et al.* 2001). The figure interestingly also suggests that the shape of the damage curve can be much different for different vulnerable systems, and that the difference between stabilizing at different levels can have implications for millions of people.

1.5   Rationale for action: adaptation and mitigation as part of a sustainable development strategy

From the previous sections it is clear that climate change is real, changes in important climate variables such as global average temperature have been observed, and there is growing evidence that observed changes in natural and managed ecosystems can, indeed, be associated with these climatic changes. Future serious and possible irreversible impacts are projected for a broad set of socioeconomic scenarios. Risks to unique and threatened ecosystems, especially from extreme climate events, and the probability that even modest climatic changes would exacerbate inequities between regions, suggest that action is warranted to prevent (further) dangerous anthropogenic interference with the climate system. Climate-change response options include, primarily, *mitigation* and *adaptation*. We define mitigation as a set of actions that will reduce the likelihood of climate change, whereas adaptation involves actions that will reduce the impacts of climate change (without necessarily altering the likelihood that it will occur).

Scientific assessment (e.g. Parry *et al.* 2001, Arnell *et al.* 2002, Swart *et al.* 1998, and Watson *et al.* 2000) suggest that stabilizing carbon dioxide concentrations indeed reduces risk of adverse impacts, more so if stabilization levels are lower. There are many remaining uncertainties, but politically it has been agreed that lack of full scientific certainty should not be used as a reason for postponing mitigation measures (see also Box 1.1). As will be discussed in Chapter 9, many no-regrets options are available, i.e. options the benefits of which equal or exceed the costs to society, even excluding the benefits of avoided climate change. Such benefits include the reduction of local and regional air pollution and associated health effects, reduced energy costs, improved local employment, and others. Assuming that climate impacts have predominantly adverse impacts, as suggested by the main scientific literature (assessed by McCarthy *et al.* 2001), reducing these impacts makes sense economically and ethically. The effects are likely to endanger sustainable development opportunities. Therefore, is has been concluded already in the Second Assessment Report (IPCC 1996a, b) that there is a rationale to go beyond no-regrets measures in mitigating greenhouse gas emissions. A key – political – question remains how far beyond no regrets one should go and how fast. The key conclusions from the IPCC Working Group III (IPCC 2001a, b, c) (see Box 1.8) suggest that there are abundant options to mitigate climate change, and that aggregate costs may not be prohibitive. Maybe more importantly, mitigation options can be made more attractive and policies and measures more effective if they are integrated within broader socioeconomic strategies aiming at sustainable development. The reason that, notwithstanding these positive opportunities internationally, co-ordinated mitigation appears to be extremely difficult, is that many financial, cultural, institutional and political barriers need to be overcome. For example, costs of mitigation may be modest at the national level, but they can be very significant for vulnerable economic sectors, which are often politically influential. However, the question does not seem to be IF mitigation action is warranted, but rather WHAT KIND of action it should be, WHEN it should be taken and by WHOM? Or: what are the most appropriate measures on the short term, taking into account the long-term risks of climate change and the many other problems competing for attention and resources?

At the same time, any package of response measures would include adaptation next to mitigation, since even aggressive mitigation would not prevent all climatic changes and associated impacts. It would also include reducing vulnerability. Many sectors (e.g. agriculture and water management) are vulnerable to natural climatic variability and extreme events. Part of the package therefore would be technologies and practices that reduce this vulnerability and increase resilience. While this would be useful from the perspective of future climatic changes, it would already be useful today to reduce vulnerability to current climate variability and to many other natural and manmade pressures on vulnerable systems. Hence, there is a strong rationale for action also in the areas of adaptation and vulnerability reduction. To some extent, ecosystems and societies will adapt naturally to climate change, but planned adaptation can facilitate this process considerably. Of course, this especially applies to human systems, but natural systems can also be 'helped' to adapt, e.g. by protective measures taking into account projected climatic changes.

Both adaptation, vulnerability reduction, and mitigation action can contribute to sustainable development, while sustainable development strategies, if appropriately designed, can help stabilize greenhouse gas concentrations and reduce the risks posed by climate change, This is the key focus of the following chapters of this book.

Box 1.8   Selected key conclusions of the Third Assessment Report
(IPCC 2001a, b, c), Working Group III: *Mitigation*

The effectiveness of climate change mitigation can be enhanced when
climate policies are integrated with the non-climate objectives:[1]

- Alternative development paths can result in very different
  greenhouse gas emissions.
- Climate change mitigation will both be affected by, and have
  impacts on, broader socioeconomic policies and trends, e.g. those
  relating to development, sustainability and equity.
- Lower emissions scenarios require different patterns of energy
  resource development.
- There is an inter-relationship between the environmental
  effectiveness of an international regime, the cost-
  effectiveness of climate policies and the equity of the
  agreement.
- The effectiveness of climate change mitigation can be enhanced
  when climate policies are integrated with the non-climate objectives
  of national and sectoral policy development and be turned into
  broad transition strategies to achieve the long-term social and
  technological changes required by both sustainable development
  and climate change mitigation.

There are many options to limit or reduce greenhouse gas emissions at the
short and long term:[1]

- Significant technical progress relevant to greenhouse gas emission
  reduction has been made since the Second Assessment
  Report (IPCC 1996a, b) in 1995 and has been faster than
  anticipated.
- Forests, agricultural lands, and other terrestrial ecosystems offer
  significant carbon mitigation potential. Although not necessarily
  permanent, conservation and sequestration of carbon may
  allow time for other options to be further developed and
  implemented.
- There is no single path to a low-emission future and countries and
  regions will have to choose their own paths. Most model results
  indicate that known technological options could achieve a broad
  range of atmospheric carbon dioxide stabilization levels (e.g. 550 or

450 ppmv or below over the next 100 years or more), but implementation would require associated socioeconomic and institutional changes.

- Social learning and innovation, and changes in institutional structure could contribute to climate change mitigation.

Costs of climate mitigation depend on many factors but in general can be considered to be modest at the aggregate level:[1]

- Some sources of greenhouse gas emissions can be limited at no or negative net social cost to the extent that policies can exploit no-regrets opportunities: removal of market imperfections, accounting for ancillary benefits and revenue recycling to achieve a double dividend.
- The cost estimates for Annex B countries to implement the Kyoto Protocol vary between studies and regions, and depend strongly upon the assumptions regarding the use of the Kyoto mechanisms, and their interactions with domestic measures: (a) Annex II countries: in the absence of emissions-trading between Annex B countries, the majority of global studies show reductions in projected GDP of about 0.2–2 per cent in 2010 for different Annex II regions – with full emissions trading between Annex B countries, the estimated reductions in 2010 are between 0.1–1.1 per cent of projected GDP; (b) economies in transition: for most of these countries, GDP effects range from negligible to a several per cent increase.
- Cost-effectiveness studies with a century timescale estimate that the costs of stabilizing carbon dioxide concentrations in the atmosphere increase as the concentration stabilization level declines. Different baselines can have a strong influence on absolute costs.
- Emission constraints in Annex I countries have well established, albeit varied, 'spillover' effects on non-Annex I countries.

But: there are many barriers to implementing the options, which however could be overcome by appropriate policies and measures:

- The successful implementation of greenhouse gas mitigation options needs to overcome many technical, economic, political, cultural, social, behavioural and/or institutional barriers that prevent the full exploitation of the technological, economic and social opportunities of these mitigation options.

- National responses to climate change can be more effective if deployed as a portfolio of policy instruments to limit or reduce greenhouse gas emissions.
- Co-ordinated actions among countries and sectors may help to reduce mitigation cost, address competitiveness concerns, potential conflicts with international trade rules, and carbon leakage. A group of countries that wants to limit its collective greenhouse gas emissions could agree to implement well-designed international instruments.
- Earlier actions, including a portfolio of emissions mitigation, technology development and reduction of scientific uncertainty, increase flexibility in moving towards stabilization of atmospheric concentrations of greenhouse gases. The desired mix of options varies with time and place.

[1] Subheadings from the authors, bullets from IPCC-WG3 SPM.

References

Arnell, N. W., Cannell, M. G. R., Hulme, M., Kovats, R. S., Mitchell, J. F. B., Nichols, R. J., Parry, M. L., Livermore, M. T. J. and White, A. (2002) The consequences of CO stabilisation for the impacts of climate change, *Climatic Change*, **53**, 143–446.

Cohen, S., Demeritt, J., Robinson, J. and Rothman, D. (1998) Climate change and sustainable development: towards dialogue, *Global Environmental Change*, **8**(4), 341–71.

Fuglestvedt, J. S., Berntsen, T. K., Godal, O., Sausen, R., Shine, K. P. and Skodvin, T. (2001) *Assessing Metrics of Climate Change: Current Methods and Future Possibilities*. CICERO Report 2001:04. Oslo.

Houghton, J. T., Callander, B. A., and Varney, S. K. eds. (1990) *Climate Change: The IPCC Scientific Assessment*. Cambridge: Cambridge University Press.

Houghton, J. T., Meira Filho, L. G., Callander, B. A., Harris, N., Kattenberg, A. and Maskell, K. eds. (1996) *Climate Change 1995: The Science of Climate Change*. Cambridge: Cambridge University Press.

Houghton, J., Ding, Y., Griggs, D. J., Noguer, M., van der Linden, P. J. and Xiaosu, D. eds. (2001) *Climate Change 2001: The Scientific Basis*. Cambridge: Cambridge University Press.

IPCC (1990) (H. T. Houghton, G. J. Jenkins and J. J. Ephraums, eds.) *Climate Change: The IPCC Assessment*. First Assessment Report of the Intergovernmental Panel on Climate Change. Cambridge: Cambridge University Press.

IPCC (1996a) (J. J. Houghton, L. G. Meiro Filho, B. A. Callander, N. Harris, A. Kattenberg and K. Maskell, eds.) *Climate Change 1995: The Science of Climate Change*. Contribution of Working Group I to the Second Assessment Report of the Intergovernmental Panel on Climate Change, Cambridge: Cambridge University Press.

IPCC (1996b) (R. T. Watson, M. C. Zinyowera and R. H. Moss, eds.) *Climate Change 1995: Impacts, Adaptation, and Mitigation of Climate Change: Scientific-Technical Analyses.* Contribution of Working Group II to the Second Assessment Report of the Intergovernmental Panel on Climate Change. Cambridge: Cambridge University Press.

IPCC (2001a) (J. T. Houghton, Y. Ding, D. J. Griggs, M. Noguer, P. J. van der Linden, X. Dai, K. Maskell and C. A. Johnson, eds.) *Climate Change 2001: The Scientific Basis.* Contribution of Working Group I to the Third Assessment Report of the Intergovernmental Panel on Climate Change. Cambridge: Cambridge University Press.

IPCC (2001b) (J. J. McCarthy, O. F. Canziani, N. A. Leary, D. J. Dokken and K. S. White, eds.) *Climate Change 2001: Impacts, Adaptation, and Vulnerability.* Contribution of Working Group II to the Third Assessment Report of the Intergovernmental Panel on Climate Change. Cambridge: Cambridge University Press.

IPCC (2001c) (B. Metz, O. Davidson, R. Swart and J. Pan, eds.) *Climate Change 2001: Mitigation.* Contribution of Working Group III to the Third Assessment Report of the Intergovernmental Panel on Climate Change. Cambridge: Cambridge University Press.

IPCC (2001d) (N. Nakicenovic and R. Swart, eds.) *Emissions Scenarios.* A Special Report of the Intergovernmental Panel on Climate Change. Cambridge: Cambridge University Press.

IPCC (2001e) (R. T. Watson, [remaining authors to be added], eds.) *Climate Change 2001: Synthesis Report.* A Contribution of Working Groups I, II, and III to the Third Assessment Report of the Intergovernmental Panel on Climate Change. Cambridge: Cambridge University Press.

Leggett, J., Pepper, W. J. and Swart, R. J. (1992) *Emissions Scenarios for IPCC : An Update.* In J. T. Houghton, B. A. Callander and S. K. T. Varney, eds., *Climate Change 1992: Supplementary Report to the IPCC.* Cambridge: Cambridge University Press.

McCarthy, J. J., Canziani, O. F., Leary, N. A., Dokken, D. J. and White, K. S. (2001) *Climate Change 2001: Impacts, Adaptation, and Vulnerability.* Cambridge: Cambridge University Press.

Mendelsohn, R. O., Morrison, W., Schlesinger, M. E. and Andronova, N. G. (2000) Country-specific market impacts of climate change. *Climatic Change,* **45**, 553–69.

Munasinghe, M. (2000) Development, equity and sustainability (DES) in the context of climate change. In R. Pachauri, K. Tanaka and T. Taniguchi, eds., *Guidance Papers on the cross cutting issues of the Third Assessment Report of the IPCC.* Geneva: Intergovernmental Panel on Climate Change.

Munasinghe, M. and Swart, R. (2000) *Climate Change and Its Linkages with Development, Equity and Sustainability.* Geneva: Intergovernmental Panel on Climate Change.

Nakicenovic, N. and Swart, R. (2000) *Special Report on Emissions Scenarios.* Cambridge: Cambridge University Press.

Nordhaus, W. D. and Boyer, J. (2000) *Warming the World: Economic Models of Climate Change.* Cambridge, MA: MIT Press.

Parry, M., Arnell, N., McMichael, T., Nicholls, R., Marttens, P., Kovats, S., Livermore, M., Rosenzweig, C., Iglesias, A., and Fisher, G. (2001) Millions at risk: defining critical climate change threats and targets. *Global Environment Change*, **11**(3), 1–3.

Smith, J. B., Schellnhuber, H-J. and Mirza, M. M. Q. (2001) Vulnerability to climate change and reasons for concern: a synthesis. In: J. J. McCarthy, O. F. Canziani, N. A. Leary, D. J. Dokken and K. S. White, eds., *Climate Change 2001: Impacts, Adaptation, and Vulnerability*. Cambridge: Cambridge University Press.

Swart, R. J., Berk, M., Janssen, M., Kreileman, E. and Leemans, R. (1998) The safe landing approach: risks and trade-offs in climate change. In: J. Alcamo, R. Leemans and E. Kreileman, eds., *Global Change Scenarios of the 21st Century: Results from the IMAGE 2.1 Model*. Oxford: Elsevier/Pergamon Press.

Tol, R. S. J. (1999) The marginal costs of greenhouse gas emissions. *The Energy Journal*, **20**(1), 61–81.

Watson, R. T., Noble, I. R., Bolin, B., Ravindranath, N. H., Verardo, D. J. and Dokken, D. J. (2000) *Land Use, Land-use Change, and Forestry*. Intergoevernmental Panel on Climate Change Special report. Cambridge: Cambridge University Press.

2

Future scenarios of development and climate change

2.1   Introduction and methodology

Climate change is a problem, which plays out over long timescales. Inertia in natural and socioeconomic systems plays an important role. If the emissions of greenhouse gases were to cease today, global temperatures would continue to increase for decades, and sea-level continue to rise for centuries. While household appliances could be replaced within years, replacement of major industrial installations and power plants and institutional structures can take decades, and changes in large-scale infrastructure, or fundamental human values, a century or more. While the evidence is mounting that climate is changing already, and impacts on natural systems have been identified, the main risks are expected to occur in the future. Exploring the options to address climate change makes the use of scenario analysis indispensable, since the future is unknowable and dependent on choices still to be made. Scenarios are images of the future and of the pathways towards such futures. They are not to be confused with predictions. Some scenarios are explorative – exploring how the future may evolve starting from the conditions of today and trends. Other scenarios are normative – describing a future to which the authors attach some value: it may be a desirable future, like a sustainable one, or it may be an undesirable one, in which the world falls into despair and chaos. It is impossible to predict the future and, hence, there is no such thing as 'business-as-usual'. In order to evaluate possible future actions in order to combat climate change in a sensible fashion, a multibaseline scenario approach is the right way to go. Response options can be explored that appear to be robust against alternative future developments.

In Section 2.2, we discuss two sets of scenarios, which have been developed in recent years. First, the Intergovernmental Panel on Climate Change (IPCC) developed the *Special Report on Emissions Scenarios* (*SRES*), (Nakicenovic & Swart 2000). The *SRES* scenarios are of the explorative kind and have been used as the basis for the assessment of climate change in the IPCC's Third Assessment Report (IPCC 2001a, b, c), and of the options to mitigate climate change sufficiently to stabilize atmospheric concentrations in accordance with Article 2 of the United Nations Framework Convention on Climate Change (UNFCCC). They do not assume any additional climate policies beyond those agreed at the time of the development of the scenarios (around 1998). We also discuss in this section the normative results of the so-called 'post-*SRES*' analysis, in which a number of modelling groups evaluated how concentrations of greenhouse gases, notably carbon dioxide, could be stabilized in the different worlds represented by the *SRES* reference scenarios. Second, the Global Scenario Group, co-ordinated by the Stockholm Environment Institute (Gallopin *et al.* 1997; Raskin *et al.* 1998) developed sets of normative scenarios, some targeting a sustainable future of the global system, including stabilization of greenhouse gas concentrations at low levels. The Global Scenario Group work complements the *SRES* analyses not only in the sense that explicit sustainability targets are set – something very difficult in the intergovernmental, IPCC, context – but also because it considers a broad set of social and environmental sustainability aspects beyond climate change. Both sets of scenarios couple a narrative description of the way the world may evolve, with a detailed quantification of the developments of the population, economy, energy system, and various other characteristics, at a regional level. In Section 1.3, the implications of the *SRES* scenarios for climate change and sea-level are presented. Although specific impact analysis on the basis of the *SRES* or Global Scenario Group scenarios has not been performed, we discuss in Section 1.4 the potential impacts of the scenarios on the basis of available information about the linkages between climatic changes and impacts, in the context of Article 2 of the UNFCCC.

2.2   Intergovernmental Panel on Climate Change scenarios

2.2.1   *Introduction and background*

How will climate develop in the future? How may human developments exacerbate or mitigate climatic changes and their impacts? To analyse such questions, long-term scenarios are needed for the release of greenhouse gases into the atmosphere. These again depend on the development of the population of the world, the size and structure of the economy, and the technologies applied to produce the goods and services needed to satisfy the demands of the population. Since such scenarios were not available in a fashion sufficiently detailed and comprehensive to allow for a comprehensive analysis of anthropogenic climate change, the IPCC has developed greenhouse gas emissions scenarios three times since its foundation in 1988. A short account of these attempts to formulate scenarios is useful to illustrate the evolution in the thinking about long-term scenarios in the climate change context. For the First Assessment Report (IPCC 1990a, b), two modelling teams – one from the USA – the Atmospheric Stabilization Framework/ASF (Lashof & Tirpak 1990) – and one from The Netherlands, using integrated assessment models – IMAGE (Rotmans *et al.* 1990; Swart *et al.* 1990) – developed four scenarios: 'business-as-usual', and three with increasing levels of climate policies. The lowest, with 'accelerated policies' was normative in the sense that it was designed to lead to stabilization of carbon dioxide concentrations at 'less than a doubling of pre-industrial concentrations'. These scenarios were criticized shortly after their publication, primarily because rapid developments in the understanding of key natural and socioeconomic processes at that time, and perceived progress in national and international climate policy development, made them outdated quickly. It was also perceived as inappropriate for the IPCC to develop normative scenarios, which were considered to be policy-prescriptive. It was increasingly recognized that a 'business-as-usual' or 'most likely' scenario is not meaningful, since societal development inherently is unpredictable, and projections or forecasts would be very speculative, the more so if the time horizon expands. Therefore in 1992, the same two groups developed a new set of six scenarios for the IPCC (Leggett *et al.* 1992) without climate policies additional to those in place at that time. These, which were more extensively reviewed than the 1990 cases, became known as the IS92 set of scenarios. Interestingly, notwithstanding the recommendation that no scenario should be singled out as the most probable one, one case – IS92a – has widely, if not almost exclusively, been used for climate and mitigation policy analysis since 1992 as 'the' reference or baseline case.

After 1992, the nature and structure of the IPCC process changed. Two of the changes that affected the emissions scenario development process were: first, all three IPCC Working Groups adopted a very strict, formalized scientific peer review procedure that previously had been used only by Working Group I on climate science, and second, there was an increasing and legitimate pressure to have a broader and more regionally balanced representation of experts in the writing teams of the IPCC reports. The 1992 IPCC scenario set was evaluated in 1994, and recommendations for new scenarios were made (Alcamo *et al.* 1994). Proposals for future work included: (a) take into account economic restructuring in the former Soviet Union and Eastern Europe; (b) evaluate the possible consequences of the General Agreement on Tariffs and Trade (GATT) negotiations; (c) explore different economic development pathways, notably those reducing regional income inequalities; (d) examine different trends in technological change, (e) and take into account the latest progress in the international climate negotiations. In 1997, a large international writing team started the development of the SRES, which developed into a very innovative process, leading to a new set of internationally agreed scenarios, comprehensively covering all greenhouse gases and addressing the concerns noted above. Some of the innovative features included: (a) openness of the process to broad participation by experts from around the world, a.o. through a web site; (b) development of narrative storylines before quantifying emissions trajectories, and (c) the diversity of participating groups, in particular the quantification of the emissions trajectories by six modelling groups from three major world regions. These innovations were instrumental in broadening the discussion, from mere quantification of emissions profiles, to a debate about plausible socioeconomic development pathways for a variety of aspects relevant to, but not limited to, climate change from various regional and disciplinary perspectives.

2.2.2   *Special Report on Emissions Scenarios: baseline scenarios and driving forces*

What could the world look like in 2100? How will people live? How much energy will they use and how will they produce it? Will there still be forest? Will there still be deep regional income differences as we see in the world today? Have they widened? Narrowed? This is the type of question addressed in the SRES. Two key variable features of the world were selected to distinguish between different development pathways: (a) the level of globalization of economies and institutions, and (b) the main goal of the development process – either an emphasis on material economic growth, or a more balanced focus including more attention to social and environmental aspects of development. In the tree shown in Figure 2.1, the branches in horizontal directions represent the former, the branches in vertical direction represent the latter. The roots feeding the tree are the driving forces of socioeconomic change: population and economic development, energy, and land use. The scenarios were only given numbers: A1, A2, B1 and B2, because of the difficulty adequately to capture the multifaceted nature of the scenarios by a short name.[1] Box 2.1 gives a short description of the scenario narratives, and

---

[1] After their publication, different users of the scenarios have given names to them to suit their needs, e.g. 'World Markets' (A1), 'Provincial Enterprise' (A2), 'Global Sustainability' (B1), and 'Local Stewardship' (B2) (Carter, Fronzek & Bärlund 2001). Although these names are more imaginative and provocative than the numbers, they do hide a number of important caveats, e.g. a 'global sustainability' scenario may be inconsistent with the absence of additional climate policies.

SRES scenarios

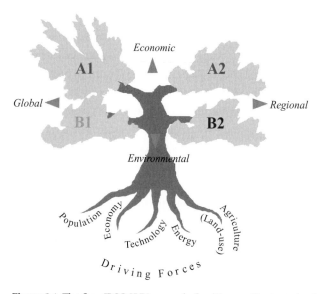

Figure 2.1. The four IPCC SRES scenario families are illustrated as branches of a two-dimensional tree. The two dimensions indicate the relative orientation of the different scenario storylines towards economic or environmental concerns and global and regional development pathways, respectively. The A1 storyline branches out into different groups of scenarios reflecting different technology development paths in one family. *Source:* Nakicenovic and Swart (2000).

Table 2.1 provides a quantitative summary. More detailed information about input assumptions and output variables can be found in Appendix 2.1

Box 2.1 Scenarios from the Intergovernmental Panel on Climate Change *Special Report on Emissions Scenarios*

A1   The A1 storyline and scenario family describes a future world of very rapid economic growth, global population that peaks in mid century and declines thereafter, and the rapid introduction of new and more efficient technologies. Major underlying themes are convergence among regions, capacity-building, and increased cultural and social interactions, with a substantial reduction in regional differences in per capita income. The A1 scenario family develops into three groups that describe alternative directions of technological change in the energy system. The three A1 groups are distinguished by their technological emphasis: fossil-intensive (A1FI), non-fossil energy sources (A1T), or a balance across all sources (A1B) (where balanced is defined as not relying too heavily on one particular energy source, on the assumption that similar improvement rates apply to all energy supply and end-use technologies).

A2   The A2 storyline and scenario family describes a very heterogeneous world. The underlying theme is self-reliance and preservation of local identities. Fertility patterns across regions converge very slowly, which results in continuously increasing population. Economic development is primarily regionally oriented and per capita economic growth and technological change more fragmented and slower than other storylines.

B1   The B1 storyline and scenario family describes a convergent world with the same global population, that peaks in mid century and declines thereafter, as in the A1 storyline, but with rapid change in economic structures toward a service and information economy, with reductions in material intensity and the introduction of clean and resource-efficient technologies. The emphasis is on global solutions to economic, social and environmental sustainability, including improved equity, but without additional climate initiatives.

B2   The B2 storyline and scenario family describes a world in which the emphasis is on local solutions to economic, social and environmental sustainability. It is a world with continuously increasing global population – at a rate lower than A2 – intermediate levels of economic development, and less rapid and more diverse technological change than in the B1 and A1 storylines. While the scenario also is oriented toward environmental protection and social equity, it focuses on local and regional levels.
An illustrative scenario was chosen for each of the six scenario groups A1B, A1FI, A1T, A2, B1, and B2. All should be considered equally sound. The *SRES* scenarios do not include additional climate initiatives, which means that no scenarios are included that explicitly assume implementation of the UNFCCC or the emissions targets of the Kyoto Protocol.

Sources: Nakicenovic and Swart (2000); Metz *et al.* (2001).

Basic driving forces: the Impact, Population, Affluence, Technology formula

The root drivers of greenhouse gas emissions are often summarised through the so-called IPAT formula: Impact (resource usage or emissions) = Population * Affluence (volume of production and consumption) * Technology (resource

Table 2.1. *Overview of main driving forces and GHG emissions at the global level in the six IPCC SRES illustrative scenarios.*
For more detail see Appendix 2.1. Source: Nakicenovic and Swart (2000).

Scenario group	1990	A1FI	A1B	A1T	A2	B1	B2
Population (billion)	5.3						
2020		7.6	7.4	7.6	8.2	7.6	7.6
2050		8.7	8.7	8.7	11.3	8.7	9.3
2100		7.1	7.1	7.0	15.1	7.0	10.4
World GDP (10^{12} 1990 US$)	21						
2020		53	56	57	41	53	51
2050		164	181	187	82	136	110
2100		525	529	550	243	328	235
Primary energy (10^{18} J yr^{-1})	351						
2020		669	711	649	595	606	566
2050		1431	1347	1213	971	813	869
2100		2073	2226	2021	1717	514	1357
Carbon dioxide (fossil fuels GtC yr^{-1})	6.0						
2020		11.2	12.1	10.0	11.0	10.0	9.0
2050		23.1	16.0	12.3	16.5	11.7	11.2
2100		30.3	13.1	4.3	28.9	5.2	13.8
Carbon dioxide (land use GtC yr^{-1})	1.1						
2020		1.5	0.5	0.3	1.2	0.6	0.0
2050		0.8	0,4	0.0	0.9	−0.4	−0.2
2100		−2.1	0.4	0.0	0.2	−1.0	−0.5
Methane (MtCH$_4$ yr^{-1})	310						
2020		416	421	415	424	377	384
2050		630	452	500	598	359	505
2100		735	289	274	889	236	597
Nitrogen oxide (MtN yr^{-1})	6.7						
2020		9.3	7.2	6.1	9.6	8.1	6.1
2050		14.5	7.4	6.1	12.0	8.3	6.3
2100		16.6	7.0	5.4	16.5	5.7	6.9
Sulphur dioxide (MtS yr^{-1})	71						
2020		87	100	60	100	75	61
2050		81	64	40	105	69	56
2100		40	28	20	60	25	48

usage or emissions per unit of production and consumption). Emissions scenario authors usually take this formula as the starting-point for their analysis. Clearly, the three factors are not independent. In the *SRES*, this has been addressed by the development of narrative storylines that would eliminate inconsistent combinations of assumptions for the three variables. For example, a combination of high affluence and a high population growth is considered to be implausible, a.o. because fertility and mortality trends depend to a large degree on the availability of education, birth control methods, health care, and other factors that are to a large extent determined by the income levels. In general, there is an inverse relationship between income and fertility rates, although this does not hold for all observed situations. Also, a combination of rapid technological improvement and low economic growth is not a likely one. In the past, technological development has gone hand in hand with wealth development and social, cultural and institutional innovation.

Population

Population projections are among the most accurate type of projections for the short to medium term because of the delays in demographic developments. For greenhouse gas emission scenarios, long-term projections are needed, and there is much less confidence in those. Therefore, and also for population, a scenario approach is appropriate. Small differences in assumptions for key indicators (e.g. fertility levels) lead to markedly different outcomes, especially beyond the first half of the twenty-first century. For example, a decrease in the long-term fertility rate of less than half a birth per woman (from 2.1 to 1.7) leads to an almost 50 per cent lower world population by 2100! Therefore, it is not surprising that population projections show large variations: from a lowest number of well below 5 billion (the current size of the population has exceeded 6 billion) to a number beyond 25 billion by 2100. The *SRES* assumes that for the affluent globalizing worlds of A1 and B1, world population would peak around the middle of the century and then decline to slightly above current levels by 2100 (see Figure 2.2). For scenario A2, with the lowest income growth and demographics more determined by slowly changing regional cultural preferences, the world population would increase to around 15 billion. In this scenario, only in Asia would there be as many people by 2100 as in the whole world in A1 and B1. For B2, in which the emphasis is on solution of local environmental and social problems, an intermediate scenario is assumed, leading to a gradual increase in world population to about 10.4 billion.

Most of the projected growth will take place in the developing regions. In terms of composition of the population, all scenarios project an ageing population. The age structure of a population and the socioeconomic development are linked. Although ageing will have implications for a wide range of issues (e.g. health care, labour force, household size, and patterns of consumer expenditures) it is unknown what the effect on economic development may be. It could be beneficial through increased savings rates, but also negative because of a reduced labour force. Similarly, the effects of another major aspect of population growth – urbanization – is not well understood at aggregate levels. Because of associated mobility

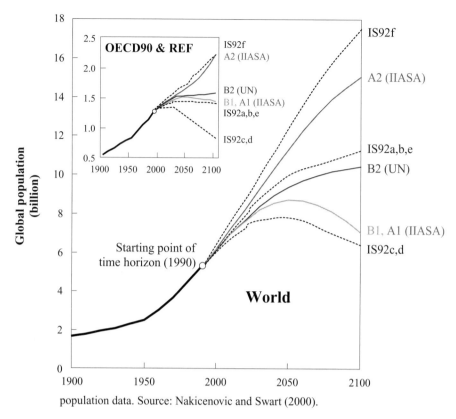

population data. Source: Nakicenovic and Swart (2000).

Figure 2.2. Population scenarios in IPCC SRES, other population projections and historical population data. *Source:* Nakicenovic and Swart (2000).

changes and income differences between urban and rural communities, there are likely to be significant effects on greenhouse gas emissions, which are not taken into account explicitly in scenario development. Another important aspect of population development is human health. While this has not been elaborated by the IPCC, others have explored the potential health implications of the *SRES* scenarios (Martens & Rotmans 2001; see also Section 2.2.4).

Economic development

The second important driving force of emissions is the growth, structure and regional distribution of the world's production and consumption system – the 'A(ffluence)' in the IPAT formula. Various indicators are used to quantify affluence, or welfare (see Box 2.2). For greenhouse gas emissions scenarios, usually gross domestic product (GDP) or GDP per capita is used as the preferred measure. The GDP quantifies only financial flows in market transactions in the formal economy and, hence, does not give an adequate indication of human welfare. Nevertheless,

Box 2.2   Indicators of human development

Gross domestic product

The GDP is used widely as one of the primary indicators of the level of human development. It represents the gross value added by all resident producers in the economy, plus any product taxes and minus any subsidies not included in the value of the products. It is calculated without making deductions for depreciation of fabricated assets or for depletion and degradation of natural resources. However, human development is not only measured in terms of economic growth, but also is concerned with creating an environment in which people can develop their full potential and lead productive, creative lives in accord with their needs and interests. Development seeks to expand the choices available to people to lead lives that they value. In this context, economic growth is only one means (albeit, a very important one), of enlarging the set of choices. However, growth alone does not guarantee human welfare (UNDP 2001). There are limitations to using only monetary measures to capture human development. They include difficulties of capturing non-marketed and non-priced goods (e.g. subsistence consumption and free social services) or other aspects important to human well-being, e.g. community resources, social relations, and the natural environment (Henninger 1998). Countries that score high on the economic dimension do not always do so on human wellbeing. Some countries experience rapid economic growth with little improvement in human welfare. Similarly, other countries enjoy relatively high levels of human welfare without high economic development by purposely allocating resources to meet the basic needs of their citizens (UNDP 1990).

To summarize, commonly used growth measures like GDP that are based on the conventional system of national accounts neglect income-distribution concerns and non-market activities (especially environmental effects). In order to overcome the latter deficiency, attempts have been made to develop a system of national accounts which will yield an environmentally adjusted GDP (or green GDP), environmentally adjusted net domestic product, and environmentally adjusted net income. In recent years, the United Nations Statistical Office (UNSO) has led the effort better to incorporate environmental concerns into a revised system of national accounts framework, through a new set of satellite accounts that encourages the compilation of relevant data on environmental and natural resources (UN 1993). These satellite accounts constitute an important step toward the eventual goal of computing the environmentally adjusted net domestic product and environmentally adjusted net income. In fact, the full incorporation of environmental effects into the system of national accounts will take many years to accomplish, even in the Organization of Economic Co-operation and Development (OECD) countries. Therefore, less comprehensive (and mainly physical) indicators of sustainability that measure air, forest, soil, and water degradation may be used in the interim (Atkinson *et al.* 1996; Munasinghe & Shearer 1995). Accounting for natural resources and environmental change will be important particularly for developing countries, with resource-dependent economies and growing environmental degradation. By expanding economic assets to include natural resources and liabilities to include stocks of pollutants, resource accounting focuses on fundamental determinants of future economic wellbeing. Assessments of resource depletion and environmental pollution can yield useful indicators of sustainability.

Human Development Index

Adding to the effort of capturing this broader concept of human development, the United Nations Development Program introduced the Human Development Index as a composite measure of human development. The Index measures the overall achievements in a country, of the basic dimensions of human development. It focuses on three essential elements of life: longevity, knowledge, and a decent standard of living. It is measured by life expectancy, educational attainment (adult literacy and combined primary, secondary and tertiary enrolment) and adjusted income per capita in purchasing power parity and US dollars, respectively. Rankings by the Human Development Index and by GDP per capita can be quite different, showing that countries do not have to wait for economic prosperity to make progress in human development.

Three more related indices have been developed by UNDP (2001):

Human Poverty Index

While the Human Development Index measures overall the achievement of human development, the Human Poverty Index reflects the distribution of progress and measures the backlog of deprivation that still exists. The Human Poverty Index measures deprivation in the same dimensions of basic human development as the Human Development Index.

Gender-related Development Index

The Gender-related Development Index measures achievements in the same dimensions and using the same indicators as the Human Development

Index, but captures inequalities in achievement between women and men. It is simply the Human Development Index adjusted downward for gender inequality. The greater the gender disparity in basic human development, the lower is a country's Gender-related Development Index compared with its Human Development Index.

The Gender Empowerment Measure
The Gender Empowerment Measure reveals whether women can take active part in economic and political life. It focuses on participation, measuring gender inequality in key areas of economic and political participation, and decision-making. It tracks the percentages of women in government, among legislators, senior officials, and managers, and among professional and technical workers – and the gender disparity in earned income, reflecting economic independence. Differing from the Gender-related Development Index, it exposes inequality in opportunities in selected areas.

Purchasing power parity
The World Bank has attempted to better capture the buying power of income by converting GDP per capita (in local currency) into US dollars at purchasing power parity, instead of simply converting local currency into dollars at the official exchange rate. Purchasing power parity rates of exchange are based on the costs of a standard basket of commodities in various countries, and are used for comparing economic measures across those nations. In the *Human Development Report* (UNDP 2001), World Bank purchasing power parities were used to provide the latest overall GDP measures covering a wide range of countries. By eliminating differences in national price levels, the method facilitates cross-country comparisons of income, poverty, inequality, and expenditure patterns, which are more realistic than those derived from using nominal foreign exchange rates. Although purchasing power parity measures have shortcomings, they are being used increasingly to supplement information based on nominal exchange rates.

Source: UNDP (2001).

it is the most widely used economic indicator and data are available for most countries in the world for a relatively long period (some decades back). A clear, internationally agreed definition of GDP makes comparisons possible, making it an easy and attractive indicator for modellers. Different from demographic development, economic development is very unpredictable, even in the short term. The assumptions made in *SRES* are summarized in Table 2.2 and in contrast with historic rates of economic growth. Economic growth is determined by demographic developments, technological innovation, social and institutional developments, trade and investment patterns, and political stability. Therefore it differs widely between different scenarios representing different socioeconomic futures.

All scenarios follow the common assumption that growth in developing countries will be faster than in developed countries, and that growth rates in developed countries will be lower than experienced in the years after the Second World War. With the exception of A2, *SRES* economic growth assumptions assume that developing countries will continue to develop faster than in the 1950–90 period. Sectors A1 and B1 explicitly explore futures in which the existing income gaps between developed and developing countries would gradually narrow, responding to criticism of the earlier IS92 scenarios. This normative assumption unfortunately does not imply that such developments towards greater income equality are considered to be more likely now. The assumptions are, however, not considered to be implausible, taking into account rates of growth differentials in the past. They show also that income convergence between the developed and the developing countries induced by rapid growth in the latter does not necessarily lead to higher greenhouse gas emissions. In A1 and B1, the relative income differences between the developed and developing countries would decrease from 15 in 1990 to less than 2 in a century, and to about 3 and 4 for A2 and B2, respectively (against 10 for IS92a). Taking into account the size of the current income gap, a further closure is considered to be implausible: because the economies of the developed and developing countries are coupled closely, it is unlikely that very high economic growth in the developing countries takes place simultaneously with very low growth in the developed countries. It should also be noted that while the relative income gap may change, this is not the case in absolute terms because of the huge size of the present gap. These normative assumptions, especially for the rapid growth with convergence assumed in scenarios A1 and B1, represent a fairly rosy picture of the economic development of the various regions. The probability of these assumptions could be questioned, e.g. because they imply long-term improvements of the labour productivity towards the high end of historical numbers, or even beyond (Nakicenovic & Swart 2000). Also, the current outlook for the assumed upturn of the African economies does not seem to be as bright as assumed in most scenarios. Such assumed improvements for the next few decades in many developing regions would be similar to the historic experience in postwar Japan and recent growth in the newly industrializing economies in East Asia. The possibilities of economic downturns such as in the former Soviet Union have not been taken into account.

Table 2.2. *Per capita GDP growth rates, historic and in IPCC SRES scenarios.*
Source: Nakicenovic and Swart (2000).

	1870–1913	1913–50	1950–80	1980–92	1990–2050				1990–2100			
					A1	A2	B1	B2	A1	A2	B1	B2
OECD90					1.6 (1.2–1.8)	1.1 (0.8–1.6)	1.5 (1.2–1.6)	1.2 (1.0–1.4)	1.6 (1.2–1.7)	1.1 (0.8–1.2)	1.2 (1.2–1.3)	1.1 (0.9–1.3)
Western Europe	1.3	0.9	3.5	1.7								
Australia, Canada, New Zealand, USA	1.8	1.6	2.2	1.3								
REF/Eastern Europe	1.0	1.2	2.9	−2.4	4.0 (2.8–4.5)	1.9 (0.5–2.2)	3.0 (2.7–3.6)	3.0 (1.9–3.3)	3.3 (2.5–3.4)	2.0 (1.5–2.0)	2.8 (2.6–2.8)	2.4 (1.6–2.6)
Asia	0.6	0.1	3.5	3.6	5.5 (5.1–5.9)	2.7 (2.7–3.6)	4.8 (4.6–5.5)	4.7 (3.3–4.8)	4.4 (3.9–4.7)	2.5 (2.4–2.9)	3.9 (3.8–4.2)	3.3 (3.1–3.4)
Africa, Middle East, Latin America (ALM)					4.0 (3.5–4.4)	1.9 (1.7–2.2)	3.5 (3.1–3.9)	2.4 (1.7–2.7)	3.3 (3.1–3.5)	1.9 (1.8–2.1)	3.0 (2.8–3.2)	2.1 (1.9–2.5)
Africa	0.5	1.0	1.8	−0.8								
Latin America	1.5	1.5	2.5	−0.6								
Industrialized countries					2.0 (1.3–2.1)	1.2 (0.8–1.8)	1.7 (1.5–1.8)	1.4 (1.1–1.6)	1.9 (1.3–2.0)	1.2 (0.9–1.4)	1.5 (1.4–1.5)	1.2 (1.0–1.4)
Developing countries					4.9 (4.4–5.2)	2.4 (2.3–3.0)	4.2 (3.9–4.8)	3.8 (2.5–3.9)	4.0 (3.6–4.1)	2.2 (2.2–2.6)	3.5 (3.4–3.7)	2.8 (2.6–3.0)
World	1.3	0.9	2.5	1.1	2.8 (2.2–2.9)	1.1 (0.7–1.5)	2.3 (2.1–2.6)	1.8 (1.1–1.9)	2.7 (2.2–2.8)	1.3 (1.3–1.5)	2.2 (2.2–2.4)	1.6 (1.4–1.7)

As to the structure of the economy, most scenarios assume a further move towards a service economy. However, the rate and extent to which this would happen varies. An important question is that of the developing countries bypassing the industrialization phase through which the industrialized countries passed, i.e. could they adopt a 'modern' economic structure, characterized by technological 'leapfrogging' and requiring rapid institutional changes (see Section 4.6.2)?

Technology development

The third key factor of the IPAT formula is the amount of resources used or emissions produced per unit of economic production. This factor is conditioned by technological development. Of the three factors in the formula, technology change is the one that can be influenced best by policies. Technological change is important for economic development in the long term, and also determines which technologies are being used to meet human needs as, for example, in the energy sector. In this way, technological change is a key factor determining greenhouse gas emissions. Through different technology assumptions, the range of possible emissions for one set of assumptions for population and economic development can be as large as the range of emissions for very different assumptions for population and economic development but similar technology development. While theoretically different technology variants could have been explored explicitly in a co-ordinated fashion for all of the SRES scenario families, this was done only for the A1 family for reasons of resource constraints. In A1, the very rapid economic growth and technology dynamism make the scenario the best choice for exploring the implication of different technology choices in a further similar socioeconomic future. Three types of choices were explored: (a) a 'fossil-intensive' future (A1FI) in which further investments in energy source development would predominantly be in fossil energy (coal, unconventional oil, and gas resources); (b) a rapid technological development case (A1T) in which rapid advances in renewable technology play a key role, and (c) a 'balanced' case (A1B) in which similar technological progress was assumed for all energy sources, leading to a more balanced energy mix. In the other families, different modelling teams explored different technology choices as well, but each team made assumptions they considered to be most consistent with the storyline.

In the SRES scenarios, the various models use different ways to simulate the development of technology in the various sectors, notably in the extraction, conversion and end-use of energy. What they have in common is that they do not assume any 'manna from the sky' in the form of efficient low-carbon, or efficient technology that is not available today (such as nuclear fusion), but only technologies that have been demonstrated at least at the stage of prototypes. Some models assume technological learning. The differences between scenarios are large. In

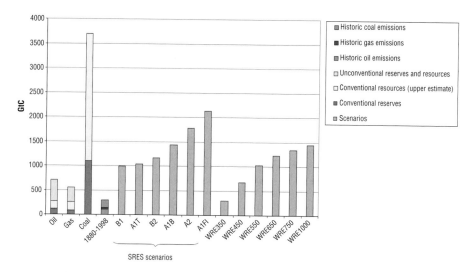

Figure 2.3. Carbon in oil, gas and coal reserves compared with historic fossil fuel carbon emissions 1860–1998, and with cumulative carbon emissions from a range of SRES and TAR stabilization scenarios up until 2100. Unconventional oil and gas includes tar sands, shale oil, other heavy oil, coal bed methane, deep geopressurized gas, gas in aquifers, etc. Gas hydrates (clathrates) that amount to an estimated 12 000 GtC are not shown. For a colour version of this graph see the original publication or website http://www.grida.no/climate/ipcc_tar/wg3/figspm2.htm. *Source:* Metz *et al.* (2001).

A1, energy-demand improvements are modest because of relatively low energy prices. In A1B, the dynamism of the scenario is assumed to lead to innovations for all energy technologies, not favouring any source above the other. For example, progress and cost reductions are assumed for both renewable technologies and for exploitation and production of natural gas. In history, there has often been a 'lock-in' into specific technologies (coal-fired steam, internal combustion engines), and therefore A1T and A1FI explore different options for further development of the energy system. Here, it is important to note that because of the large coal resources and also unconventional oil and gas resources, fossil fuel scarcity is not going to limit carbon dioxide emissions (Figure 2.3 shows the carbon locked in fossil fuel reserves and resources[2] next to the amount of carbon that would be emitted in this century in the *SRES* scenarios and a set of stabilization scenarios). However, conventional oil and gas reserves are limited, and in the coming decades there seem to be two broad directions in which the energy system could move. The first would be the fossil-intensive future, in which future investments would be in

---

[2] Reserves are the economically recoverable part of the resource base, with current energy prices and technologies.

Table 2.3. *Technology assumptions in the energy sector for the IPCC SRES scenarios.*
Source: *Nakicenovic and Swart (2000).*

Scenario	Technology improvement rates			
	Coal	Oil	Gas	Non-fossil
A1B	High	High	High	High
A2[a]	Medium	Low	Low	Low
B1[b]	Medium	Medium	Medium	Moderate-high
B2[c]	Low	Low-medium	Moderate-high	Medium
A1G[d]	Low	Very high	Very high	Medium
A1C[d]	High	Low	Low	Low
A1T	Low	High	High	Very high

[a] Technology improvement rates in the A2 scenario are heterogeneous among the world regions.

[b] B1: The assumed time-dependent learning coefficients range from 0.9 (i.e. a 10 per cent reduction in the capital:output ratio on a doubling of cumulated production) for oil, 0.9–0.95 for gas, and 0.9–0.95 for surface coalmining to about 0.94–0.96 for non-fossil electric power generation options, and 0.9–0.95 for commercial biofuels.

[c] In the specific model implementations, 'inconvenience costs' of energy-enduse, including social externalities costs, are expected to be important particularly for traditional coal technologies, e.g. underground mining, cooking with coal stoves.

[d] A1G (gas).

further developing unconventional oil and gas resources, leading to high greenhouse gas emissions (as in A1FI, A2). If climate change is to be abated in this fossil-energy based scenario, carbon could be removed from point sources (e.g. as power plants) and disposed of underground or in the deep ocean (see also Chapter 9). This last option is not explored in the *SRES*, since it covers only scenarios without climate policies (see Section 2.2.3 for stabilization scenarios based on *SRES*). A second option is investing in alternative, non-fossil technologies (as in B1 and A1T). Such technologies could include renewables, or nuclear energy.

Energy supply mix

The energy mix in the scenarios is determined by energy demand and supply dynamics. As for the demand side, in A1, low energy prices are disincentives for energy efficiency improvements, while high incomes give consumers opportunities to live a comfortable – and energy-intensive – life. Structural economic changes, rapid technology development, diffusion, and transfer lead to intensity decreases in B1. In A2 and B2, energy intensity changes are relatively slow due to slower exchange of technologies across regions and slower advances.

Table 2.3 summarizes the assumptions with respect to the extraction, distribution and conversion technologies in the various *SRES* scenarios. These assumptions

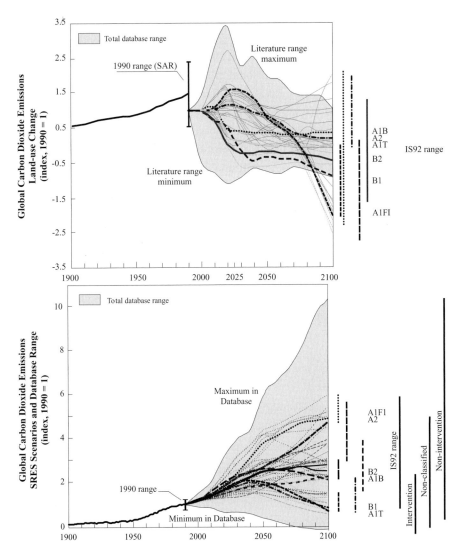

Figure 2.4. Global CO_2 emissions related to energy and industry (a) and land-use changes (b) from 1900 to 1990, and for the 40 SRES scenarios from 1990–2100, shown as an index (1990 = 1). The shaded area depicts the even wider range range of scenarios in the literature. For a colour version of the graphs see the original publication or Figure 2 on website http://www.ipcc.ch/pub/srese.pdf. *Source:* Nakicenovic and Swart (2000).

are derived from the narrative storylines and are driving the large differences in the resulting greenhouse gas emissions from the energy sector. The energy mix for the various scenarios evolves in very different ways (see Figure 2.4). Another factor explaining the differences between the energy mixes in the various scenarios is the assumed resource use. While the total resource base (see Figure 2.3) is similar for all scenarios, the usage of these resources over time is dependent on the relative prices of the various energy sources, and the technological development of exploiting them. In A1, abundant availability of all sources is derived from the narrative, including unconventional oil and gas resources, but also renewable energy. In A1B, A1T, and A1FI, the different possible developments in this rapidly changing world are explored. Also in B1, large resources are supposed to become available, but the advances in non-fossil resource development driven by learning are more rapid than in oil, gas and coal, driving the fuel mix towards more non-fossil resources. In A2, additional conventional oil and gas resources are assumed to become available only very slowly, and unconventional sources not at all. Also new renewable sources do not penetrate significantly due to slow technological development, leaving coal as the main source. In B2, oil and gas availability expand slowly, along with the development of alternative sources of energy, thereby limiting the energy options in this scenario.

As a result, all scenarios show a distinctly different development of the energy mix. While A1C (coal variant) and A2 assume a return to a coal-intensive energy sector due to resource constraints and slow development in the other sectors, in A1T, B1 and, to a lesser extent A1B, the use of non-fossil technologies increases rapidly.

Land-use change

After energy, land-use and agriculture form the second important source of greenhouse gases. The SRES scenarios explore this sector too, albeit less thoroughly than for energy due to the poor availability of information and modelling capabilities. In A1, initially the push by market growth, and an income-driven enhanced consumption of animal products, lead to an initial continuation of forest conversion. However, this trend is slowed down and, after some decades reversed by opposing trends, i.e. a rapid increase of agricultural productivity and a decreasing population size. Rapid diffusion of agricultural technologies contributes as well. In A2, the balance of these developments is different. Because of slower technological development and continuous population growth, land conversion is more sustained. In B1, trends are more favourable to reduction of land conversion than in A1 because, in addition to the rapid development and diffusion of environmentally friendly technologies enhancing productivity growth, dietary changes lead to a lower increase and eventually a decline in the demand for animal products. A key uncertainty in this scenario is to what extent the demand for renewable biomass energy may counteract these positive developments. The direction of developments in B2 is similar to A1, but for different reasons: technological development is slower and population growth higher in B2 than in A1, but increases in demands for agricultural products are lower while there is a more

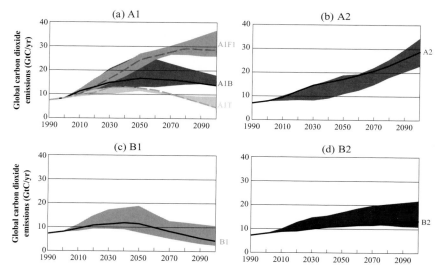

Figure 2.5. Total annual global CO_2 emissions from all sources (energy, industry and land-use change) from 1990–2100 in GtC/yr for the 4 IPCC SRES scenario families and six scenario groups. For a colour version of the graphs, see the original publication or Figure 2 on website http://www.ipcc.ch/pub/srese.pdf. *Source:* Nakicenovic and Swart (2000).

pronounced effort to protect natural resources in the more environmentally oriented B2 scenario. This also leads to continued land use changes in the first half of the century followed by the reversal of this trend in the second half.

Special Report on Emissions Scenarios greenhouse gas emissions

The *SRES* demonstrates that greenhouse gas emissions are not only greatly dependent on views of how the world develops as a result of autonomous developments, but also of human choice. Earlier scenario studies often varied variables from low to medium to high, but the *SRES* approach adds richness and consistency to the assumptions. Nevertheless, the range of possible *SRES* emissions is very large, from an eventual decline of emissions below the current level, to a fivefold increase during this century. It captures a large part of the emissions scenarios reported in the literature to date (see Figure 2.4 for carbon dioxide). The emissions ranges show that for one scenario family (i.e. one particular set of socioeconomic developments) emissions can still vary widely, depending on the technological choices made, e.g. A1 in Figure 2.5. They show also that very different worlds can have very similar emissions, e.g. the many, relatively poor, people in A2 and the fewer, but much richer, people in A1T, or the rapidly globalizing world of A1B and the local solutions scenario B2. From the perspective of sustainable development, B1 is particularly interesting, since this scenario suggests that internationally

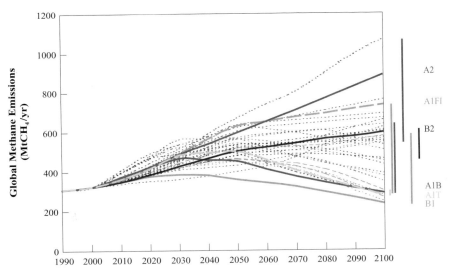

Figure 2.6. Global annual methane emissions for the IPCC SRES scenarios (in MtCH$_4$/yr). Illustrative scenarios are highlighted. *Source:* Nakicenovic and Swart (2000).

co-ordinated efforts to address social and environmental problems can lead to low and decreasing greenhouse gas emissions even without explicit climate policies. As shown in Chapter 1, this scenario may not yet lead to stabilization of carbon dioxide concentrations in the twenty-first century, but comes close to this, at levels around 550 ppmv.

A similar wide range as for carbon dioxide and a similar distribution over scenario families applies to some other greenhouse gases, notably to methane (see Figure 2.6). In scenarios like B1, A1B, and A1T, methane emissions are decreasing even without climate policies, a.o. because it makes sense economically to use landfill gas and stop natural gas leakage in oil and gas production and distribution. In addition, dietary changes play a role in B1, reducing beef and dairy consumption and, hence, emissions from cattle. Trends similar to methane are projected in the scenarios for the ozone precursors: nitrogen oxides, non-methane volatile organic compounds and, to some extent, carbon monoxide. Tropospheric ozone is an important greenhouse gas, but also causes local and regional air-quality problems. For this reason, emissions are being controlled in the scenarios even without explicit climate policies. Greenhouse gas mitigation here could be considered a side-effect of other environmental policies.

For nitrous oxide (see Figure 2.7), the range is also wide, but there is a large overlap between the various scenario families, and declining emissions in the latter half of the century are represented less prominently in the scenario

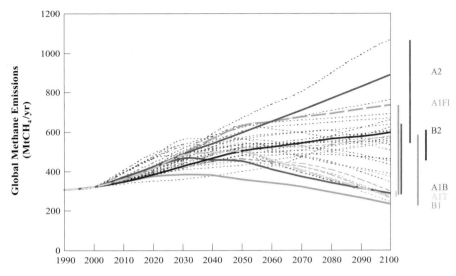

Figure 2.7. Global annual nitrous oxide emissions for the IPCC SRES scenarios (in MtN/yr). Illustrative scenarios are highlighted. *Source:* Nakicenovic and Swart (2000).

range. A reason for this is that much of the nitrous oxide emissions are tied to diverse agricultural production activities requiring nitrogen fertilizers, needed to provide an increasing amount of food and feed in all scenarios. The range of fertilizer-driven agricultural production assumptions is smaller than the range of energy production, explaining partly the narrower range for overall nitrous oxide emissions.

From the perspectives of both climate change and sustainable development, emissions of sulphur dioxide play an interesting role. The *SRES* scenarios are different from earlier scenarios in the sense that they assume an important decline of global emissions after some decades, while earlier non-climate policy scenarios assumed that sulphur emissions would continue to grow with the ongoing industrialization in developing countries. The reason for this is that, since the early 1990s, in key developing countries (e.g. China and India) government policies have successfully brought about a slowing down of sulphur emissions (see Figure 2.8), and it is now generally believed that this process will continue and spread to other developing countries. The main motivation is that these countries suffer serious problems with local air pollution and expect problems with regional pollution (acidification), while abatement technologies are available at apparently affordable prices. For climate change, this has the perverse effect that the 'masking' cooling effect of sulphur aerosols would be decreased (see Chapter 1).

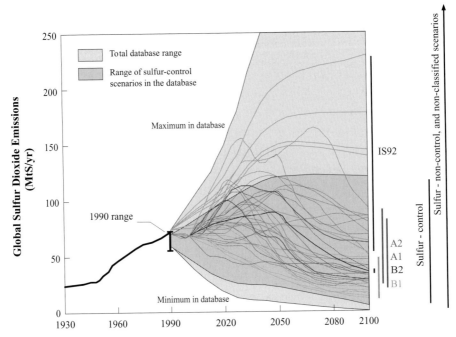

Figure 2.8. Global anthropogenic SO_2 emisisons (MtS/yr) – historical development
from 1930–1990 and in the SRES scenarios. For a colour version of the graphs see the
original publication or Figure 2 on website http://www.ipcc.ch/pub/srese.pdf. *Source:*
Nakicenovic and Swart (2000).

2.2.3   *Stabilizing atmospheric greenhouse gas concentrations in the Special Report on
         Emissions Scenarios*

As we have seen in Chapter 1, the SRES scenarios lead to concentrations of
greenhouse gases that will still be increasing by 2100, so in that sense they are not
consistent with the ultimate objective of the UNFCCC. However, in scenarios such
as B1 and A1T, carbon emissions by 2100 are on their way down, and concentration
increases are levelling off at about 550 and 600 ppmv, respectively. This implies
that, according to this IPCC scenario analysis, it would be possible for carbon
concentrations to be stabilized without 'additional climate initiatives'. This seems
surprising. However, it does not imply that nothing has to be changed and that no
environmental policies in general would be needed. On the contrary, no scenario
is policy-free, and scenarios with low greenhouse gas emissions are characterized
by vigorous policies to abate local and regional air pollution and increase resource
use efficiency that have lowering of greenhouse gas emissions as a side-effect.

One of the reasons for declining emissions in the second half of the next century
in the *SRES* A1 and B1 scenarios is the projected decrease in the population of the

world, which would peak at about 8.7 billion before declining to 7.1 billion by the end of the century.[3] More importantly, these scenarios assume a rapid vigorous move toward environmentally sound technologies to improve local and regional air quality. Not only is new technology being developed rapidly in these scenarios, but also is being adapted and transferred swiftly to all regions in this globalizing world. Also in B1, structural economic changes towards a service-oriented economy and a gradual shift towards more resource-lean lifestyles play important roles. In other words, moving society in a sustainable direction would offer possibilities eventually to reduce greenhouse gases without stringent climate policy. But if the economy as a whole would not move in such a direction, as short-term trends seem to suggest, stabilization of carbon dioxide concentrations would need additional climate policies, the more so as trends are moving more into resource and fossil-fuel-intensive directions. Evidently, also this applies for a situation in which the seriousness of climate change risk leads the world community to adopt long-term goals below 550–650 ppmv stabilization. As will be discussed later, stabilization at levels of 450 ppmv or even below may be more expensive than stabilizing at higher levels, but risks of climate impacts, especially for natural ecosystems, would be reduced.

Thus, in most future worlds, represented by the four *SRES* families of scenarios, emissions and associated concentrations of greenhouse gases continue to increase, and additional efforts are needed to achieve the UNFCCC objective of stabilizing atmospheric greenhouse gas concentrations – at levels that would prevent dangerous interference with the climate system. But how? Which technologies could be used, at which scales? Is behavioural change needed in addition to technological advances? And which policies and measures may be needed? Morita, Nakicenovic and Robinson (2000) explore this in a programme of co-ordinated modelling analyses (see also Morita *et al.* 2001). Taking the baseline assumptions for the primary driving forces (e.g. demographic and economic change from *SRES*, the matter of how emissions can be stabilized at different levels is investigated, assuming additional climate policies. The latter have to be more stringent as the baseline emissions are larger and the stabilization goal lower (see Figure 2.9). It is evident that for fossil-fuel-intensive scenarios such as A1FI and A2, it is much more difficult, if not impossible, to stabilize at low levels, while it is much easier to stabilize at such levels when baseline emissions are already low, such as in B1 or A1T. The macroeconomic costs vary similarly: GDP loss for stabilizing at 550 ppmv from a B1 baseline would be about one-tenth from the loss incurred when stabilizing at the

---

[3] The 'low' UN long-range projections indicate 5.6 billion by 2100, so the *SRES* scenarios do not seem implausibly low.

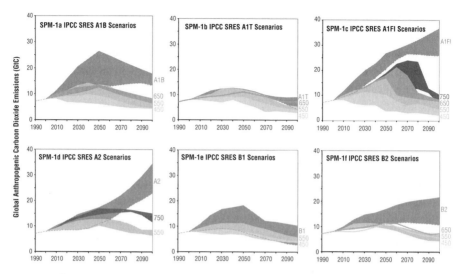

Figure 2.9. Comparison of reference scenarios (top shaded range, see also Figure 2.6) and scenarios in which CO_2 concentrations stabilize at 450, 550, 650 and 750 ppm. The lower the stabilization level and the higher the baseline emissions, the wider the gap. The difference between emissions in different scenario groups can be as large as the gap between reference and stabilization scenario within one group. For a colour version of the graphs see the original publication or Figure 2 on website http://www.grida.no/climate/ipcc_tar/wg3/figspm-1.htm. *Source:* Metz *et al.* (2001).

same level from A1, and one-twentieth as compared to A2.[4] The IPCC finds that stabilizing carbon dioxide concentrations at levels of 550, 450 ppmv, or below would be possible with known technology (albeit not necessarily economically feasible today, and with associated socioeconomic and institutional changes).

There are generic differences between the actions needed to stabilize carbon dioxide concentrations in the various baseline worlds. In general, more and more stringent measures are needed in the A1 and A2 worlds[5] than in the B1 and B2 worlds, e.g. higher carbon tax rates have to be assumed in order to meet the targets, and developing countries would have to start controlling their emissions to below the baseline earlier. Similarly, it is easier to stabilize at a particular level in A1 as compared to A2, and in B1 as compared to B2. From a intergenerational equity perspective, it should be noted that in scenarios such as A1B and A2, emissions

---

[4] Note that the direct costs per unit of reduction or the marginal costs do not necessarily follow the same pattern. Since many of the cheaper options have already been implemented for reasons unrelated to climate change in B1, such costs may actually be higher in B1, but because the necessary total emissions reduction from the baseline is so much smaller, aggregate costs are smaller.

[5] Morita, Nakicenovic and Robinson (2000) did not analyse the low-emissions A1T case.

would need to be reduced soon in order to avoid necessary rapid technological and social changes later. Across the scenarios, there are some 'robust' options for reducing carbon dioxide emissions (i.e. included in the scenarios of all or most participating modelling teams), and some options that vary across the participating modelling groups and reference–stabilization combinations. Robust technological options include energy efficiency improvements in both demand and supply sectors, and increased usage of natural gas and biomass fuels. Biological carbon sequestration (e.g. reforestation) is relevant for all scenarios, but the more so for more stringent stabilization targets. Natural gas and biomass, a.o. for combined-cycle technology, and capturing carbon from centralized fossil fuel combustion and storing it underground or in the deep ocean, can help bridge the period towards a mostly carbon-free energy economy. For the longer-term, different options were preferred by the analysts to achieve stabilization of carbon dioxide concentrations, from a renewable/solar-based hydrogen economy, in which fuel cells play a key role, to a future with an important role for nuclear energy, to an energy future in which fossil fuels would still be used, but carbon captured and disposed in underground aquifers or the deep ocean.

A more detailed description of the role of technologies in the stabilization scenarios, is provided in Table 8.4.

2.2.4   *Broadening the Special Report on Emissions Scenarios to capture issues beyond climate change*

While earlier greenhouse gas scenarios concentrated fully on the narrowly defined quantification of greenhouse gas emissions, the *SRES*, through their narrative approach, provide a basis for a wider analysis of the future in the context of sustainable development. For example, Martens and Rotmans (2001) explore the possible implications of the *SRES* worlds for water resources, biodiversity, human health, and tourism.

Fresh water resources

About one-third of the population of the world is living in countries experiencing moderate to high water stress.[6] This fraction could increase to two-thirds by 2025 (Watson *et al.*, 1998); 1.3 billion people do not have access to an adequate supply of safe drinking water, and 2 billion are without adequate sanitation facilities. Evidently, this has serious implications for health (see also below) and wider sustainable development opportunities. Also, rising water stress increasingly affects natural ecosystems, e.g. wetlands. Many of the underlying causes are

---

[6] Defined as when 20–40 per cent of available fresh water is being used already (Watson *et al.* 1998).

similar to those of greenhouse gas emissions: population growth, and the rate and type of economic growth. Hence, the qualitative *SRES* storylines provide a basis for a rough assessment of the scarcity of fresh water. Hoekstra and Huynen (2001) explore the future of the fresh water problem for the four *SRES* scenario families. They introduce the idea of a water transition. In the first stages of such a transition, increasing demand for water leads to increasing pressure on water resources, which would slow down water demand and eventually stabilize it at sustainable levels. In order to achieve this effect, increasing 'water literacy' among consumers, farmers and industrialists is needed, in addition to adequate pricing of water. Such a smooth transition depends on socioeconomic circumstances and technological developments.

In the A1 scenario, the rapid income growth and technological development would be favourable to increasing the access of people to an adequate supply of safe water. Not only can increases in demand be slowed by efficiency increases, but supplies also could be enhanced technologically, e.g. through reuse of water, salt or brackish water desalination and increased storage in reservoirs. However, a critical uncertainty would arise if the developments were not fast enough – especially in developing countries – to keep pace with increasing water demand in agriculture to boost food supplies, and in industry to support the rapid production growth. Climatic changes associated with A1 may further stress water supplies, but this will be even more so in A2. Scenario A2 paints a much more pessimistic picture in terms of fresh-water scarcity. Technological development and diffusion is slow, population growth fast and concentrated in urban areas, economic growth slow and unevenly distributed, interest in environmental problems low, and high climatic changes would affect regional water availability in some important regions. In this scenario, water crisis, especially in many developing countries, appears to be a very serious possibility, with dramatic consequences for human health and international security. The scenario that appears to come closest to the idealized water scenario sketched above is B1, which combines low population growth with rapid technological developments and diffusion in an environmentally friendly direction, and rapid exchange of technologies and practices across regions. In B1, the implicit tensions that may put water security at risk in A1 are eased. But also in B2, where the emphasis is on local solutions to environmental and social problems, adequate solutions may be found to avoid a deterioration of the water problems. The analysis shows that the kind of future socioeconomic development not only influences greenhouse gas emissions, but also offers different challenges and solutions for the pressing water scarcity problems.

Biological diversity

It is well recognized that biological diversity is seriously threatened worldwide. Conservatively, the current extinction rate for animals and plants is estimated to be 50–100 times the average 'natural' rate. Since 1600, more than 484 animal and 654 plant species become extinct (Watson *et al.* 2001). Again, main driving forces are similar to those of climate change, especially as they lead to changes in land use and land cover, and affect the quality of natural ecosystems. In simple terms, what are important here are the size reductions and fragmentation of natural ecosystems, and the changing environmental conditions of the remaining natural ecosystems. Not only are terrestrial systems at risk, but also fresh water and marine systems, e.g. 60 per cent of coral reefs are threatened.

Pressure on biodiversity resources comes also from both poverty and wealth. In the early stages of economic development, basic needs such as food and fuel are of overriding importance for the poor. Meeting these needs increases pressure on remaining natural areas, through the extraction of fuel wood and conversion of forests to agricultural lands. But also the rich contribute to the pressure: (a) directly, through conversion of land for urban development, development of infrastructure or water extraction, and (b) indirectly, through climate change, decreasing air quality and acid deposition from industry. At the same time, technological improvements, notably through the increasing productivity in agricultural production, relieve some of the pressure. These complexities make projections of biodiversity changes extremely hazardous and it is no surprise that the land-use change projections in the *SRES* scenarios (Nakicenovic & Swart, 2000) show wide variations in the model runs even for one scenario family.

Rotmans *et al.* (2001) evaluate the extent to which biodiversity risks would be exacerbated or ameliorated in the *SRES* scenarios. They relate biodiversity risk to the storylines and modelled land-use conversions (see above). Not surprisingly, B1 offers the lowest risks for biodiversity losses and A2 the highest. Both A1 and B2 have an intermediate risk, for different reasons. In most scenarios, a land-use transition is realized: after an initial continuation of forest conversion, eventually the relaxation of the pressure on the forests leads to increasing forest area, notably because of slowing population growth or even population decline, and continued productivity increases in agriculture. It can be concluded that the choice of socioeconomic development pathway impacts directly on the opportunities to preserve biological diversity. From a perspective of sustainable development, and in addition to its intrinsic value, loss of biodiversity affects the wellbeing and development possibilities of the local and global populations by reducing the availability of the goods and services provided by natural ecosystems.

Human health

What would be the health situation of the world population in the *SRES* worlds? Martens and Hilderink (2001) suggest that health conditions may move towards three 'ages': (a) of emerging infectious diseases; (b) of medical technology, and (c) of sustained health. Evidently, health developments in the real world may show different combinations of these 'ages' in different places and over time, but the kinds of development represented by the *SRES* allow for a different emphasis. The 'age of medical technology' may be typical for the A1 scenario, characterized by rapid material economic growth and technology development. Technology focused on life extension could counteract increased health risks due to lifestyle and environmental changes in this scenario. The regionalized A2 scenario, with more modest economic and technological developments and high population densities, may see a re-emergence of infectious diseases or emergence of new ones: the 'age of emerging infectious diseases'. Increased pressure on the environment and movement of people in and between regions could contribute to these undesirable developments, which is not countered by sufficiently rapid medical technology developments or reduced health risks through behavioural changes. The B1 world appears to come closest to something that may be called the 'sustainable health age', in which investments in social services and behavioural changes will reduce lifestyle-related diseases, and infectious diseases can be controlled through a globalized surveillance network. Transfer of knowledge and technology in this scenario leads to a diminishing difference between developing and developed countries. The B2 scenario, which combines regionalized development with local solutions for social and environmental problems, eventually may lead to a similar situation, but especially in the early decades (re-)emergence of infectious diseases may pose a serious risk.

Tourism

Tourism will be affected by changes in precipitation patterns, severely affecting income-generating activities (IPCC 1998). The impact in any particular area depends on whether the tourist activity is summer- or winter-oriented and, for the latter, the elevation of the area and the impact of climate on alternative activities and destinations is important. For example, in spring 1997, when conditions in alternative destinations in the Alps were poor, the number of skiers in the High Atlas in Morrocco increased significantly (Parish & Funnell 1999).

The impacts of sea-level rise on coastal tourism affects tourist development planning and execution in many cases, leading to environmental problems such as water shortages (Wong 1998). Furthermore, tourism businesses, which usually are location-specific, have a lower potential to adapt to climate change than tourists themselves (who have a wide variety of options) (Wall 1998). Small island states are especially vulnerable to changes in the tourism due to their often high economic dependence on tourism, concentration of assets and infrastructure in the coastal zone, and poor resident population.

2.3   Sustainability scenarios: the Global Scenario Group

2.3.1   Introduction

The scenarios developed by the Global Scenario Group[7] extend the range of issues beyond climate change into the domain of not only other environmental, but also economic and social dimensions. Rather than as an afterthought, as in the additions to the *SRES* work discussed above, these scenarios explore from the start how the future may evolve into unsustainable directions, but also analyse how economic, social, and environmental sustainability could be achieved as a next step in the development of mankind.

Three broad categories, or 'classes' of scenarios are developed – 'Conventional Worlds', 'Barbarization', and 'Great Transitions' – all by extrapolating a distinct set of trends visible in the world today. The intention is to transcend the often-used practice in scenario analysis of simply varying key socioeconomic variables from low, through medium, to high assumptions. 'Conventional Worlds' envision the world to evolve without major surprises, sharp discontinuities, or fundamental transformations in human civilization, at least as seen through the eyes of today's prominent priorities. Material values and the free market system dominate this class of scenarios. Two examples are elaborated: (a) a reference 'Market Forces' case, in which free markets and little attention for socioeconomic equity and environmental quality dominate, and (b) a 'Policy Reform' case, in which sustainability goals are pursued within the same dominant paradigm. By contrast, 'Barbarization' scenarios assume that social, economic, and moral underpinnings of societies deteriorate and overwhelm the coping capacity of both markets and policy reform (Gallopin *et al.* 1997). Gated communities abound, and outside the gates poverty and environmental degradation determine the lives of the poor. Again, two possible futures are elaborated. In 'Fortress Worlds' the rich and the poor live more and more in separate worlds, the rich protecting themselves effectively from the poor infringing on their affluent lifestyles. In the 'Breakdown' scenario the rich cannot sustain this position and a conflict-ridden breakdown occurs, leaving famine, wars, and loss of technological and institutional achievements in its wake. 'Great Transitions' intends to demonstrate that alternative futures are possible, or maybe even probable. Sustainability values increase in importance, stimulated

---

[7] The Global Scenario Group was established in 1995, initially in support of the Global Environmental Outlook project of the UN Environment Programme, e.g. UNEP (1999). The Global Scenario Group consists of roughly a dozen professionals from different regional and disciplinary backgrounds, with a long-term commitment to examining the requirements for sustainability and the possible strategies to achieve it. This ongoing process of global and regional scenario analysis is supported through a quantification of main trends with the help of the PoleStar accounting framework, developed at SEI.

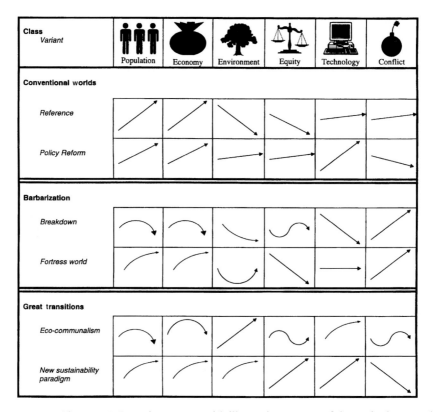

Figure 2.10. Scenario structure with illustrative patterns of change in the scenarios of the Global Scenario Group. *Source:* Raskin *et al.* (1998).

by increasingly influential nongovernmental organization networks. Technology development is rechannelled in the direction of exceedingly high resource efficiency supporting the development of a more environmentally friendly and equitable society. One example, 'Ecocommunalism', paints a world in which people return to a sustainable living patterns on a large scale and self-reliance at small scales, possibly following major disastrous events due to the pressures building up on the environment and social systems today. The New Sustainability Paradigm assumes that a paradigmatic shift gradually moves priorities away from short-term material economic growth interests towards a caring, humane, global civilization with full appreciation of long-term environmental, economic, and social stresses at all scales. The main patterns of change are summarized in Figure 2.10. In Figure 2.11, we show the relationship between economic output and population in the three classes of scenarios. Economic growth is greatest in 'Conventional Worlds', because of the economic stagnation and breakdown in the 'Barbarization' scenarios, and the shift from material economic growth to the improvement of

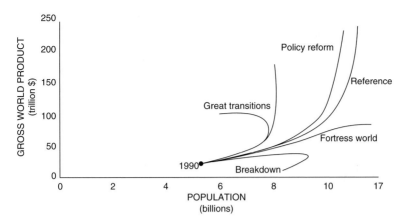

Figure 2.11. Relationship between economic output and population in the scenarios of the Global Scenario Group. *Source:* Raskin *et al.* (1998).

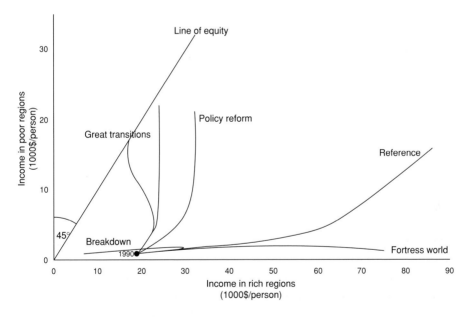

Figure 2.12. Relationship between income in rich and poor regions in the scenarios of the Global Scenario Group. *Source:* Raskin *et al.* (1998).

all aspects of welfare in the 'Great Transitions' scenarios, including strict environmental protection and sharp reductions in income inequalities between regions. In Figure 2.12, we summarize the relationship between rich and poor regions for all the scenarios. The different scenarios are discussed below in somewhat more detail.

Table 2.4. *Sustainability goals in the scenarios of the Global Scenario Group.* Source: *Raskin* et al. *(1998).*

Indicator		1995	2025	2050
Hunger	Millions of people	900	445	220
	% of 1995 value	–	50%	25%
	% of population	16%	6%	2%
Unsafe water	Millions of people	1360	680	340
	% of 1995 value	–	50%	25%
	% of population	24%	9%	4%
Illiteracy	Millions of people	1380	690	345
	% of 1995 value	–	50%	25%
	% of population	24%	9%	4%
Life expectancy	Years	66	70 in all countries	

'Two Conventional Worlds': reference/market forces and policy reform

The Global Scenario Group explores the possibilities of reaching sustainability goals within different paradigms. In the 'Conventional Worlds' scenarios, the 'Market Forces' case is contrasted with the 'Policy Reform' case, very much following the currently dominant thinking of changing unsustainable developments gradually through incremental policy interventions, steered through governments. In the world today, many trends appear to be unsustainable, millions of people live in poverty, and their numbers are increasing, the environment is degraded in many places, and the global biogeochemical cycles are being changed by human activities with often unknown risks. In the 'Reference' scenario, these trends are exacerbated. From the perspective of sustainable development, a number of sustainability goals are proposed, some of them following some of the recommendations of world conferences held during the 1990s. Main social goals imply halving the number of hungry and illiterate people and those without access to safe water by 2025, and again cutting this number in half by 2050. Decreasing income inequalities are instrumental to achieve these challenging goals. Also, environmental goals are set, including stabilizing atmospheric carbon dioxide concentrations at 450 ppmv and limiting the rate of global average temperature change to 0.10 °C per decade, sharply increasing the ecoefficiency (the amount of material resource needed for a unit of economic production) by a factor of 4 by 2025, and a factor of 10 by 2050, sharply reducing releases of persistent organic pollutants and heavy metals, reducing water scarcity, and converting net forest loss into forest increase within the next decades (see Tables 2.4 and 2.5 for details).

Table 2.5. *Environmental Indicators and targets*

Region	Indicator	1995	2025	2050
Climate				
World	Carbon dioxide concentration Warming rates Carbon dioxide emissions	380 ppmv	Stabilize at < 450 ppmv by 2100 average 0.1 °C decade 1990–2100 < 700 GIC cumulative, 1990–2100	
OECD	Carbon dioxide emissions rate	Various and rising	<65% of 1990 (<90% of 1990 by 2010)	<35% of 1980
non-OECD	Carbon dioxide emissions rates	Various and rising	Increases slowing, energy efficiency rising	Reach OECD per capita rates by 2075
Resource use				
OECD	Eco-efficiency Materials use/capita	$100 GDP/300 kg 80 t	4-fold increases ($100 GDP/75 kg) <60 t	10-fold increases ($100 GDP/30 kg) <30 t
non-OECD	Eco-efficiency Materials use/capita	Various but low Various but low	Converge toward OECD practices Converge toward OECD per capita values	
Toxics				
OECD	Releases of persistent organic pollutants & heavy metals	Various but high	<50% of 1996	<10% of 1995
non-OECD	Releases of persistent organic pollutants & heavy metals	Various and rising	Increases slowing	converge to OECD per capita values
Freshwaterr				
World	Use-to-resource radio	Various and rising	Reaches peak values	0.2–0.4 maximum (in countries >4 in 1995, less than 1995 values)
	Population in water stress	1.9 billion (34%)	Less than 3 billion (<40%)	Less than 3.5 billion, Begins decreasing (<40%)
Ecosystem pressure				
World	Deforestation Land degradation Marine overfishing	Various but high Various but high Fish stocks declining	No further deforestation No further degradation Over-fishing stopped	Net reforestation Net restoration Healthy fish-stocks

The main question here is: in a world in which values, lifestyles, economic structures, and governance systems remain essentially the same as today, can sustainability goals be met through phasing-in policies and measures worldwide and adjustments to market-driven growth? Raskin *et al.* (1998) suggest they can. But this would require a number of conditions to be met, including: (a) awareness of the risk of conventional development; (b) adequately responding to institutions; (c) successful co-ordination of policies and technological development, and (d) sufficient political will to accept the costs of carrying out the required actions. Since many key actors on the world scene today have priorities other than sustainability, these latter conditions may be difficult to meet, although not beyond imagination. However, Raskin *et al.* (1998) acknowledge these difficulties and explore alternative ways in which the, possibly even tighter, sustainability goals may be achieved (Raskin *et al.* 2001).

'Great Transitions': new sustainability paradigm

In the 'Market Forces' scenario dominated by private sector interests, social, economic, and environmental developments move the world beyond sustainable development goals. There are internal inconsistencies and tensions in this scenario which may not only make this future implausible, but also undesirable. On the one hand, globilization and rapid economic growth carry a lot of promise. But on the other hand, the side-effects in terms of income inequalities and increasing environmental pressures are likely to be unsustainable. In 'Policy Reform', governments address these unsustainable developments through policy interventions. However, in such a dynamic world, if these policy interventions would not go far enough, the sustainability goals may not be achieved, and in case they would go too far, development may be stalled. That is why 'Great Transitions' builds upon the current emergence of civil society as a third force. In this scenario, a gradually changing value and knowledge system of peoples around the globe is instrumental in facilitating the social and technological changes needed to meet the sustainable development goals. Civil society, the private sector, and governments all have equally important roles to play.

2.4   Other scenarios

Many other scenarios exploring sustainability issues have been developed, in addition to the scenarios of the IPCC and the Global Scenario Group. Although it is beyond the scope of this book to discuss such other scenarios in detail, we mention a few that interested readers may pursue further. In UNEP's Global

Environmental Outlook (UNEP 2002), the scenarios of the Global Scenario Group have been taken as the starting-point for discussion and regional elaboration, among a worldwide network of collaborating centres. The UNEP distinguishes four possible future worlds: (a) in the 'Markets First' world, current dominant trends such as globalization, liberalization, and privatization expand; (b) the 'Security First' future emphasizes heightened security concerns and leads to a polarized world; (c) in the 'Policy First' scenario, negative social and environmental side-effects of market forces are addressed through government policies, and (d) the 'Sustainability First' world is one in which the apparent insufficient effects of policies are complemented by more fundamental value changes, making development more sustainable. Regional implications of these scenarios have been analysed in detail, e.g. for Europe (UNEP/RIVM 2003). Also, sustainability scenarios have been developed for particular sectors, e.g. for the world's water problems (Gallopin & Rijsberman 2000), and for land use and food futures (Rosegrant et al. 2001). Gallopin and Rijsberman explore two types of future scenarios as alternatives for current unsustainable trends: (a) water problems would be resolved through economic and technological advances relying on the market (economics, technology and private sector), and (b) sustainable water management is achieved through a revival of human values, strengthened international co-operation, heavy emphasis on education, international mechanisms, international rules, increased solidarity and changes in lifestyles and behaviour. Similar to the analysis in the IPCC, Global Scenarios Group and Global Environmental Outlook scenarios we discussed above, the UNEP scenario analysis suggests that the most promising ways to make development more sustainable rely on changes which require gradual social and institutional transitions that go beyond reliance on purely market mechanisms.

Appendix 2.1

Table A2.1. *Overview of main primary driving forces in 1990, 2020, 2050, and 2100. Bold numbers show the value for the illustrative scenario and the numbers between brackets show the value for the range a across all forty SRES scenarios in the six scenario groups that constitute the four families. Units are given in the table. Technological change is not quantified in the table.*

Family		A1			A2	B1	B2
Scenario group	1990	A1F1	A1B	A1T	A2	B1	B2
Population (billion)							
1990	5.3						
2020		**7.6**	**7.5**	**7.6**	**8.2**	**7.6**	**7.6**
		(7.4–7.6)	(7.2–7.6)	(7.4–7.6)	(7.5–8.2)	(7.4–7.6)	(7.6–7.8)
2050		**8.7**	**8.7**	**8.7**	**11.3**	**8.7**	**9.3**
			(8.3–8.7)		(9.7–11.3)	(8.6–8.7)	(9.3–9.8)
2100		**7.1**	**7.1**	**7.0**	**15.1**	**7.0**	**10.4**
		(7.0–7.1)	(7.0–7.7)		(12.0–15.1)	(6.9–7.1)	(10.3–10.4)
World GDP (10^{12} 1990 US $ per year)							
1990	21						
2020		**53**	**56**	**57**	**41**	**53**	**51**
		(53–7)	(48–61)	(52–7)	(38–45)	(46–57)	(41–51)
2050		**164**	**181**	**187**	**82**	**136**	**110**
		(163–87)	(120–81)	(177–87)	(59–111)	(110–66)	(76–111)
2100		**525**	**529**	**550**	**243**	**328**	**235**
		(522–50)	(340–536)	(519–50)	(197–249)	(328–50)	(199–525)
Per capita income ratio: developed countries and economies in transition (Annx-I) to developing countries (non-Annex-I)							
1990	16.1						
2020		**7.5**	**6.4**	**6.2**	**9.4**	**8.4**	**7.7**
		(6.2–7.5)	(5.2–9.2)	(5.7–6.4)	(9.0–12.3)	(5.3–10.7)	(7.5–12.1)
2050		**2.8**	**2.8**	**2.8**	**6.6**	**3.6**	**4.0**
			(2.4–4.0)	(2.4–2.8)	(5.2–8.2)	(2.7–4.9)	(3.7–7.5)
2100		**1.5**	**1.6**	**1.6**	**4.2**	**1.8**	**3.0**
		(1.5–1.6)	(1.5–1.7)	(1.6–1.7)	(2.7–6.3)	(1.4–1.9)	(2.0–3.6)

For some driving forces, no range is indicated because all scenario runs have adopted exactly the same assumptions.

Table A2.2. *Overview of main secondary scenario driving forces in 1990, 2020, 2050, and 2100. Bold numbers show the value for the illustrative scenario and the numbers between brackets show the value for the range across all forty SRES scenarios in the six scenario groups that constitute the four families. Units are given in the table.*

Family		A1		A2	B1	B2
Scenario group	1990 A1F1	A1B	A1T	A2	B1	B2
Final energy intensity (10^6 J per US \$)[a]						
1990	16.7					
2020	**9.4**	**9.4**	**8.7**	**12.1**	**8.8**	**8.5**
	(8.5–9.4)	(8.1–12.0)	(7.6–8.7)	(9.3–12.4)	(6.7–11.6)	(8.5–11.8)
2050	**6.3**	**5.5**	**4.8**	**9.5**	**4.5**	**6.0**
	(5.4–6.3)	(4.4–7.2)	(4.2–4.8)	(7.0–9.5)	(3.5–6.0)	(6.0–8.1)
2100	**3.0**	**3.3**	**2.3**	**5.9**	**1.4**	**4.0**
	(2.6–3.2)	(1.6–3.3)	(1.8–2.3)	(4.4–7.3)	(1.4–2.7)	(3.7–4.6)
Primary energy (10^{18} J per year)[a]						
1990	351					
2020	**669**	**711**	**649**	**595**	**606**	**566**
	(653–752)	(573–875)	(515–649)	(485–677)	(438–774)	(506–633)
2050	**1431**	**1347**	**1213**	**971**	**813**	**869**
	(1377–601)	(968–1611)	(913–1213)	(679–1059)	(642–1090)	(679–966)
2100	**2073**	**2226**	**2021**	**1717**	**514**	**1357**
	(1988– 2737)	(1002– 2683)	(1255– 2021)	(1304– 2040)	(514–1157)	(846–1625)
Share of coal in primary energy (%)[a]						
1990	24					
2020	**29**	**23**	**23** (8–23)	**22**	**22**	**17**
	(24–42)	(8–28)		(18–34)	(8–27)	(14–31)
2050	**33**	**14**	**10** (2–13)	**30**	**21**	**10**
	(13–56)	(3–42)		(24–47)	(2–37)	(10–49)
2100	**29** (3–48)	**4**	**1** (1–3)	**53**	**8**	**22**
		(4–41)		(17–53)	(0–22)	(12–53)
Share of zero carbon in primary energy (%)[a]						
1990	18					
2020	**15**	**16**	**21**	**8**	**21**	**18**
	(10–20)	(9–26)	(15–22)	(8–16)	(7–22)	(7–18)
2050	**19**	**36**	**43**	**18**	**30**	**30**
	(16–31)	(21–40)	(39–43)	(14–29)	(18–40)	(15–30)
2100	**31**	**65**	**85**	**28**	**52**	**49**
	(30–47)	(27–75)	(64–85)	(26–37)	(33–70)	(22–49)

[a] 1990 values include non-commercial energy consistent with IPCC (2001b) but with *SRES* accounting conventions. Note that ASF, MiniCAM, and IMAGE scenarios do not consider non-commercial renewable energy. Hence, these scenarios report lower energy use.

Table A2.3. *Overview of greenhouse gas, sulphur dioxide, and ozone precursor emissions a in 1990, 2020, 2050, and 2100, and cumulative carbon dioxide emissions to 2100. Bold numbers show the value for the illustrative scenario and the numbers between brackets show the value for the range across all forty SRES scenarios in the six scenario groups that constitute the four families. Units are given in the table.*

Family		A1			A2	B1	B2
Scenario group	1990	A1FI	A1B	A1T	A2	B1	B2
Carbon dioxide, fossil fuels (Gt carbon per year)	6.0						
2020		**11.2**	**12.1**	**10.0**	**11.0**	**10.0**	**9.0**
		(10.7–14.3)	(8.7–14.7)	(8.4–10.0)	(7.9–11.3)	(7.8–13.2)	(8.5–11.5)
2050		**23.1**	**16.0**	**12.3**	**16.5**	**11.7**	**11.2**
		(20.6–26.8)	(12.7–25.7)	(10.8–12.3)	(10.5–18.2)	(8.5–17.5)	(11.2–16.4)
2100		**30.3**	**13.1**	**4.3**	**28.9**	**5.2**	**13.8**
		(27.7–36.8)	(12.9–18.4)	(4.3–9.1)	(17.6–33.4)	(3.3–13.2)	(9.3–23.1)
Carbon dioxide, land use (Gt carbon per year)	1.1						
2020		**1.5**	**0.5**	**0.3**	**1.2**	**0.6**	**0.0**
		(0.3−1.8)	(0.3−1.6)	(0.3−1.7)	(0.1−3.0)	(0.0−1.3)	(0.0−1.9)
2050		**0.8**	**0.4**	**0.0**	**0.9**	**−0.4**	**−0.2**
		(0.0−0.9)	(0.0−1.0)	(−0.2−0.5)	(0.6−0.9)	(−0.7−0.8)	(−0.2−1.2)
2100		**−2.1**	**0.4**	**0.0**	**0.0**	**−1.0**	**−0.5**
		(−2.1−0.0)	(−2.4−2.2)	(0.0−0.1)	(−0.1−2.0)	(−2.8−0.1)	(1.7−1.5)
Cumulative carbon dioxide, fossil fuels (Gt carbon)							
1990–2100		**2128**	**1437**	**1038**	**1773**	**989**	**1160**
		(2079−478)	(1220−989)	(989−1051)	(1303−860)	(794−306)	(1033−627)
Cumulative carbon dioxide, land use (Gt carbon)							
1990–2100		**61**	**62**	**31**	**89**	**−6**	**4** (4−153)
		(31−69)	(31−84)	(31−62)	(49−181)	(−22−84)	
Cumulative carbon dioxide, total (Gt carbon)							
1990–2100		**2189**	**1499**	**1068**	**1862**	**983**	**1164**
		(2127–538)	(1301–2073)	(1049–113)	(1352–938)	(772–1390)	(1164–686)

Table A2.3. (*cont.*)

Sulphur dioxide,	70.9					
(Mt sulphur per year)						
2020		87	100	60	100	75	61
		(60–134)	(62–117)	(60–101)	(66–105)	(52–112)	(48–101)
2050		81	64	40	105	69	56
		(64–139)	(47–120)	(40–64)	(78–141)	(29–69)	(42–107)
2100		40	28	20	60	25	48
		(27–83)	(26–71)	(20–27)	(60–93)	(11–25)	(33–48)
Methane (Mt per year)	310					
2020		416	421	415	424	377	384
		(415–479)	(400–444)	(415–466)	(354–493)	(377–430)	(384–469)
2050		630	452	500	598	359	505
		(511–636)	(452–636)	(492–500)	(402–671)	(359–546)	(482–536)
2100		735	289	274	889	236	597
		(289–735)	(289–640)	(274–291)	(549–1069)	(236–579)	(465–613)
Nitrous oxide (Mt per year)	6.7					
2020		9.3	7.2	6.1	9.6	8.1	6.1
		(6.1–9.3)	(6.1–9.6)	(6.1–7.8)	(6.3–12.2)	(5.8–9.5)	(6.1–11.5)
2050		14.5	7.4	6.1	12.0	8.3	6.3
		(6.3–14.5)	(6.3–14.3)	(6.1–6.7)	(6.8–13.9)	(5.6–14.8)	(6.3–13.2)
2100		16.6	7.0	5.4	16.5	5.7	6.9
		(5.9–16.6)	(5.8–17.2)	(4.8–5.4)	(8.1–19.3)	(5.3–20.2)	(6.9–18.1)
Chlorofluorocarbons,	1672					
hydrofluorocarbons,						
hydrochlorofluorocarbons						
Polyfluorocarbons						
(Mt carbon equivalent						
per year)						
2020		337	337	337	292	291	299
2050		566	566	566	312	338	346
2100		614	614	614	753	299	649
Polyfluorocarbon	32.0					
(Mt carbon equivalent						
per year)[b]						
2020		42.7	42.7	42.7	50.9	31.7	54.8
2050		88.7	88.7	88.7	92.2	42.2	106.6
2100		115.3	115.3	115.3	178.4	44.9	121.3
Sulphur hexafluoride	37.7					
(Mt carbon equivalent						
per year)[b]						
2020		47.8	47.8	47.8	63.5	37.4	54.7
2050		119.2	119.2	119.2	104.0	67.9	79.2
2100		94.6	94.6	94.6	164.6	42.6	69.0

(*cont.*)

Table A2.3. (*cont.*)

Family	A1			A2	B1	B2
Scenario group	1990 A1F1	A1B	A1T	A2	B1	B2
Carbon monoxide (Mt per year)	879					
2020	1204	1032	1147	1075	751	1022
	(1123–552)	(978–1248)	(1147–60)	(748–1100)	(751–1162)	(632–1077)
2050	2159	1214	1770	1428	471	1319
	(1619–2307)	(949–1925)	(1244–770)	(642–1585)	(471–1470)	(580–1319)
2100	2570	1663	2077	2326	363	2002
	(2298–3766)	(1080–2532)	(1520–2077)	(776–2646)	(363–1871)	(661–2002)
Non-methane volatile organic compounds (Mt per year)	139					
2020	192	222	190	179	140	180
	(178–230)	(157–222)	(188–190)	(166–205)	(140–193)	(152–180)
2050	322	279	241	225	116	217
	(256–322)	(158–301)	(206–41)	(161–242)	(116–237)	(147–217)
2100	420	194	128	342	87	170
	(167–484)	(133–552)	(114–28)	(169–342)	(58–349)	(130–304)
Nitrous oxide (Mt nitrogen per year)	30.9					
2020	50 (46–51)	46	46	50	40	43
		(46–66)	(46–49)	(42–50)	(38–59)	(38–52)
2050	95	48	61	71	39	55
	(49–95)	(48–100)	(49–61)	(50–82)	(39–72)	(42–66)
2100	110	40	28	109	19	61
	(40–151)	(40–77)	(28–40)	(71–110)	(16–35)	(34–77)

[a] The uncertainties in the *SRES* emissions for non-carbon dioxide greenhouse gases are generally greater than those for energy carbon dioxide. Therefore, the ranges of non-carbon dioxide greenhouse gas emissions provided in the Report may not reflect fully the level of uncertainty compared to carbon dioxide e.g. only a single model provided the sole value for halocarbon emissions.

[b] In the SPM the emissions of chlorofluorocarbons, hydrofluorocarbons, hydrochlorofluorocarbons, perfluorocarbons, and sulphur hexafluoride are presented as carbon-equivalent emissions. This was done by multiplying the emission by weight of each substance by its global warming potential and subsequent summation. The results were then converted from carbon dioxide-equivalents (reflected by the GWPs) into carbon-equivalent. Note that the use of GWP is less appropriate for emission profiles that span a very long period. It is used here, in the interest of readability of the SPM in preference to a more detail breakdown by the twenty-seven substances. The method here is also preferred over the even less desirable option to display weighted numbers for the aggregate categories in this table.

References

Alcamo, J., Bouwman, A., Edmonds, J., Gruebler, A., Morita T. and Sugandhy, A. (1994) An evaluation of the IPCC IS92 emissions scenarios. In J. T. Houghton, L. G. Meira Filho, J. Bruce, Hoesung Lee, B. A. Callander, E. Haites, N. Harris and K. Maskell, eds., *Climate Change*. Cambridge: Cambridge University Press.

Carter, T. R., Fronzek, S. and Bärlund, I. (2004) FINSKEN: a framework for developing consistent global change scenarios for Finland in the 21st century. *Boreal Environment Research*, **9**, 91–107.

Gallopin, G., Hammond, A., Raskin, P. and Swart, R. (1997) Branch Points: Global Scenarios and Human Choice. Boston: Global Scenario Group/Stockholm Environment Institute.

Gallopin, G. C. and Rijsberman, F. (2000). Three global water scenarios. In W. J. Cosgrove, and F. Rijsberman: *World Water Vision: Making Water Everybody's Business*. London: Earthscan.

Henninger, N. (1998) Mapping and Geographical analysis of human welfare and poverty – Review and assessment. Washington: World Resources Institute.

Hoekstra, A. and Huynen, M. (2001) Balancing the world water demand and supply. In P. Martens, and J. Rotmans, eds., *Transitions in a Globalising World*. Maastricht: International Centre for Integrative Studies, University of Maastricht.

IPCC (1990) *Climate change: The IPCC Response Strategies*. Washington: Island Press.

IPCC (1998) The Regional Impacts of Climate Change: An Assessment of Vulnerability. In R. T., Watson, M. C. Zinyowera and R. H. Moss, eds., In *Special Report of IPCC Working Group II*. Cambridge: Cambridge University Press.

Lashof, D. and Tirpak, D. (1990) *Policy Options for Stabilising Global Climate, 21P-2003*. Washington: US Environmental Protection Agency.

Leggett, J., Pepper, W. J. and Swart, R. J. (1992) Emissions scenarios for the IPCC: an update. In J. T. Houghton, B. A. Callander and S. K. Varney, eds., *IPCC climate change 1992: the supplementary report to the IPCC scientific assessment*. Cambridge: Cambridge University Press.

Martens, P. and Hilderink, H. (2001) Human health in transition: towards more disease or sustained health? In P. Martens, and J. Rotmans, eds., *Transitions in a Globalising World*. Maastricht: International Centre for Integrative Studies, University of Maastricht.

Martens, P. and Rotmans, J. eds., (2001) *Transitions in a Globalising World*. Maastricht: International Centre for Integrative Studies, University of Maastricht.

Metz, B., Davidson, O., Swart, R. and Pan, J. (2001) *Climate Change 2001: Mitigation*. Cambridge: Cambridge University Press.

Morita, T., Nakicenovic, N. and Robinson, J. (2000) Overview of mitigation scenarios for global climate stabilisation based on new IPCC emissions scenarios, *Environmental Economic and Policy Studies*, **3**, 65–88.

Morita, T., Robinson, J., Adegbulugbe, A., Alcama, J., Herbert, D., Lebre La Rovere, E., Nakicenovic, N., Pitcher, H., Raskin, P., Riahi, K., Sankovski, A., Sokolov, B., de Vries, B. and Zhou, D. (2001) Greenhouse emissions mitigation scenarios and implications. In B. Metz, O. Davidson, R. Swart and J. Pan, eds., *Climate Change 2001:*

Mitigation. Contribution of Working Group III to the Third Assessment Report of the Intergovernmental Panel on Climate Change. Cambridge: Cambridge University Press.

Munasinghe, M. and Shearer, W. eds. (1995) *Defining and Measuring Sustainability: The Biogeophysical Foundations*. Tokyo and Washington: UN University and World Bank.

Nakicenovic, N. and Swart, R. eds. (2000) *IPCC Special Report on Emissions Scenarios*. Cambridge: Cambridge University Press.

Parish, R., and Funnell, D. C. (1999) Climate change in mountain regions: some possible consequences in the Moroccan High Atlas. *Climate Change*, 9(1), 15–58.

Raskin, P., Gallopin, G., Gutman, P., Hammond A. and Swart, R. (1998) *Bending the Curve: Towards Global Sustainability*. Boston: Global Scenario Group/Stockholm Environment Institute.

Raskin, P., Gallopin, G., Gutman, P., Hammond, A., Kates R. and Swart, R. (2001) *Great Transitions: Global Scenario Group*. Boston: Stockholm Environment Institute.

Rosegrant M. W., Paisner, M. S., Meijer, S. and Witcover, J. (2001) *Global Food Projections to 2020, Emerging Trends and Alternative Futures*. Washington: IFPRI.

Rotmans, J. (1990) *IMAGE: An Integrated Model to Assess the Greenhouse Effect*. Dordrecht: Kluwer.

Rotmans, J., de Groot, D., and van Vliet, A. (2001) *Biodiversity: luxury or necessity?* In P. Martens, and J. Rotmans, eds., *Transitions in a Globalising World*. Maastricht: International Centre for Integrative Studies. Maastricht of University.

Swart, R. J., den Elzen, M. G. J. and Rotmans, J. (1990) *Emission scenarios for the Intergovernemental Panel on Climate Change, Milieu*, 3, 82–9.

UNDP (1990) *Human Development Report*. New York: Oxford University Press.

UNDP (2001) *Human Development Report*. New York: Oxford University Press.

UNEP (1999) Global environmental outlook (2000) *Millennium Report on the Environment*. Nairobi: UNEP.

UNEP (2002) *Global Environmental Outlook*. London: Earthscan.

UNEP/RIVM (2003) *Four Scenarios for Europe*. Nairobi/Bilthoven: UNEP/RIVM.

Wall, G. (1998) Implication of global climate change for tourism and recreation in wetland areas. *Climatic Change*. 40(2), 371–89.

Watson R. T., Dixon, J. A., Hamburg, S. P., Janetos, A. C. and Moss, R. H. (1998) *Protecting our Planet, Securing our Future: Linkages among Global Environmental Issues and Human Needs*. Washington: UNEP/USNASA/World Bank.

IPCC (2001) (R. T. Watson, [other editors to be added], eds.) *Climate Change 2001: Synthesis Report*. A Contribution of Working Groups I, II, and III to the Third Assessment Report of the Intergovernmental Panel on Climate Change. Cambridge: Cambridge University Press.

Wong, P. P. (1998) Coastal tourism development in Southeast Asia: relevance and lessons for coastal zone management. *Oceanand Coastal Management*, 38, 89–109.

3

Framework for making development more sustainable (MDMS): concepts and analytical tools

3.1 Preliminary ideas

World decision-makers are looking for new solutions to many critical problems, including traditional development issues (e.g. economic stagnation, persistent poverty, hunger, malnutrition, and illness), as well as newer challenges, e.g. worsening environmental degradation and accelerating globalization. One key approach that is receiving growing attention is based on the concept of sustainable development or 'development which lasts'. Following the 1992 Earth Summit in Rio de Janeiro and the adoption of the UN's Agenda 21, sustainable development has become well accepted worldwide (UN 1993; WCED 1987).

A key question for policy-makers is: how can we make development more sustainable? In order to help them address this question, analysts have a number of concepts and tools at their disposal. In this chapter, we discuss generic approaches, including the sustainable development triangle, integrative methods (e.g. optimality and durability) and other elements, e.g. indicators, cost–benefit analysis, and multicriteria analysis. Externalities, valuation techniques, and discounting are explained. In Chapter 5, we expand on the use of cost–benefit analysis for adaptation to climate change at the project and more aggregate level. In addition to cost–benefit analysis and multicriteria analysis, we also discuss other techniques in Chapter 5. The latter techniques include methods that have been applied specifically in the context of mitigation options, notably the so-called 'safe-landing' and 'tolerable windows' approaches, and cost-effectiveness analysis.

Both development and sustainable development are wide-ranging topics that have been researched thoroughly in past decades, and boast an extensive literature. The scope of this book does not permit a comprehensive review of that literature. A more fruitful and effective approach would be to pursue a more modest goal, i.e. defining a framework for the identification of those elements of sustainable development that are relevant to climate change, and to analyse the linkages between them. In this chapter, we set out such a framework, and in Chapter 4, we begin to apply it. The key issues to be investigated are: (a) the implications of climate change and climate change responses (including adaptation and mitigation) for sustainable development, and (b) the implications of development strategies and policies in both developing and developed countries for climate change and climate-change-response options, and vulnerability to climate-change impacts, including associated synergies and trade-offs.

For our purposes, sustainable development may be defined as an approach that will, *inter alia*, permit continuing improvements in the present quality of life at a lower intensity of resource use, thereby leaving behind for future generations an undiminished stock of productive assets (i.e. manufactured, natural, and social capital) that will enhance opportunities for improving their quality of life. While no universally acceptable practical definition of sustainable development exists as yet, the concept has evolved to encompass three major points of view – economic, social, and environmental – as represented by the triangle in Figure 3.1(a), (see for example, Munasinghe 1992). Each viewpoint corresponds to a domain (and system) that has its own distinct driving forces and objectives. The economy is geared mainly towards improving human welfare, primarily through increases in the consumption of goods and services. The environmental domain focuses on protection of the integrity and resilience of ecological systems. The social domain emphasizes the enrichment of human relationships and achievement of individual and group aspirations.

We have outlined in the previous chapters the unprecedented challenge that the threat of climate change poses to humanity. Nevertheless, from the developing country perspective, it is crucial to recognize also that, while climate change is important in the long run, there are a number of other sustainable development issues that affect human welfare more immediately, e.g. hunger and malnutrition, poverty, health, and pressing local environmental issues. Meanwhile, in the industrialized world, high levels of per capita consumption, material throughput, and greenhouse gas emissions threaten future sustainable development prospects, not only directly, but also indirectly by providing an inappropriate example that many developing countries seek to follow.

The wide-ranging potential interactions between climate change and sustainable development suggest that the linkages between these two topics need to be analysed critically. Accordingly, we sketch out in this chapter a transdisciplinary meta-framework (named 'sustainomics') that may be applied to the nexus of sustainable development and climate change.

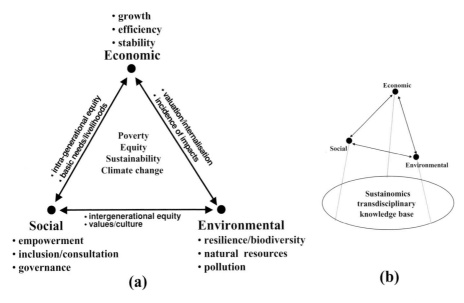

Figure 3.1. (a) Elements of sustainable development; (b) sustainable development triangle supported by the sustainomics framework *Source:* adapted from Munasinghe (*1992, 1994*)

3.1.1   Introduction to sustainomics: a framework for making development more sustainable

Although no widely accepted approach or framework exists that attempts to define, analyse, and implement sustainable development, Munasinghe (1992; 1994) proposed the term *sustainomics* to describe 'a transdisciplinary, integrative, comprehensive, balanced, heuristic and practical meta-framework for making development more sustainable'. The multiplicity and complexity of issues involved cannot be covered by a single discipline. Hitherto, multidisciplinary approaches involving teams of specialists from different disciplines have been applied to sustainable development issues. A step further has also been taken through interdisciplinary work, which seeks to break down the barriers among various disciplines. However, what is required now is a truly transdisciplinary meta-framework, that would weave the knowledge from existing disciplines into new concepts and methods that could address the many facets of sustainable development – from concept to actual practice. Thus, sustainomics would provide a comprehensive and eclectic knowledge base to support sustainable development efforts (see Figure 3.1(b)). This approach is related to, and complements, the recently propagated 'sustainability science' (Kates *et al.* 2001), which seeks to develop the knowledge base better to understand the linkages between the various natural and socioeconomic systems.

The precise definition of sustainable development remains an elusive (and perhaps unreachable) goal. Thus, one key element of the sustainomics approach is its reliance on a less ambitious, but more promising, strategy that first seeks *to make development more sustainable* (MDMS). Such an incremental (or gradient-based) method is more practical, because many unsustainable activities may be easier to recognize and eliminate. In particular, it will help us avoid sudden catastrophic ('cliff edge') outcomes.

The sustainomics approach seeks to synthesize and integrate key elements drawn from core disciplines like ecology, economics, and sociology, as well as anthropology, botany, chemistry, demography, ethics, geography, law, philosophy, physics, psychology, zoology, etc., and apply this knowledge to practical development questions. Thus, sustainomics is complementary to sustainability science, building on its results and applying them. Technological skills such as engineering, biotechnology (e.g. to enhance food production), and information technology (e.g. to improve the efficiency of natural resource use), also play key roles. Methods that bridge the economy–society–environment interfaces are especially important. For example, environmental and resource economics attempts to incorporate environmental considerations into traditional neoclassical economic analysis (Freeman 1993; Teitenberg 1992). The growing field of ecological economics goes further in combining ecological and socioeconomic methods to address environmental problems, and emphasizes the importance of key concepts like the scale of economic activities (for a good introduction, see Costanza *et al.* 1997). Newer areas related to ecological science (e.g. conservation ecology, ecosystem management, and political ecology) have originated alternative approaches to the problems of sustainability, including crucial concepts like system resilience and integrated analysis of ecosystems and human actors (Holling 1992). Recent workers in sociology have explored ideas about the integrative glue that binds societies together, while drawing attention to the concept of social capital and the importance of social inclusion (Putnam 1993). The literature on energetics and energy economics has focused on the relevance of physical laws, e.g. the first and second laws of thermodynamics (covering mass–energy balance, and entropy, respectively). This research has yielded valuable insights into how energy flows link physical, ecological, and socioeconomic systems, and analysed the limits placed on ecological and socioeconomic processes by laws governing the transformation of 'more available' (low entropy) to 'less available' (high entropy) energy (Georgescu-Roegen 1971; Hall 1995; Munasinghe 1990). Recent work on sociological economics, environmental sociology, cultural economics, economics of sociology, and sociology of the environment, also are relevant. The literature on environmental ethics has explored many issues, including the weights to be attached to values and human motivations, decision-making processes, consequences of decisions, intra- and intergenerational equity, the 'rights' of animals and the rest of nature, and human responsibility for the stewardship of the environment (Andersen 1993; Environmental Ethics; Sen 1987; Westra 1994). In the same vein, Siebhuner (2000) has recently defined '*Homo sustinens*' as a moral, co-operative individual with social, emotional and nature-related skills, as opposed to the conventional '*Homo economicus*' motivated primarily by economic self-interest and competitive instincts.

While seeking to build on such earlier work, the term '*sustainomics*' projects a more neutral image. The neologism is necessary in order to focus attention explicitly on sustainable development, and avoid the implication of any disciplinary bias or hegemony. We acknowledge that researchers in areas such as ecological economics and sustainability science have made strong efforts to address sustainable development issues. Nevertheless, many development practitioners share a concern that such disciplines focus more on the ecological and scientific aspects rather than the human dimensions of development. In fact, the substantive transdisciplinary framework underlying sustainomics could help to make the rather knowledge-oriented sustainability science more relevant and accessible to decision-makers at various levels.

The sustainomics approach should lead to the balanced and consistent treatment of the economic, social, and environmental dimensions of sustainable development (as well as other relevant disciplines and paradigms). Balance is needed also in the relative emphasis placed on traditional development versus sustainability. For example, much of the mainstream literature on sustainable development which originates in the North tends to focus on pollution, the unsustainability of growth, and population increase. These ideas have far less resonance in the South, the priorities of which include continuing development, consumption and growth, alleviation of poverty, and equity.

Many disciplines contribute to the sustainomics framework, while sustainable development itself involves every aspect of human activity, including complex interactions among socioeconomic, ecological, and physical systems. The scope of analysis needs to extend from the global to the local scale, cover time-spans extending to centuries (e.g. in the case of climate change), and deal with problems of uncertainty, irreversibility, and non-linearity. The sustainomics framework seeks to establish an overarching design for analysis and policy guidance, while the constituent components (or disciplines) provide the 'reductionist' building blocks and foundation. The heuristic element underlines the need for continuous rethinking based on new research, empirical findings, and current best practice, because reality is more complex than our models, our understanding is incomplete, and we have no consensus on the subject.

In this chapter, we identify some of the key constituent elements of sustainomics, and how they might be used to analyse issues arising from the nexus of climate change and sustainable development. Also, we illustrate some of these concepts, by applying them to case studies involving a variety of climate change problems. The current state of knowledge is inadequate to provide a comprehensive definition of sustainomics. Furthermore, sustainomics must provide a heuristic, dynamically evolving framework, in order to address rapidly changing sustainable development issues – a goal outside the scope of this book. Therefore, the intention here is to sketch out several preliminary ideas, which would serve as a starting-point, better to understand the interaction between climate change and sustainable development.

3.2   Key elements of sustainomics

Current approaches to sustainable development draw on the experience of several decades of development efforts. Historically, the development of the industrialized world focused on material production. Not surprisingly, most industrialized and developing nations have pursued the economic goal of increasing output and growth during the twentieth century. Thus, the traditional approach to development was associated strongly with economic growth, but has important social dimensions as well (see Section 3.2.4 below).

By the early 1960s, the large and growing numbers of poor in the developing world, and the lack of 'trickle-down' benefits to them, resulted in greater efforts to improve income distribution directly. The development paradigm shifted towards equitable growth, where social (distributional) objectives, especially poverty alleviation, were recognized to be as important as economic efficiency, and distinct from the latter (see Section 3.2.4 below).

Protection of the environment now has become the third major objective of sustainable development. By the early 1980s, a large body of evidence had accumulated for environmental degradation being a major barrier to development, and new proactive safeguards were introduced, e.g. the environmental assessments.

The main elements of sustainomics outlined below are as follows. Issues are analysed first through the prism of the sustainable development triangle – from the economic, social, and environmental viewpoints. Integrated analysis is facilitated by a joint optimality–durability approach. Development and growth may be restructured more sustainably, using a 'tunnelling' perspective that internalizes externalities. Sustainable development assessments are important, especially at the subnational and project levels. A mapping model facilitates the implementation of sustainable development assessments, by incorporating environmental and social assessments, into the conventional economic decision-making process, with economic valuation of environmental and social impacts serving as the bridge to cost–benefit analysis. Multicriteria analysis plays a key role in making trade-offs among diverse objectives, especially when economic valuation is difficult. The action impact matrix approach and comprehensive, multisector models (e.g. computable general equilibrium models) based on an expanded set of national accounts help to integrate economic, social, and environmental issues at the macroeconomic decision-making level (Chapter 4). Integrated assessment models play key roles in analysing global level problems, e.g. climate change. A range of sustainable development indicators help to measure progress and make choices at various levels of aggregation.

3.2.1 Economic aspects

Economic progress often is evaluated in terms of welfare (or utility) – measured as willingness to pay for goods and services consumed. Thus, many economic policies typically seek to enhance income, and induce more efficient production and consumption of (mainly marketed) goods and services. The stability of prices and employment are among other important objectives. At the same time, the equation of welfare with monetary income and consumption has been challenged for centuries by philosophers and religious leaders (Narada 1988). More recently, Maslow (1970) and others have identified hierarchies of needs that provide psychic satisfaction beyond mere goods and services.

The degree of economic efficiency is measured in relation to the ideal of Pareto optimality, which encourages actions that will improve the welfare of at least one individual without worsening the situation of anyone else. The idealized, perfectly competitive, economy is an important (Pareto-optimal) benchmark, where (efficient) market prices play key roles in both allocating productive resources to maximize output and ensuring optimal consumption choices that maximize consumer utility. If significant economic distortions are present, appropriate shadow prices need to be used. The well-known cost–benefit criterion accepts all projects, the net benefits of which are positive, i.e. aggregate benefits exceed costs (Munasinghe 1992). It is based on the weaker 'quasi' Pareto condition, which assumes that such net benefits could be redistributed from the potential gainers to the losers, so that no one is worse off than before. More generally, interpersonal comparisons of (monetized) welfare are fraught with difficulty – both within and across nations, and over time, e.g. the value of human life.

Economic sustainability

The modern concept underlying economic sustainability seeks to maximize the flow of income that could be generated, while at least maintaining the stock of assets (or capital) that yield these beneficial outputs (Maler 1990; Solow 1986). This approach is based on the pioneering work of Hicks (1946) who implied that people's maximum sustainable consumption is 'the amount that they can consume without impoverishing themselves'. Much earlier, Fisher (1906) had defined *capital* as 'a stock of instruments existing at an instant of time', and *income* as 'a stream of services flowing from this stock of wealth'. Economic efficiency continues to play a key role – in ensuring both efficient allocation of resources in production, and efficient consumption choices that maximize utility. Problems of interpretation arise in identifying the kinds of capital to be maintained (e.g. manufactured, natural, and human resource stocks, as well as social capital have been identified) and their substitutability (see Section 3.2.2). Often, it is difficult to value these assets and the services they provide, particularly in the case of ecological and social resources (Munasinghe 1992). Even key economic assets may be overlooked, for example, in informal or subsistence economies where non-market based transactions are important. The issues of uncertainty, irreversibility, and catastrophic collapse pose additional difficulties in determining dynamically efficient development paths (Pearce & Turner 1990). Many commonly used microeconomic approaches rely heavily on marginal analysis based on small perturbations, e.g. comparing incremental costs and benefits of economic activities. From the viewpoint of resilience theory (see Section 3.2.2), this type of system soon returns to its dominant stable equilibrium and thus there is little risk of instability. Such methods assume smoothly changing variables, and therefore are rather inappropriate for analysing large changes, discontinuous phenomena, and sudden transitions among multiple equilibria. More recent work (especially at the cutting edge of the economics–ecology interface) has begun to explore the behaviour of large, non-linear, dynamic and chaotic systems, as well as newer concepts, e.g. system vulnerability and resilience.

3.2.2   *Environmental aspects*

Development in the environmental sense is a rather recent concern relating to the need to manage scarce natural resources in a prudent manner, because human welfare depends ultimately on ecological services. Ignoring safe ecological limits will increase the risk of undermining long-term prospects for development. Dasgupta and Maler (1997) point out that until the 1990s, the mainstream development literature hardly mentioned the topic of environment (see, for example, Chenery & Srinivasan 1988; 1989; Dreze & Sen 1990; Stern 1989). An even more recent review paper on economic growth mentions the role of natural resources only in passing (Temple 1999). Examples of the growing literature on the theme of environment and sustainable development include books by Faucheux, Pearce and Proops (1996) describing models of sustainable development, and Munasinghe, Sunkal and de Miguel (2001) addressing the links between growth and environment.

Environmental sustainability

The environmental interpretation of sustainability focuses on the overall viability and health of ecological systems, defined in terms of a comprehensive, multiscale, dynamic, hierarchical measure of resilience, vigour, and organization (Costanza 2000). The classic definition of resilience was provided by Holling (1973) in terms of the ability of an ecosystem to persist despite external shocks. Resilience is determined by the amount of change or disruption that will cause an ecosystem to switch from one system state to another. An ecosystem state is defined by its internal structure and set of mutually reinforcing processes. Petersen, Allen and Holling (1998) argue that the resilience of a given ecosystem depends on the continuity of related ecological processes at both larger and smaller spatial scales (see Box 3.1). Further discussion of resilience may be found in Pimm (1991), and Ludwig, Walker and Holling (1997). Vigour is associated with the primary productivity of an ecosystem. It is analogous to output and growth as an indicator of dynamism in an economic system. Organization depends on both complexity and structure of an ecological or biological system. For example, a multicellular organism, such as a human being, is more highly organized (having more diverse subcomponents and interconnections among them), than a single-celled amoeba. Higher states of organization imply lower levels of entropy. Thus, the second law of thermodynamics requires that the sustainability of more complex organisms depends on the use of low-entropy energy derived from their environment, which is returned as (less useful) high-entropy energy. The ultimate source of this energy is solar radiation.

In this context, natural resource degradation, pollution, and loss of biodiversity are detrimental because they increase vulnerability, undermine system health, and reduce resilience (Munasinghe & Shearer 1995; Perrings & Opschoor 1994). The notion of a safe threshold (and the related concept of carrying capacity) are important, often to avoid catastrophic ecosystem collapse (Holling 1986). It is useful also to think of sustainability in terms of the normal functioning and longevity of a nested hierarchy of ecological and socioeconomic systems, ordered according to scale, e.g. a human community would consist of many individuals, who themselves are composed of a large number of cells (see Box 3.1 for details). Gunderson and Holling (2001) use the term 'panarchy' to denote such a hierarchy of systems and their adaptive cycles across scales. A system at a given level is able to operate in its stable (sustainable) mode, because it is protected by the slower and more conservative changes in the supersystem above it, while being simultaneously invigorated and energized by the faster cycles taking place in the subsystems below it. In brief, both conservation and continuity from above, and innovation and change from below, play a useful role in the panarchy.

Box 3.1   Spatial and temporal aspects of sustainability

An operationally useful concept of sustainability must refer to the persistence, viability, and resilience of organic or biological systems, over their 'normal' life-span (see the main text for a discussion of resilience). In this ecological context, sustainability is linked with both spatial and temporal scales, as shown in Figure 3.2. The x axis indicates lifetime in years and the y axis shows linear size (both in logarithmic scale). The central O represents an individual human being, having a longevity and size of the order of 100 years and 1.5 m, respectively. The diagonal band shows the expected or 'normal' range of life-spans for a nested hierarchy of living systems (both ecological and social), starting with single cells and culminating in the planetary ecosystem. The bandwidth accommodates the variability in organisms as well as longevity.

Environmental changes that reduce life-spans below the normal range imply that external conditions have made the systems under consideration unsustainable. In short, the regime above and to the left of the normal range denotes premature death or collapse. Examples are international military conflicts or deforestation activities affecting resilience of large ecosystems. At the same time, it is unrealistic to expect any system to last for ever. Indeed, each subsystem of a larger system (such as single cells within a multicellular organism) generally has a shorter life-span than the larger system itself. If subsystem life-spans increase too much, the system above it is likely to lose its plasticity and become 'brittle', as indicated by the region below and to the right of the normal range (Holling 1973). In other words, it is the timely death and replacement of subsystems that facilitate successful adaptation, resilience, and evolution of larger systems, e.g. like individual cells in the body that have shorter life-spans than the human being. Similarly, if the scale of a system becomes small, it may become non-viable. An example is provided by threatened species, where too-low numbers threaten the viability and resilience of the species.

Gunderson and Holling (2001) use the term 'panarchy' to denote such a nested hierarchy of systems and their adaptive cycles across scales. A system at a given level is able to operate in its stable (sustainable) mode, because it is protected by the slower and more conservative changes in the supersystem above it, while being invigorated simultaneously and energized by the faster cycles taking place in the subsystems below it. In brief, both conservation and continuity from above, and innovation and change from below, play useful roles in the panarchy.

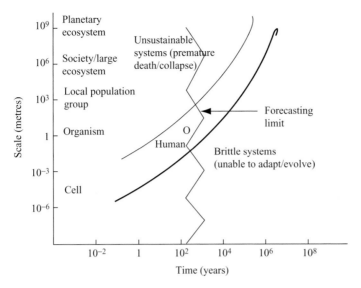

Figure 3.2. Spatial and temporal norms for sustainable biological and social systems organized as a nested hierarchy across scales. *Source:* adapted from Munasinghe (1994)

We may argue that sustainability requires biological systems to be able to enjoy a normal life-span and function normally, within the range indicated in Figure 3.2. Thus, leftward and downward movements would be especially undesirable. For example, the horizontal arrow might represent a case of infant death, indicating an unacceptable deterioration in human health and living conditions. In this specific case, extended longevity involving a greater than normal life-span would not be a matter for particular concern. On the practical side, forecasting up to a timescale of even several hundred years is rather imprecise. Thus, it is important to improve the accuracy of scientific models and data, in order to make very long-term predictions of sustainability (or its absence) more convincing, especially in the context of persuading decision-makers to spend large sums of money to reduce unsustainability. One way of dealing with uncertainty, especially if the potential risk is large, relies on a precautionary approach, i.e. avoiding unsustainable behaviour using low-cost measures, while studying the issue more carefully.

To conclude, sustainable development of ecological systems requires both adaptive capacity and opportunities for improvement. Improving adaptive capacity will increase resilience and sustainability. Expanding the set of opportunities for system improvement will give rise to development. Heuristic system behaviour facilitates learning, the testing of new processes, adaptation, and improvement.

In the context of ecological and social systems, sustainable development requires both adaptive capacity and opportunities for improvement. Improving adaptive capacity will increase resilience and sustainability. Expanding the set of opportunities for system improvement will give rise to development. Heuristic system behaviour facilitates learning, the testing of new processes, adaptation, and improvement.

Sustainable development is not necessarily synonymous with the maintenance of the ecological status quo. From an economic perspective, a coupled ecological–socioeconomic system should evolve so as to maintain a level of biodiversity that will guarantee the resilience of the ecosystems on which human consumption and production depend. Sustainable development demands compensation for the opportunities foregone by future generations because today's economic activity changes the level or composition of biodiversity in a way that will affect the flow of vital future ecological services, and narrow the options available to unborn generations. This holds true even if positive rates of economic growth indicate an increase in the instrumental (or use) values of options currently available.

3.2.3   Social aspects

Social development usually refers to improvements in both individual wellbeing and the overall welfare of society (more broadly defined), that result from increases in social capital – typically, the accumulation of capacity for individuals and groups of people to work together to achieve shared objectives. The institutional component of social capital refers mainly to the formal laws as well as traditional or informal understandings that govern behaviour, while the organizational component is embodied in the entities (both individuals and social groups) that operate within these institutional arrangements. The quantity and quality of social interactions that underlie human existence, including the level of mutual trust and extent of shared social norms, help to determine the stock of social capital. Thus, social capital tends to grow with greater use and erodes through disuse, unlike economic and environmental capital, which are depreciated or depleted by use. Furthermore, some forms of social capital may be harmful, e.g. co-operation within criminal gangs may benefit them, but impose far greater costs on the larger community.

There is an important element of equity and poverty alleviation as well (see Section 3.2.4). Thus, the social dimension of development includes protective strategies that reduce vulnerability, improve equity, and ensure that basic needs are met. Future social development will require sociopolitical institutions that can adapt to meet the challenges of modernization, which often destroy traditional coping mechanisms that have evolved in the past (especially to protect disadvantaged groups).

Social sustainability

Social sustainability is able to draw on the ideas discussed earlier regarding environmental sustainability, since habitats may be interpreted broadly to include also manmade environments such as cities and villages (UNEP, IUCN, and WWF 1991). Reducing vulnerability and maintaining the health (i.e. resilience, vigour, and organization) of social and cultural systems, and their ability to withstand shocks, is also important (Bohle, Downing & Watts 1994; Chambers 1989; Ribot, Najam & Watson 1996). Enhancing human capital (through education) and strengthening social values and institutions (e.g. trust and behavioural norms) are key aspects. Weakening social values, institutions, and equity will reduce the resilience of social systems and undermine governance. Preserving cultural diversity and cultural capital across the globe, strengthening social cohesion and networks of relationships, and reducing destructive conflicts, are integral elements of this approach. Better participation in decision-making and stakeholder empowerment within new institutional frameworks are key elements of social sustainability, especially since climate change potentially will affect all human beings, while the poorest and most marginalized groups are likely to suffer the worst impacts. An important aspect of empowerment and broader participation is subsidiarity, i.e. decentralization of decision-making to the lowest (or most local) level at which it is still effective. In summary, for both ecological and socioeconomic systems, the emphasis is on improving system health and its dynamic ability to adapt to change across a range of spatial and temporal scales, rather than the conservation of some 'ideal' static state (see also Box 3.1).

3.2.4   *Equity and poverty*

Equity and poverty are two important issues for making development more sustainable, and have social, economic, and environmental dimensions (see Figure 3.1(a)). Recent worldwide statistics are compelling. Over 2.8 billion people (almost half the global population) live on less than US $2 per day, and 1.2 billion barely survive on under US $1 per day. The top 20 percentile of the world's population consumes about 83 per cent of total output, while the bottom 20 percentile consumes only 1.4 per cent. Income disparities are worsening – the per capita ratio between the richest and the poorest 20 percentile groups was 30 : 1 in 1960 and over 80 : 1 by 1995. In poor countries, up to half the children under 5 years of age are malnourished, whereas the corresponding figure in rich countries is less than 5 per cent.

Equity is an ethical and usually people-oriented concept with primarily social, and some economic and environmental, dimensions. It focuses on the basic fairness of both the processes and outcomes of decision-making. The equity of any action may be assessed in terms of a number of generic approaches, including parity, proportionality, priority, utilitarianism, and Rawlsian distributive justice. For example, Rawls (1971) stated that 'Justice is the first virtue of social institutions, as truth is of systems of thought.' Societies normally seek to achieve equity by balancing and combining several of these criteria.

Poverty alleviation, improved income distribution and intragenerational (or spatial) equity are key aspects of economic policies seeking to increase overall human welfare (Sen 1981; 1984). Brown (1998) points out shortcomings in utilitarianism, which underlies much of the economic approach to equity. Broadly speaking, economic efficiency provides guidance on producing and consuming goods and services more efficiently, but is unable to provide a means of choosing (from a social perspective) among alternative patterns of consumption that are efficient. Equity principles provide better tools for making judgements about such choices.

Social equity is linked also to sustainability, because highly skewed or unfair distributions of income and social benefits are less likely to be acceptable or lasting in the long run. Equity is likely to be strengthened by enhancing pluralism and grass-roots participation in decision-making, as well as by empowering disadvantaged groups (defined by income, gender, ethnicity, religion, caste, etc.), (Rayner & Malone 1998). In the long term, considerations involving intergenerational equity and safeguarding the rights of future generations are key factors. In particular, the economic discount rate plays a key role with respect to both equity and efficiency aspects (Arrow *et al.* 1995). Further details of equity–efficiency interactions that need to be reconciled in order to make development more sustainable are reviewed in Box 3.2.

3.3   Integration of economic, social, and environmental considerations

As a prelude to integration, it is useful to compare the concepts of ecological, social, and economic sustainability. One useful idea is that of the maintenance of the set of opportunities, as opposed to the preservation of the value of the asset base (Githinji & Perrings 1992). In fact, if preferences and technology vary through successive generations, merely preserving a constant value of the asset base becomes less meaningful. By concentrating on the size of the opportunity set, the importance of biodiversity conservation becomes more evident, for the sustainability of an ecosystem. The preservation of biodiversity allows the system to retain resilience by protecting it from external shocks, in the same manner that

Box 3.2   Interactions between social equity and economic efficiency

Conflicts between economic efficiency and equity may arise due to assumptions about the definition, comparison, and aggregation of the welfare of different individuals or nations. For example, efficiency often implies maximization of output subject to resource constraints. The common assumption is that increases in average income per capita will make most or all individuals better off. However, this approach potentially can result in a less equitable income distribution. Overall welfare could drop, depending on how welfare is defined in relation to the distribution of income. Conversely, total welfare might increase if policies and institutions can ensure appropriate resource transfers – typically from the rich to the poor.

In the same context, aggregating and comparing welfare across different countries is a disputable issue. Gross national product (GNP) is simply a measure of the total measurable economic output of a country, and does not represent welfare directly. Aggregating GNP across nations is not necessarily a valid measure of global welfare. Box 2.2 in Chapter 2 gives various indicators that could be used to quantify welfare. However, national economic policies frequently focus more on the growth of GNP rather than its distribution, indirectly implying that additional wealth is equally valuable to rich and poor alike, or that there are mechanisms to redistribute wealth in a way that satisfies equity goals. Attempts have been made to incorporate equity considerations within a purely economic framework, by the weighting of costs and benefits so as to give preference to the poor. Although systematic procedures exist for determining such weights, often the element of arbitrariness in assigning weights has caused many practical problems.

At the same time, it should be recognized that all decision-making procedures do assign weights (arbitrarily or otherwise). For example, progressive personal income taxes are designed to take proportionately more from the rich. On the other hand, traditional cost–benefit analysis based on economic efficiency (which seeks to maximize net benefits) assigns the same weight of unity to all monetary costs and benefits, irrespective of income levels. More pragmatically, in most countries the tension between economic efficiency and equity is resolved by keeping the two approaches separate, e.g. by maintaining a balance between maximizing GNP and establishing institutions and processes charged with redistribution, social protection, and provision of various social goods to meet basic needs. The interplay of equity and efficiency at the international level is illustrated later, in the climate change case study (Chapter 4).

Equity in the environmental sense has received more attention recently, because of the disproportionately greater environmental damages suffered by disadvantaged groups. In the same vein, poverty alleviation efforts (which traditionally focused on raising monetary incomes), are being broadened to address the degraded environmental and social conditions facing the poor.

In summary, both equity and poverty have not only economic, but also social and environmental, dimensions, and therefore they need to be assessed using a comprehensive set of indicators (rather than income distribution alone). From an economic policy perspective, emphasis needs to be placed on expanding employment and gainful opportunities for poor people through growth, improving access to markets, and increasing both assets and education. Social policies would focus on empowerment and inclusion, by making institutions more responsive to the poor, and removing barriers that exclude disadvantaged groups. Environmentally related measures to help poor people might seek to reduce their vulnerability to disasters and extreme weather events, crop failures, loss of employment, sickness, economic shocks, etc. Thus, an important objective of poverty alleviation is to provide poor people with assets (e.g. enhanced physical, human and financial resources) that will reduce their vulnerability. Such assets increase the capacity for both coping (i.e. making short-run changes) and adapting (i.e. making permanent adjustments) to external shocks (Moser 1998). The foregoing ideas merge quite naturally with the sustainable livelihoods approach, which focuses on access to portfolios of assets (social, natural, and manufactured), the capacity to withstand shocks, gainful employment, and social processes, within a community or individual-oriented context.

An even broader non-anthropocentric approach to equity involves the concept of fairness in the treatment of non-human forms of life or even inanimate nature. One view asserts that humans have the responsibility of prudent 'stewardship' (or 'trusteeship') over nature, which goes beyond mere rights of usage (see, for example, Brown 1998).

preservation of the capital stock protects economic assets for future consumption. Differences emerge because economists tend to rely on the Hicks–Lindahl income test, in which a society that consumes its fixed capital without replacement is considered not sustainable, whereas by using an ecological approach, unsustainability could arise from a loss of resilience due to a reduction in the self-organization of the system, without reference to any change in productivity. In the case of social systems, resilience depends to a certain extent on the capacity of human societies to adapt and continue functioning in the face of stress and shocks. Thus, linkages between sociocultural and ecological sustainability emerge through the organizational similarities between human societies and ecological systems, and the parallels between biodiversity and cultural diversity. From a longer-term perspective, the concept of co-evolution of social, economic and ecological systems provides useful insights into the harmonious integration of the various elements of sustainable development (see Figure 3.1(a)), (Costanza *et al.* 1997; Munasinghe 1994).

One may conclude that the exact definition of sustainable development paths is likely to be extremely difficult at this stage, and may be considered a long-run or ideal objective. However, a more promising and practical shorter-run goal that is consistent with the sustainomics approach to make development more sustainable is to seek strategies that might make future development prospects more sustainable. In such an approach, one key step would be to begin by eliminating the many unsustainable activities that are readily identifiable.

It is important and possible to integrate and reconcile the economic, social, and environmental aspects in a holistic and balanced manner. Economic analysis has a special role in contemporary national policy-making, since some of the most important decisions fall within the economic domain. While mainstream economics used for practical policy-making often has ignored many crucial aspects of the environmental and social dimensions of sustainable development, there is a small but growing body of literature that seeks to address such shortcomings (see, for example, recent issues of the journals *Ecological Economics* and *Conservation Ecology* (published on the Internet)).

Two broad approaches are relevant for integrating the economic, social, and environmental dimensions of sustainable development. They are distinguished by the degree to which the concepts of *optimality* and *durability* are emphasized. While there are overlaps between the two approaches, the main thrust is somewhat different in each case. Uncertainty often plays a key role in determining which approach would be preferred. For example, relatively steady and well-ordered conditions may encourage optimizing behaviour that attempts to control and even fine-tune outcomes, whereas a subsistence farmer facing chaotic and unpredictable circumstances might opt for a more durable response that simply enhances survival prospects.

3.3.1   *Optimality*

The optimality-based approach has been widely used in economic analysis generally to maximize welfare (or utility), subject to the requirement that the stock of productive assets (or welfare itself) is non-decreasing in the long term. This assumption is common to most sustainable economic growth models – for useful reviews, see Islam (2001) and Pezzey (1992). The essence of the approach is illustrated by the simple example of maximization of the flow of aggregate welfare, W, cumulatively discounted over infinite time, t, as represented by the expression

$$\max \int_0^\infty W(C,Z) \cdot e^{-rt} dt.$$

Here, W is a function of C (the consumption rate), and Z (a set of other relevant variables), while r is the discount rate. Further side constraints may be imposed to satisfy sustainability needs, e.g. non-decreasing stocks of productive assets (including natural resources).

Some ecological models also optimize variables such as energy use, nutrient flow, or biomass production, giving more weight to system vigour as a measure of sustainability. In economic models, utility often is measured mainly in terms of the net benefits of economic activities, i.e. the benefits derived from development activities minus the costs incurred to carry out those actions (for more details about valuation, see Box 3.4 below, and Freeman (1993) or Munasinghe 1992). More sophisticated economic optimization approaches seek to include environmental and social variables, e.g. by attempting to value environmental externalities, system resilience, etc. However, given the difficulties of quantifying and valuing many such 'non-economic' assets, the costs and benefits associated with market-based activities tend to dominate in most economic optimization models.

Basically, the optimal growth path maximizes economic output, while the sustainability requirement is met (within this framework) by ensuring non-decreasing stocks of assets (or capital). Some analysts support a 'strong sustainability' constraint, which requires the separate preservation of each category of critical asset (e.g. manufactured, natural, sociocultural and human capital), assuming that they are complements rather than substitutes. One version of this rule might correspond roughly to maximizing economic output, subject to side constraints on environmental and social variables that are deemed critical for sustainability, e.g. biodiversity loss or meeting the basic needs of the poor. Other researchers have argued in favour of 'weak sustainability', which seeks to maintain the aggregate monetary value of the total stock of assets, assuming that the various asset types may be valued and that there is some degree of substitutability among them (see, for example, Nordhaus and Tobin 1972).

Side constraints are often necessary, because the underlying basis of economic valuation, optimization and efficient use of resources may not be applied easily to ecological objectives such as protecting biodiversity and improving resilience, or to social goals such as promoting equity, public participation and empowerment. Thus, such environmental and social variables cannot easily be combined into a single valued objective function with other measures of economic costs and benefits (see Sections 3.4.2 and 3.4.3 below). Moreover, the price system (which has time lags) might fail to anticipate reliably irreversible environmental and social harm, and non-linear system responses that could lead to catastrophic collapse.

Nevertheless, even if it may be difficult to do so, in principle the optimality approach could incorporate multiple objectives, thereby making it relevant and applicable within the holistic sustainomics approach. For example, non-economic indicators of environmental and social status could be included (e.g. area under forest cover) and incidence of conflict (see, for example, Hanna and Munasinghe 1995; Munasinghe and Shearer 1995; UNDP 1998; World Bank 1998). The constraints on critical environmental and social indicators are proxies representing safe thresholds, and help to maintain the viability of those systems. In this context, techniques like multicriteria analysis may be required, to facilitate trade-offs among a variety of non-commensurable variables and objectives (see, for example, Meier and Munasinghe 1994). Risk and uncertainty will also necessitate the use of decision-analysis tools (for a concise review of climate change decision-making frameworks, see Toth 1999). Recent work has underlined the social dimension of decision science, by pointing out that risk perceptions are subjective and depend on the risk measures used, as well as other factors, e.g. ethnocultural background, socioeconomic status, and gender (Bennet 2000).

3.3.2   *Durability*

The second broad integrative approach would focus primarily on sustaining the quality of life, e.g. by satisfying environmental, social, and economic sustainability requirements. Such a framework favours 'durable' development paths that permit growth, but are not necessarily economically optimal. There is more willingness to trade-off some economic optimality for the sake of greater safety, in order to stay within critical environmental and social limits – especially among increasingly risk-averse and vulnerable societies or individuals who face chaotic and unpredictable conditions (see the discussion on the precautionary principle in Section 4.2). The economic constraint might be framed in terms of maintaining consumption levels (from the durability perspective, usually defined broadly to include environmental services, leisure, and other 'non-economic' benefits) – i.e. per capita consumption that never falls below some minimum level, or is non-declining. The environmental and social sustainability requirements may be expressed in terms of indicators of 'state' that seek to measure the durability or health (resilience, vigour, and organization) of complex ecological and socioeconomic systems. As an illustrative example, consider a simple durability index, D, for an ecosystem measured in terms of its expected life-span (in a healthy state), as a fraction of the normal life-span (see also Box 3.1). We might specify: $D = D(R,V,O,S)$ to indicate the dependence of durability on resilience, R, vigour, V, organization, O, and the state of the external environment, S, especially in relation to potentially damaging shocks. There is the likelihood of further interaction here due to linkages between the sustainability of social and ecological systems, e.g. social disruption and conflict could exacerbate damage to ecosystems, and vice versa. For example, longstanding social norms in many traditional societies have helped to protect the environment (Colding & Folke 1997).

Durability encourages a holistic systemic viewpoint, which is important in sustainomics analysis. The self-organizing and internal structure of ecological and socioeconomic systems makes 'the whole more durable (and valuable) than the sum of the parts'. A narrow definition of efficiency based on marginal analysis of individual components may be misleading (Schutz 1999). For example, it is more difficult to value the integrated functional diversity in a forest ecosystem than the individual species of trees and animals. Therefore, the former is more likely to fall victim to market failure (as an externality). Furthermore, even where correct environmental shadow prices prevail, some analysts point out that cost minimization could lead to homogenization and consequent reductions in system diversity (Daly & Cobb 1989; Perrings, Maler & Folke, 1995). Systems analysis also helps to identify the benefits of co-operative structures and behaviour, which a more partial analysis may neglect.

In the durability approach, constraints based on sustainability could be represented also by the approach discussed earlier, which focuses on maintaining stocks of assets. Here, the various forms of capital are viewed as bulwarks decreasing vulnerability to external shocks and reducing irreversible harm, rather than mere accumulations of assets that produce economic outputs. System resilience, vigour, organization, and ability to adapt will depend dynamically on the capital endowment as well as the magnitude and rate of change of a shock.

3.3.3   *Complementarity and convergence of optimal and durable approaches*

National economic management often provides good examples of how the two approaches complement one another. For example, economywide policies involving both fiscal and monetary measures (e.g. taxes, subsidies, interest rates, and foreign exchange rates) might be optimized on the basis of quantitative macroeconomic models. Nevertheless, decision-makers inevitably modify these economically 'optimal' policies before implementing them, to take into account other sociopolitical considerations based more on durability (such as protection of the poor, regional factors, etc.), which facilitate governance and stability. The determination of an appropriate target trajectory for future global greenhouse gas emissions (and corresponding target greenhouse gas concentration) provides another useful illustration of the interplay of the durability and optimality approaches (for details see IPCC 1996 and Munasinghe 1998, and the case study in Chapter 4).

The practical potential for convergence of the two approaches may be realized in several ways. First, wastes ought to be generated at rates less than, or equal to, the assimilative capacity of the environment, e.g. emissions of greenhouse gases and ozone-depleting substances into the global atmosphere. Second, renewable resources, especially if they are scarce, should be utilized at rates less than, or equal to, the natural rate of regeneration. Third, non-renewable resource use should be managed in relation to the substitutability between these resources and technological progress. Both wastes and natural resource input use might be minimized by moving from the linear throughput to the closed-loop mode. Thus, factory complexes could be designed in clusters – based on the industrial ecology concept – to maximize the circular flow of materials and recycling of wastes among plants. Finally, inter- and intragenerational equity (especially poverty alleviation), pluralistic and consultative decision-making, and enhanced social values and institutions, are important additional aspects that should be considered (at least in the form of safe limits or constraints).

Greenhouse gas mitigation provides an interesting example of how such an integrative framework could help to incorporate climate change-response measures within a national sustainable development strategy. The rate of total greenhouse gas emissions, G, may be decomposed by means of the identity

$$G = [Q/P] \times [Y/Q] \times [G/Y] \times P,$$

where $[Q/P]$ is quality of life per capita; $[Y/Q]$ is the material consumption required per unit of quality of life, $[G/Y]$ is the greenhouse gas emission per unit of consumption, and P is the population. A high quality of life can be consistent with low total greenhouse gas emissions, provided that each of the other three terms on the right-hand side of the identity could be minimized (see also the discussion in Section 4.6.2 on 'restructuring growth'). Reducing $[Y/Q]$ implies mainly 'social decoupling' (or 'dematerialization') whereby satisfaction becomes less dependent upon material consumption through changes in tastes, behaviour, and social values. Similarly, (G/Y) may be reduced by 'technological decoupling' (or 'decarbonization') that reduces the intensity of greenhouse gas emissions in consumption and production. Finally, population growth can be reduced, especially where emissions per capita are already high. The linkages between social and technological decoupling need to be explored (see, for example, IPCC (1999)). For example, changes in public perceptions and tastes could affect the directions of technological progress, and influence the effectiveness of mitigation and adaptation policies. The scope and need for emissions reduction is illustrated in Box 3.3.

Box 3.3   Making world consumption and production patterns more sustainable

As stated in Agenda 21, 'the major cause of the continued deterioration of the global environment is the unsustainable pattern of consumption and production, particularly in industrialized countries'.

Under the UNFCCC, developed countries are requested to take the lead in combating climate change, for several reasons. First, they are responsible for 83 per cent of the rise in cumulative carbon dioxide concentrations since 1800 (Loske 1996), and are responsible for 62 per cent of global emissions (UNDP 1998). Second, the adverse effects of global warming will be unequally distributed between North and South, since those who have caused the problem are (in relative terms) likely to be the gainers, and those who have been the bystanders are likely to be the victims. Finally, developed countries possess more capabilities for responding to climate change, in terms of financial and technical capacity to both mitigate and adapt to climate change.

The pressure the human economy exerts on the environment depends on the levels, patterns, and flows of extracted raw materials, production, consumption, and waste disposal between the economy and the biosphere. Sustainability implies that humanity keeps the utilization of nature within the (flexible) boundaries of global environmental space. However, dangerous infringement of some of these boundaries (including the climate system) cannot reliably be determined beforehand. Thus, it is a matter of foresight and precaution to embark upon a path of reducing resource and material throughput. Several studies have suggested that industrial economies reduce the current level of resource flow by a factor of 10 in the coming 40–50 years (Factor 10 Club 1995; McLaren et al., 1997; Schmidt-Bleek 1994). The concept of resource productivity calls for a reduction in resource use while aiming at maintaining economic and social wellbeing at the same time.

Higher population growth and expanding economies in the South could increase emissions from developing countries to nearly half of global emissions by 2010 (Tarnoff 1997). At the same time, a study by Austin, Goldemberg and Parker (1998) revealed the following.

1. While the annual industrial emissions from developing countries reach those of industrialized nations by 2015, the stock contribution to the two groups will become equal only in 2055.
2. By including emissions from land use change and forestry (with developing countries accounting for 77 per cent of the build-up during 1850–1990), contributions will be equal in 2038.
3. By adjusting for population, contributions equal in 2100.

Harrison (1992) revealed that population accounted for 35.6 per cent of the annual carbon dioxide emission growth between 1965 and 1989, while per capita consumption increases contributed 64.4 per cent. Rahman, Robins and Roncerel (1993) established that 'even immediate stabilization of the world population would still result in a doubling of carbon dioxide in 30 years, if all people achieved US per capita consumption levels'. Nevertheless, a reduction in population growth in developing countries, that today account for 80 per cent of the world population, is desirable as much for making national level development more sustainable, as for climate considerations.

While greenhouse gas emissions from developing countries must grow, their ability to contain these emissions is limited. A long-term solution to enhance mitigative capacity (i.e. the potential ability to undertake mitigation efforts) of developing countries, would be to invest in building social and economic infrastructure in these countries.

There is considerable scope for examining how both the optimality and durability approaches might be applied in a consistent manner to the various submodels within an integrated assessment model, where appropriate. Integrated assessment models are large and complex modelling frameworks currently being explored by climate change researchers; they contain coupled submodels that represent a variety of ecological, geophysical, and socioeconomic systems (IPCC 1997).

3.4   Making development more sustainable: decision criteria and analytical tools

In the context of climate change analysis, the sustainomics approach seeks to use a number of familiar criteria and analytical tools. Several generic approaches that are particularly relevant for climate change analysis are discussed below, while their specific applications to climate change are described in Chapters 5 and 8.

3.4.1   *Indicators of sustainable development*

In view of the importance of asset stocks to both the optimal and durable approaches, the practical implementation of sustainomics principles will require the identification of specific economic, social, and environmental indicators, at different levels of aggregation ranging from the global/macro to local/micro, that are relevant. It is important that the indicators be comprehensive in scope, multidimensional in nature (where appropriate), and account for spatial differences. A wide variety of indicators are described already in the literature (Adriaanse 1993; Alfsen & Saebo 1993; Azar, Homberg & Lindgren 1996; Bergstrom 1993; CSD 1998; Gilbert & Feenstra 1994; Holmberg & Karlsson 1992; Kuik & Verbruggen 1991; Liverman *et al.* 1988; Munasinghe & Shearer 1995; Moffat 1994; OECD 1994; Opschoor & Reijnders 1991; UN 1996; UNDP 1998; World Bank 1997; 1998).

Measuring economic, environmental (natural), human, and social capital also raises various problems. Manufactured capital may be estimated using conventional neoclassical economic analysis. As described later in Section 3.4.2. on cost–benefit analysis, market prices are useful when economic distortions are relatively low, and shadow prices could be applied in cases where market prices are unreliable (see, for example, Squire and van der Tak 1975). Natural capital needs to be quantified first in terms of key physical attributes. Typically, damage to natural capital may be assessed by the level of air pollution (e.g. concentrations of suspended particulate, sulphur dioxide or greenhouse gases), water pollution (e.g. biological oxygen demand or chemical oxygen demand), and land degradation, e.g. soil erosion or deforestation. Then the physical damage could be valued using a variety of techniques based on environmental and resource economics (see, for example, Freeman 1993, Munasinghe 1992 and Teitenberg 1992). Human resource stocks often are measured in terms of the value of educational levels, productivity, and earning potential of individuals. Social capital is the one that is most difficult to assess (Grootaert 1998). Putnam (1993) described it as 'horizontal associations' among people, or social networks and associated behavioural norms and values, which affect the productivity of communities. A somewhat broader view was offered by Coleman (1990), who viewed social capital in terms of social structures, which facilitate the activities of agents in society – this permitted both horizontal and vertical associations (like firms). An even wider definition is implied by the institutional approach espoused by North (1990) and Olson

(1982), that includes not only the mainly informal relationships implied by the earlier two views, but also the more formal frameworks provided by governments, political systems, legal and constitutional provisions, etc. Recent work has sought to distinguish between social and political capital, i.e. the networks of power and influence that link individuals and communities to the higher levels of decision-making. Box 2.2 in Chapter 2 provides an overview of various human development indicators that could be used to quantify wellbeing.

3.4.2   *Cost–benefit analysis*

Conventional cost–benefit analysis is one well-known example of a single valued approach, which seeks to assign economic values to the various consequences of an economic activity. The resulting costs and benefits are combined into a single decision-making criterion like the net present value (NPV), internal rate of return (IRR), or benefit–cost ratio (BCR).[1] The basic criterion for accepting a project is that the NPV of benefits is positive. Typically, NPV = PVB – PVC,

$$\text{where PVB} = \sum_{t=0}^{T} B_t/(1+r)^t; \quad \text{and} \quad \text{PVC} = \sum_{t=0}^{T} C_t/(1+r)^t,$$

where PVB is the present value of benefits, and PVC is the present value of costs, B_t and C_t are the project benefits and costs in year t, r is the discount rate, and T, the time horizon. Both benefits and costs are defined as the difference between what would occur *with and without* the project being implemented.

When two projects are compared, the one with the higher NPV is deemed superior. Furthermore, if both projects yield the same benefits (PVB), then it is possible to derive the least cost criterion, where the project with the lower PVC is preferred. The IRR is defined as that value of the discount rate for which PVB = PVC, while BCR = PVB/PVC. Further details of these criteria, as well as their relative merits in the context of sustainable development, are provided in Munasinghe (1992).

If a purely financial analysis is required from the private entrepreneur's viewpoint, then B, C, and r are defined in terms of market or financial prices, and NPV yields the discounted monetary profit. This situation corresponds to the economist's ideal world of perfect competition, where numerous profit-maximizing producers and utility-maximizing consumers achieve a Pareto-optimal outcome. However, conditions in the real world are far from perfect,

---

[1] Munasinghe *et al.* (1996) interpret cost–benefit analysis more broadly, encompassing the traditional project-level cost–benefit discussed here, cost-effectiveness cost–benefit analysis (determining least-cost options to meet a desired level of benefits), multicriteria analysis, and decision analysis.

due to monopoly practices, externalities (e.g. environmental impacts that are not internalized in the private market), and interference in the market process, e.g. taxes. Such distortions cause market (or financial) prices for goods and services to diverge from their economically efficient values. Therefore, the economic efficiency viewpoint usually requires that shadow prices (or opportunity costs) be used to measure B, C, and r. In simple terms, the shadow price of a given scarce economic resource is given by the change in value of economic output caused by a unit change in the availability of that resource. In practice, there are many techniques for measuring shadow prices, e.g. removing taxes, duties, and subsidies from market prices (Munasinghe 1992 and Squire & van der Tak 1975). Issues arising from the choice of a discount rate, r, for practical application to climate problems are discussed in Section 8.1

The incorporation of environmental considerations (especially externalities) into the economist's single valued cost–benefit analysis criterion requires special adjustments. Externalities refer to the costs and benefits that arise from the activity of one person, which does not take full account of its impacts on others. For example, emissions of particulate pollution from a power station will affect the health of people living downwind. However, such impacts often are not considered, or inadequate weight is given to them when costs and benefits of the power plant are estimated. Such external effects cannot be valued directly from market data as there are no 'prices' for these resources, e.g. clean air or water.

Ideally, all significant environmental impacts need to be valued as economic benefits and costs. As explained earlier in Section 3.4.1, environmental assets may be quantified in physical or biological units. Recent concepts for economically valuing environmental assets and impacts, and practical application techniques, are summarized in Box 3.4 (especially Figure 3.3 and Table 3.1). However, many environmental assets (e.g. biodiversity) cannot be valued accurately in monetary terms, despite the progress that has been made in recent years (Bateman & Willis, 1999; Freeman 1993; Hanley *et al.*, 1997; Markandya *et al.*, 2000; Munasinghe 1992). Therefore, criteria like the NPV often fail adequately to represent the environmental aspect of sustainable development (see Section 3.4.3 on multicriteria analysis).

Benefit transfer is a cost-effective alternative to new non-market valuation methods summarized in Box 3.4 (Desvousges, Naughton & Parsons 1992; McConnell 1992). The term 'benefit transfer' reflects the transfer of the estimated value from one environmental good or site to another. This reduces the need to design and implement a new potentially expensive valuation exercise for another site. For example, the estimated willingness to pay for a given risk reduction from contaminated water in Wyoming could be transferred to a reduced risk of poor water quality in Mongolia, as long as the transfer protocol is satisfied.

Table 3.1. *Techniques for economically valuing environmental impacts.*

	Type of market		
Behaviour type	Conventional market	Implicit market	Constructed market

Actual behaviour	Effect on production
	Effect on health
	Defensive or
	  preventive costs
		Travel cost
		Wage differences
		Property values
		
		Proxy marketed
		  goods
			Artificial market

Intended behaviour	Replacement cost shadow project		Contingent valuation

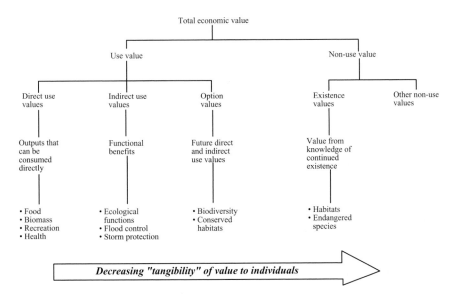

Figure 3.3. Categories of economic value attributed to environmental assets – with examples from a tropical rainforest). *Source:* Adapted from Munasinghe (1992) and Pearce (1992).

Capturing the social dimension of sustainable development within cost–benefit analysis is even more problematic. Some attempts have been made to attach 'social weights' to costs and benefits so that the resultant NPV favours poorer groups (see also Box 3.2). However, such adjustments (or preferential treatment for the poor) are rather arbitrary, and have weak foundations in economic theory. Other key social considerations (e.g. empowerment and participation) are hardly represented within cost–benefit analysis. In summary, the conventional cost–benefit

Box 3.4 Economic valuation of the environment

Economic valuation of environmental assets and services is an important input to the decision-making process. There has been some modest progress in recent years, in both the theory and application of valuation methods. The conceptual basis for valuation and various practical techniques are summarized briefly below (for details, see Munasinghe 1992).

Valuation concepts

The basic purpose of valuation is to determine the *total economic value* TEV of a resource (see Figure 3.3). The TEV consists of two broad categories: (a) use value, UV, and (b) non-use value, NUV, i.e. TEV = UV + NUV.

Use values may be broken down further into: (a) direct use value, DUV; (b) indirect use value, IUV, and (c) potential use value or option value, OV. The DUV is the immediate contribution an environmental asset makes to production or consumption, e.g. food or recreation. The IUV includes the benefits derived from functional services that the environment provides to support production and consumption, e.g. recycling nutrients or breaking down wastes. Option value is the willingness to pay now for the future benefit to be derived from an existing asset.

Non-use values are based generally on altruistic, non-utilitarian motives (Schechter & Freeman 1992), and occur even though the individual may have no intention of using a resource – one important category called *existence value* arises from the satisfaction of merely knowing that the asset exists, e.g. a rare and exotic species. As indicated in Figure 3.3, there is decreasing tangibility of the valuation concept, as one moves from left to right.

For the practitioner, what is important is not necessarily the precise conceptual breakdown of economic value, but rather the various empirical techniques derived from the above valuation concepts. Such valuation techniques, discussed below, permit us to estimate a monetary value for environmental assets and impacts. However, the results derived from some of these techniques are uncertain even in developed economies, and therefore their use in developing countries should be tempered by caution and sound judgment.

The willingness to pay of individuals for an environmental service or resource is the economic basis for deriving a variety of useful valuation techniques (Kolstad & Braden 1991). Willingness to pay is defined strictly as the area under the compensated or Hicksian demand curve, which indicates how demand varies with price while keeping the user's utility level constant. Equivalently, the difference between the values of two expenditure (or cost) functions could be used to measure the change in value of an environmental asset. The former are the minimum amounts required to achieve a given level of utility for a household (or output for a firm) before and after varying the quality of, price of, and/or access to, the environmental resource in question. All other aspects are kept constant. However, the commonly estimated demand function is the Marshallian one – that indicates how demand varies with the price of the environmental good, while keeping the user's income level constant. In practice, it has been shown that the Marshallian and Hicksian estimates of willingness to pay are comparable under normal conditions. Furthermore, in a few cases, once the Marshallian demand function has been estimated, the equivalent Hicksian function may be derived in turn. The payments people are willing to accept as compensation for environmental damage, as another measure of economic value that is related to willingness to pay. Willingness to accept and willingness to pay could diverge significantly (Cropper & Oates 1992; Shogren et al. 1994). In practice, either or both measures are used for valuation. A frequent criticism of this approach is that it is inequitable because a richer person is likely to have a greater willingness to pay or willingness to accept than a poorer person and, hence, the former will have greater weight when choices are made. This valid criticism is addressed better by dealing with equity issues separately from valuation methods based on economic valuation (see Section 3.2.4. above, and Box 3.2).

Valuation techniques

Valuation methods may be categorized according to which type of market on which they rely, and by considering how they make use of actual or potential behaviour (see Table 3.1). The most useful methods are based on how environmental quality changes affect directly observable actions, with the consequences valued in conventional markets.

Effect on production. An investment decision often has environmental impacts, which in turn affect the quantity, quality, or production costs of a range of productive outputs that may be valued readily in economic terms.

Effect on health. This approach is based on health impacts caused by pollution and environmental degradation. One practical measure related to the effect on production is the value of human output lost due to ill health or premature death. The loss of potential net earnings (called the human capital technique) is one proxy for foregone output, to which the costs of health care or prevention may be added.

Defensive or preventive costs. Often, costs may be incurred to mitigate the damage caused by an adverse environmental impact. For example, if the drinking water is polluted, extra purification may be needed. Then, such additional defensive or preventive expenditures (ex-post) could be taken as a minimum estimate of the benefits of mitigation.

Replacement cost and shadow project. If an environmental resource that has been impaired is likely to be replaced in the future by another asset that provides equivalent services, then the costs of replacement may be used as a proxy for the environmental damage, assuming that the benefits from the original resource are at least as valuable as the replacement expenses. A shadow project usually is designed specifically to offset the environmental damage caused by another project. For example, if the original project was a dam that inundated some forest land, then the shadow project might involve the replanting of an equivalent area of forest elsewhere.

Travel cost. This method seeks to determine the demand for a recreational site (e.g. number of visits per year to a park), as a function of variables like price, visitor income, and socioeconomic characteristics. The price is usually the sum of entry fees to the site, costs of travel, and opportunity cost of time spent. The consumer surplus associated with the demand curve provides an estimate of the value of the recreational site in question.

Property value. In areas for which relatively competitive markets exist for land, it is possible to decompose real estate prices into components attributable to different characteristics like house and lot size, air, and water quality. The marginal willingness to pay for improved local environmental quality is reflected in the increased price of housing in cleaner neighbourhoods. This method has limited application in developing countries, since it requires a competitive housing market, as well as sophisticated data and tools of statistical analysis.

Wage differences. As in the case of property values, the wage differential method attempts to relate changes in the wage rate to environmental conditions, after accounting for the effects of all factors other than environment (e.g. age, skill level, job responsibility, etc.) that might influence wages.

Proxy marketed goods. This method is useful when an environmental good or service has no readily determined market value, but a close substitute exists that does have a competitively determined price. In such a case, the market price of the substitute may be used as a proxy for the value of the environmental resource.

Artificial market. Such markets are constructed for experimental purposes, to determine consumer willingness to pay for a good or service. For example, a home water purification kit might be marketed at various price levels, or access to a game reserve may be offered on the basis of different admission fees, thereby facilitating the estimation of values.

Contingent valuation. This method puts direct questions to individuals to determine how much they might be willing to pay for an environmental resource, or how much compensation they would be willing to accept if they were deprived of the same resource. The contingent valuation method is more effective when the respondents are familiar with the environmental good or service (e.g. water quality) and have adequate information on which to base their preferences. Recent studies indicate that the contingent valuation method, cautiously and rigorously applied, could provide rough estimates of value of non-market items that would be helpful in economic decision-making, especially when other valuation methods are unavailable.

Source: Munasinghe (1992).

analysis methodology would tend to favour the market-based economic viewpoint, although environmental and social considerations might be introduced in the form of side constraints.

3.4.3   *Multicriteria analysis*

Multicriteria analysis, or multiobjective decision-making, is particularly useful in situations when a single-criterion approach such as monetized, project-level cost–benefit analysis falls short. In multicriteria analysis, desirable objectives are specified, usually within a hierarchical structure. The highest level represents the broad overall objectives (e.g. improving the quality of life or making development more sustainable), which often are vaguely stated. However, they can be broken down usually into more operationally relevant and easily measurable lower-level objectives, e.g. increased income, or indicators of the economic, social, and environmental dimensions of sustainable development. Sometimes only proxies are available, e.g. if the objective is to preserve biological diversity in a rainforest, the practically available attribute may be the number of hectares of rainforest remaining. Although value judgements may be required in choosing the proper attribute (especially if proxies are used), actual measurement does not have to be in monetary terms – unlike cost–benefit analysis. More explicit recognition is

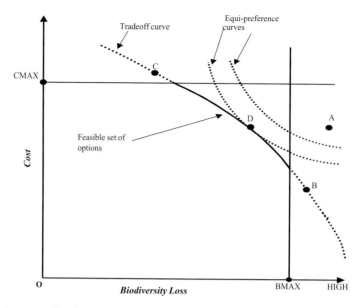

Figure 3.4. Simple two-dimensional example of multicriteria analysis. Adapted from Munasinghe (1993).

given to the fact that a variety of objectives and indicators may influence planning decisions.

Figure 3.4 is a two-dimensional representation of the basic concepts underlying multicriteria analysis. Consider an electricity supplier evaluating a hydroelectric project that could potentially cause biodiversity loss. Objective Z_1 is the additional project cost required to protect biodiversity, and Z_2 is an index indicating the loss of biodiversity. The points A, B, C, and D in Figure 3.4 represent alternative projects, e.g. different designs for the dam. In this case, project B is superior to (or dominates) A in terms of both Z_1 and Z_2, because B exhibits lower costs as well as less biodiversity loss relative to A. Thus, alternative A may be discarded. However, when we compare B and C, the choice is more complicated because the former is better than the latter with respect to costs, but worse with respect to biodiversity loss. Proceeding in this fashion, a trade-off curve (or locus of best options) may be defined by all the non-dominated feasible project alternatives such as B, C, and D. Such a curve implicitly places both economic and environmental attributes on a more equal footing, in the spirit of sustainomics.

Further ranking of alternatives is not possible without the introduction of value judgements (for an unconstrained problem). Typically, additional information may be provided by a family of equipreference curves that indicate the way in which the decision-maker or society trades-off one objective against the other (see Figure 3.4). Each such equipreference curve indicates the locus of points along which society is indifferent to the trade-off between the two objectives. The preferred alternative is the one that yields the greatest utility, i.e. at the point of tangency D of the trade-off curve with the best equipreference curve (the one closest to the origin).

Since equipreference curves usually are not measurable, other practical techniques may be used to narrow down the set of feasible choices on the trade-off curve. One approach uses limits on objectives or 'exclusionary screening'. For example, the decision-maker may face an upper bound on costs (i.e. a budgetary constraint), depicted by Cmax in the Figure 3.4. Similarly, ecological experts might set a maximum value of biodiversity loss Bmax, e.g. a level beyond which the ecosystem suffers catastrophic collapse. These two constraints may be interpreted in the context of durability considerations, mentioned earlier. Thus, exceeding Cmax is likely to threaten the viability of the electricity supplier, with ensuing social and economic consequences, e.g. jobs, incomes, returns to investors, etc. Similarly, violating the biodiversity constraint will undermine the resilience and sustainability of the forest ecosystem. In a more practical sense, Cmax and Bmax help to define a more restricted portion of the trade-off curve (darker line), thereby narrowing and simplifying the choices available to the single alternative D.

This type of analysis may be expanded to include other dimensions and attributes. For example, in our hydroelectric dam case, the number of people displaced (or resettled) could be represented by another social variable Z_3.

3.5   Improving the sustainability of conventional growth

3.5.1   Restructuring development and growth for greater sustainability

Growth is a major objective of almost all developing countries, especially the poorest ones. It cannot be fulfilled unless economic growth is sustained into the long term. The developing countries need to ensure that their endowments of natural resources are not taken for granted and squandered. If valuable resources such as air, forests, soil, and water are not protected, development is unlikely to be sustainable, not just for a few years, but for many decades. Furthermore, on the social side, it is imperative to reduce poverty, create employment, improve human skills, and strengthen our institutions.

Next, let us examine the alternative growth paths available, and the role of sustainomics principles in choosing options. Lovelock (1975) made a pioneering contribution with his Gaia hypothesis. He proposed that the totality of life on Earth might be considered an integrated web that works to create a favourable

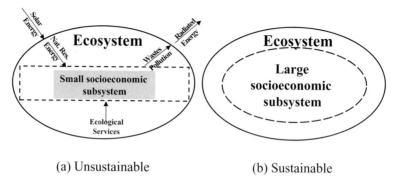

(a) Unsustainable   (b) Sustainable

Figure 3.5. Restructuring development to make the embedded socio-economic subsystem more sustainable within the larger ecosystem.

environment for survival. As a corollary, unregulated expansion of human activity might threaten the natural balance. In this spirit, Figure 3.5(a) shows how the socioeconomic subsystem (solid rectangle) has always been embedded in a broader ecological system (large oval). National economies are linked inextricably to, and are dependent upon, natural resources, since everyday goods and services are derived from natural resource inputs that originate from the larger ecological system. We extract oil from the ground and timber from trees, and we use water and air freely. At the same time, such activities have continued to expel polluting waste into the environment quite liberally. The broken line in Figure 3.5(a) shows symbolically that in many cases, the scale of human activity has increased to the point at which it is now impinging on the underlying ecosystem. This is evident today, if we consider that forests are disappearing, water resources are being polluted, soils are being degraded, and even the global atmosphere is under threat. Consequently, the critical question is: how might human society contain or manage this problem?

One traditional view that has caused confusion among world leaders is the assumption that concern for the environment is not necessarily good for economic activity. Often, they view the environmental system as a subsystem of a larger economic system, in contrast to the view shown in Figure 3.5. Thus, until recently, the conventional wisdom held that it was not possible to have economic growth and a good environment at the same time, because they were mutually incompatible goals. However, the more modern viewpoint (embodied also in sustainomics), indicates that growth and environment are indeed complementary. One key underlying assumption is that it is often possible to devise so-called 'win–win' policies that lead to economic, as well as environmental, gains (Munasinghe, Sunkel & de Miguel 2001). As illustrated earlier in Figure 3.5(a), the traditional approach to development certainly would lead to a situation in which the economic system would impinge upon the boundaries of the ecosystem in a harmful manner. On the other hand, Figure 3.5(b) summarizes the modern approach that would allow us to have the same level of prosperity without damaging the environment severely. In this case, the oval outer curve is matched by an oval inner one in which economic activities have been restructured in a way that is more harmonious with the ecosystem.

It would be fruitful to seek specific interventions that might help to make the crucial change in mindset, where the emphasis would be on the structure of development, rather than the magnitude of growth (conventionally measured). Policies that promote environmentally and socially friendly technologies that use natural resource inputs more frugally and efficiently, reduce polluting emissions, and facilitate public participation in decision-making, are important. One example is the information technology (IT) revolution, which might facilitate desirable restructuring from an environmental perspective, by making modern economies more services-oriented, and shifting activities away from highly polluting and material-intensive types of manufacturing and extractive industries (Munasinghe 1994; 1989). If managed properly, IT also could make development more socially sustainable by improving access to information, increasing public participation in decision-making, and empowering disadvantaged groups. The correct blend of market forces and regulatory safeguards is required.

3.5.2   *Linking climate change impacts and sustainable development with conventional decision-making*

Sustainomics helps in identifying practical economic, social, and natural resource management options that facilitate sustainable development. It serves as an essential bridge between the traditional techniques of (economic) decision-making and modern environmental and social analysis. In this context, the assessment of climate change impacts and vulnerability is a key concern. Sustainable development assessment is an important tool to ensure balanced analysis of climate change impacts, and development and sustainability concerns. The 'economic' component of sustainable development assessment is based on conventional economic and financial analysis (including cost–benefit analysis), as described earlier. The other two key components are environmental and social assessments (see, for example, World Bank 1998). Poverty assessment is often interwoven with sustainable development assessment. Economic, environmental, and social analyses need to be integrated and harmonized within it. Since traditional decision-making relies heavily on economics, a first step towards such an integration would be the systematic incorporation of environmental and social concerns into the policy framework of human society.

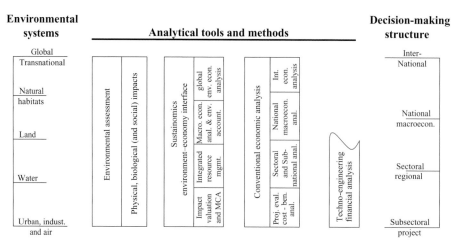

Figure 3.6. Incorporating environmental concerns in sustainable development into conventional economic decision making. This diagram may be readily adapted to incorporate social concerns (see text). *Source:* adapted from Munasinghe (1992).

Figure 3.6 provides an example of how environmental assessment is integrated into economic analysis. The right-hand side indicates the hierarchical nature of conventional decision-making in a modern society. The global and transnational level consists of sovereign nation states. In the next level are individual countries, each having a multisectored macroeconomy. Various economic sectors (e.g. industry and agriculture) exist in each country. Finally, each sector consists of different subsectors and projects. The usual decision-making process on the right side of Figure 3.6 relies on technoengineering, financial and economic analyses of projects and policies. In particular, conventional economic analysis has been well developed in the past, and uses techniques such as project evaluation/cost–benefit analysis, sectoral/regional studies, multisectoral macroeconomic analysis, and international economic analysis (finance, trade, etc.) at the various hierarchic levels.

Unfortunately, environmental and social analysis cannot be carried out readily using the above decision-making structure. We examine how environmental issues might be incorporated into this framework (with the understanding that similar arguments may be made with reference to social issues). The left side of Figure 3.6 shows one convenient environmental breakdown in which the issues are:

1.   Global and transnational environmental changes, e.g. climate change, ozone layer depletion, regional air pollution.
2.   Changing natural habitats, e.g. forests and other ecosystems.

3. Environmental problems of land management, e.g. in agricultural zones.
4. Threatened water resources, e.g. river basins, aquifers, and watersheds.
5. Local urban–industrial pollution, e.g. wastes in metropolitan areas, local air, water and soil pollution.

In each case, a holistic environmental analysis would seek to study a physical or ecological system in its entirety. Complications arise when such natural systems cut across the structure of human society. For example, a large and complex forest ecosystem (e.g. the Amazon) could span several countries, and also interact with many economic sectors within each.

The causes of environmental degradation arise from human activity (ignoring natural disasters and other events of non-human origin), and therefore we begin on the right side of the figure. The ecological effects of economic decisions must then be traced through to the left side. The techniques of environmental assessment have been developed to facilitate this difficult analysis (World Bank 1998). For example, destruction of a primary moist tropical forest may be caused by hydroelectric dams (energy sector policy), roads (transport sector policy), 'slash and burn' farming (agriculture sector policy), mining of minerals (industrial sector policy), land clearing encouraged by land tax incentives (fiscal policy), and so on. Disentangling and prioritizing these multiple causes (right side) and their impacts (left side) will involve a complex analysis.

Figure 3.6 also shows how sustainomics could play its bridging role at the ecology–economy interface, by mapping the environmental assessment results (measured in physical or ecological units) on to the framework of conventional economic analysis. A variety of environmental economic techniques, including valuation of environmental impacts (at the local/project level), integrated resource management (at the sector/regional level), environmental macroeconomic analysis and environmental accounting (at the economy-wide level), and global/transnational environmental economic analysis (at the international level), facilitate this process of incorporating environmental issues into traditional decision-making. Since there is considerable overlap among the analytical techniques described above, this conceptual categorization should not be interpreted too rigidly. Furthermore, when economic valuation of environmental impacts is difficult, techniques such as multi-criteria analysis would be useful (see Figure 3.4 and Section 3.4.3).

Once the foregoing steps are completed, projects and policies must be redesigned to reduce their environmental impacts and shift the development process towards a more sustainable path. Clearly, the formulation and implementation of such policies is itself a difficult task. In the deforestation example described earlier, protecting this ecosystem is likely to raise problems of co-ordinating policies in a large number of disparate and (usually) non-co-operating ministries and line institutions, i.e. energy, transport, agriculture, industry, finance, forestry, etc.

Analogous reasoning may be applied readily to social assessment at the society–economy interface, in order to incorporate social considerations more effectively into the conventional economic decision-making framework. In this case, the 'environmental systems' on the left side of Figure 3.6 would be replaced by relevant 'social systems', e.g. affected groups and communities at different scales. The Environmental Assessment box would be replaced by a Social Assessment one including key elements such as asset distribution, inclusion, cultural considerations, values, and institutions. Finally, a sustainomics box representing the society–economy interface would be substituted for the environment–economy interface.

Impacts on human society (i.e. beliefs, values, knowledge, and activities), and on the biogeophysical environment (i.e. both living and nonliving resources), are often interlinked via second and higher order paths, requiring integrated application of social and environmental assessments. This insight reflects current thinking on the co-evolution of socioeconomic and ecological systems.

In the framework of Figure 3.6, the right side represents a variety of institutional mechanisms (ranging from local to global) which would help to implement policies, measures, and management practices to achieve a more sustainable outcome. Implementation of sustainable development strategies and good governance would benefit from the transdisciplinary approach advocated in sustainomics. For example, economic theory emphasizes the importance of pricing policy to provide incentives that will influence rational consumer behaviour. However, cases of seemingly irrational or perverse behaviour abound, which might be better understood through findings in areas like behavioural and social psychology, and market research. Such work has identified basic principles that help to influence society and modify human actions, including reciprocity (or repaying favours), behaving consistently, following the lead of others, responding to those we like, obeying legitimate authorities, and valuing scarce resources (Cialdini 2001).

References

Adriaanse, A. (1993) *Environmental Policy Performance Indicators*. SDU, Den Haag.
Alfsen, K. H., and Saebo, H. V. (1993) Environmental quality indicators: background, principles and examples from Norway. *Environmental and Resource Economics*, **3**, 415–35.
Andersen, E. (1993) *Values in Ethics and Economics*. Cambridge MA: Harvard University Press.
Arrow, K. J., Cline, W., Maler, K. G., Munasinghe, M. and Stiglitz, J. (1995) Intertemporal equity, discounting, and economic efficiency. In M. Munasinghe, ed., *Global Climate Change: Economic and Policy Issues*. World Bank, Washington.

Austin, D. G., Goldenberg, J. and Parker, G. (1998) *Contributions to Climate Change: Are Conventional Metrics Misleading the Debate*. Washington: World Resources Institute.

Azar, C., Homberg, J. and Lindgren, K. (1996) Socio-ecological indicators for sustainability. *Ecological Economics*, **18**, 89–112.

Bateman, I. J. and Willis, K. G. (1999) *Valuing Environmental Preferences: Theory and Practice of the Contingent Valuation Method in the US, EC and Developing Countries*. Oxford: Oxford University Press.

Bennet, R. (2000) Risky business. *Science News*, **158**, 190–91.

Bergstrom, S. (1993) Value standards in sub-sustainable development: on limits of ecological economics. *Ecological Economics*, **7**, 1–18.

Bohle, H. G., Downing, T. E. and Watts, M. J. (1994) Climate change and social vulnerability: toward a sociology and geography of food insecurity. *Global Environmental Change*, **4**(1), 37–48.

Brown, P. G. (1998) Towards an economics of stewardship: the case of climate. *Ecological Economics*, **26**, 11–21.

Chambers, R. (1989) Vulnerability, coping and policy. *Institute of Development Studies Bulletin*, **20**(2), 1–7.

Chenery, H. and Srinivasan, T. N., eds. (1988) *Handbook of Development Economics*, Vol. I. Amsterdam: North-Holland.

Chenery, H. and Srinivasan, T. N. (eds) (1989) *Handbook of Development Economics*, Vol. II. Amsterdam: North-Holland.

Cialdini, R. B. (2001) *Influence: Science and Practice*, 4th edn. London: Allyn and Bacon.

Colding, J. and Folke, C. (1997) The relations among threatened species, their protection, and taboos. *Conservation Ecology*, **1**(1), 6 (available from: http://www.consecol.org/vol1/iss1/art6).

Coleman, J. (1990) *Foundations of Social Theory*. Cambridge MA: Harvard University Press.

CSD (1998) *Indicators of Sustainable Development*. New York: Commission on Sustainable Development.

Costanza, R. (2000) Ecological sustainability, indicators and climate change. In M. Munasinghe and R. Swart, eds., *Climate Change and its Linkages with Development, Equity and Sustainability*. Geneva: Intergovernmental Panel on Climate Change.

Costanza, R., Cumberland, J., Daly, H., Goodland, R. and Norgaard, R. (1997) *An Introduction to Ecological Economics*. Boca Raton: St Lucia's Press.

Cropper, M. L. and W. E. Oates (1992) Environmental economics: a Survey. *Journal of Economic Literature*, **XXX**, 675–740.

Daly, H. E. and Cobb, J. B. Jr. (1989) *For the Common Good*. Boston: Beacon Press.

Dasgupta, P. and Maler, K. G. (1997) The resource basis of production and consumption: an economic analysis. In P. Dasgupta and K. G. Maler, eds., *The Environment and Emerging Development Issues*, vol. 1. Oxford: Clarendon Press.

Desvousges, W. H., Naughton, M. C. and Parsons, G. R. (1992) Benefits transfer: conceptual problems in estimating water quality benefits using existing studies. *Water Resources Research*, **28**.

Dreze, J. and Sen, A. (1990) *Hunger and Public Action*. Oxford: Clarendon Press.

Ecological Economics (various issues). Amsterdam: Elsevier.

Environmental Ethics (various issues). Amsterdam: Elsevier.

Factor 10 Club (1995) *Carnoules Declaration*. Carnoules.

Faucheux, S., Pearce, D. and Proops, J., eds. (1996) *Models of Sustainable Development*. Cheltenham: Edward Elgar.

Fisher, I. (1906, reprinted 1965) *The Nature of Capital and Income*. New York: Augustus M. Kelly.

Freeman, A. M., (1993) *The Measurement of Environmental and Resource Values: Theory and Methods, Resources for the Future*. Washington: Resources for the Future.

Georgescu-Roegen, N. (1971) *The Entropy Law and the Economic Process*. Cambridge MA: Harvard University Press.

Gilbert, A. and Feenstra, J. (1994) Sustainability indicators for the Dutch environmental policy theme 'diffusion' cadmium accumulation in soil. *Ecological Economics*, **9**, 253–65.

Githinji, M. and Perrings, C. (1992) Social and ecological sustainability in the use of biotic resources in sub-Saharan Africa: rural institutions and decision making in Kenya and Botswana, Mimeograph, Beijer Institute and University of California, Riverside.

Grootaert, C. (1998) Social capital: the missing link, *Social Capital Initiative Working Paper No. 3*, Washington: World Bank.

Gunderson, L. and Holling, C. S. (2001) *Panarchy: Understanding Transformations in Human and Natural Systems*. New York: Island Press.

Hall, C. (ed.) (1995) *Maximum Power: The Ideas and Applications of H. T. Odum*. Niwot CO: Colorado University Press.

Hanley, N. F., Shogren, J. F. and White, B. (1997) *Environmental Economics in Theory and Practice*. Oxford: Oxford University Press.

Hanna, S. and Munasinghe, M. (1995) *Property Rights in Social and Ecological Context*. Stockholm and Washington: Beijer Institute and World Bank.

Harrison, P. (1992) *The Third Revolution: Population, Environment and a Sustainable World*. London: Penguin Books.

Hicks, J. (1946) *Value and Capital* (2nd edn). Oxford: Oxford University Press.

Holling, C. S. (1973) Resilience and stability of ecological systems, *Annual Review of Ecology and Systematics*, **4**, 1–23.

Holling, C. S. (1986) The resilience of terrestrial ecosystems: local surprises and global change. In W. C. Clark and R. E. Munn, eds., *Sustainable Development of the Biosphere*. Cambridge: Cambridge University Press, pp. 292–317.

Holling, C. S. (1992) Cross scale morphology, geometry and dynamics of ecosystems, *Ecological Monographs*, **62**, 447–502.

Holmberg, J., and Karlsson, S. (1992) On designing socio-ecological indicators. In U. Svedin and Bhagerhall-Aniansson, eds., *Society and Environment: A Swedish Research Perspective*. Boston: Kluwer Academic.

IPCC (1996) Climate Change 1995: Economic and Social Dimensions of Climate Change. In J. P. Bruce, H. Lee and E. F. Haites (eds.) Contribution of Working Group

III to the *Second Assessment Report of the Intergovernmental Panel on Climate Change*. Cambridge: Cambridge University Press.

IPCC (1997) *Climate Change and Integrated Assessment Models (IAMs)*. Geneva: Intergovernmental Panel on Climate Change.

IPCC (2000) *Methodological and technological Issues in Technology Transfer* – Intergovernmental Panel on Climate Change Special Report. Cambridge: Cambridge University Press.

Islam, Sardar M. N. (2001) Ecology and optimal economic growth: an optimal ecological economic growth model and its sustainability implications. In M. Munasinghe, O. Sunkel and C. de Miguel, eds., *The Sustainability of Long Term Growth*. Cheltenham: Edward Elgar.

Kates, R. W., W. C. Clark, R. Corell, J. M. Hall, C. C. Jaeger, I. Lowe, J. J. McCarthy, H. J. Schellnhuber, B. Bolin, N. M. Dickson, S. Faucheux, G. C. Gallopin, A. Gr bler, B. Huntley, J. J ger, N. S. Jodha, R. E. Kasperson, A. Mabogunje, P. Matson, H. Mooney, B. Moore III, T. O'Riordan, U. Svedin (2001) Environment and development: Sustainability science, *Sceince,* **292** (5517): 641–642.

Kolstad, C. D. and Braden, J. B. (eds.), (1991) *Measuring the Demand for Environmental Quality*. New York: Elsevier.

Kuik, O. and Verbruggen, H. eds. (1991) *In Search of Indicators of Sustainable Development*. Boston: Kluwer.

Liverman, D., Hanson, M., Brown, B. J., and Meredith, R. Jr. (1988) Global sustainability: towards measurement. *Environmental management*, **12**, 133–43.

Loske, R. (1996) *Klimapolitik. Im Spannungsfeld von Kurzzeitinteressen und Langzeiterfordernissen*. Marburg: Metropolis.

Lovelock, J. (1979) *Gaia: A New Look at Life on Earth*. Oxford: Oxford University Press.

Ludwig, D., Walker, B. and Holling, C. S. (1997) Sustainability, stability, and resilience. *Conservation Ecology* (online) **1**(1), 7 (http://www.consecol.org/vol1/iss1/art7).

Maler, K. G. (1990) Economic theory and environmental degradation: a survey of some problems. *Revista de Analisis Economico*, **5**, 7–17.

Markandya, A., Mason, P. and Taylor, T. (2000) *Dictionary of Environmental Economics*. London: Earthscan.

Maslow, A. H. (1970) *Motivation and Personality*. New York: Harper and Row.

McConnell, K. (1992) Model building and judgment: implications for benefits transfers with travel cost models. *Water Resources Research*, **28**, 695–700.

Mc Laren, D., Bullock, S. and Yousuf, N. (1997) *Tomorrow's World: Britain's Share in a Sustainable Future*. Earthscan: London.

Meier, P. and Munasinghe, M. (1994) *Incorporating Environmental Concerns into Power Sector Decision Making*. Washington: World Bank.

Moffat, I. (1994) On measuring sustainable development indicators. *International Journal of Sustainable Development and World Ecology*, **1**, 97–109.

Moser, C. (1998) The asset vulnerability framework: reassessing urban poverty reduction strategies. *World Development*, **26**(1), 1–19.

Munasinghe, M. (1990) *Energy Analysis and Policy*. Butterworth London: Heinemann.

Munasinghe, M. (1992) *Environmental Economics and Sustainable Development.* Environment Paper No. 3. UN Earth Summit, Rio de Janeiro. Washington: World Bank.

Munasinghe, M. (1994) Sustainomics: a transdisciplinary framework for sustainable development. In *Proceedings*, 50th Anniversary Colombo: Sessions of the Sri Lanka Association for the Advancement of Science.

Munasinghe, M. (1998). Climate change decision-making: science, policy and economics. *International Journal of Environment and Pollution*, **10**(2), 188–239.

Munasinghe, M., ed. (1989) *Computers and Informatics in Developing Countries*. London: Butterworth, for the Third World Academy of Sciences, Trieste.

Munasinghe, M. and Shearer, W., eds., (1995) *Defining and Measuring Sustainability: The Biogeophysical Foundations*. Tokyo and Washington: UN University and World Bank.

Munasinghe, M., P. Meier, M. Hoel, S. Wong, and A. Aaheim (1996) Applicability of techniques of cost–benefit analysis to climate change. In J. P. Bruce, H. Lee, and E. H. Haites, eds., *Climate Change 1995: Economic and Social Dimensions*, Chap. 5. Geneva: Intergovernmental Panel on Climate Change, Cambridge: Cambridge University Press.

Munasinghe, M., O. Sunkel and C. de Miguel, eds., (2001) *The Sustainability of Long Term Growth*. London: Edward Elgar.

Narada, The Venerable (1988) *The Buddha and His Teachings*, 4th edn. Kuala Lumpur: Buddhist Missionary Society.

Nordhaus, W. and Tobin, J. (1972) 'Is growth obsolete?' In Milton, M. ed., *The Measurement of Economic and Social Performance*. Studies in Income and Wealth, Vol. 38. National Bureau of Economic Research. New York: Columbia University Press.

North, D. (1990) *Institutions, Institutional Change and Economic Performance*. Cambridge: Cambridge University Press.

OECD (1994) *Environmental Indicators*. Paris: Organization for Economic Co-operation and Development.

Olson, M. (1982) *The Rise and Decline of Nations*. New Haven: Yale University Press.

Opschoor, H. and Reijnders, L. (1991) Towards sustainable development indicators. In O. Kuik and H. Verbruggen, eds., *In Search of Indicators of Sustainable Development*. Boston: Kluwer.

Pearce, D. W. and Turner, R. K. (1990) *Economics of Natural Resources and the Environment*. London: Harvester Wheatsheaf.

Perrings, C. and Opschoor, H. (1994) *Environmental and Resource Economics*. Cheltenham: Edward Elgar.

Perrings, C., Maler, K. G. and Folke, C. (1995) *Biodiversity Loss: Economic and Ecological Issues*. Cambridge: Cambridge University Press.

Petersen, G. D., Allen, C. R. and Holling, C. S. (1998) Diversity, ecological function, and scale: resilience within and across scales, *Ecosystems*, **1**, 6–18.

Pezzey, J. (1992) Sustainable development concepts: an economic analysis. *Environment Paper No. 2*. Washington: World Bank.

Pimm, S. L. (1991) *The Balance of Nature?*. Chicago: University of Chicago Press.

Putnam, R. D. (1993) *Making Democracy Work: Civic Traditions in Modern Italy*. Princeton: Princeton University Press.

Rahman, A., Robins, N. and Roncerel, A. eds. (1993) *Consumption Versus Population: Which is the Climate Bomb? Exploding the Population Myth*. Paris: Climate Network Europe.

Rawls, J. A. (1971) *Theory of Justice* Cambridge MA: Harvard University Press.

Rayner, S. and Malone, E., eds. (1998) *Human Choice and Climate Change*. Columbus OH: Batelle Press, pp. 1–4.

Ribot, J. C., Najam, A. and Watson, G. (1996) Climate variation, vulnerability and sustainable development in the semi-arid tropics. In J. C. Ribot, A. R. Magalhaes and S. S. Pangides, eds., *Climate Variability, Climate Change and Social Vulnerability in the Semi-Arid Tropics*. Cambridge: Cambridge University Press.

Schechter, M. and S. Freeman (1992) *Some Reflections on the Definition and Measurement of Non-Use Value*. University of Haifa: Draft mimeograph, Natural Resources and Environmental Research Center and Department of Economics.

Schmidt-Bleek, F. (1994) *Wieviel Umwelt braucht der Mensch?* Birkhauser: Berlin–Basel.

Schutz, J. (1999) The value of systemic reasoning. *Ecological Economics*, **31**(1), 23–9.

Sen, A. K. (1981) *Poverty and Famines: An Essay on Entitlement and Deprivation*. Oxford: Clarendon Press.

Sen, A. K. (1984) *Resources, Values and Development*. Oxford: Basil Blackwell.

Sen, A. K. (1987) *On Ethics and Economics*. Cambridge MA: Basil Blackwell.

Shogren, J., S. Shin, D. Hayes, and J. Kliebenstein (1994) Resolving differences in willingness to pay and willingness to accept. *American Economic Review*, **84**, 255–70.

Siebhuner, B. (2000) Homo sustinens – towards a new conception of humans for the science of sustainability. *Ecological Economics*, **32**, 15–25.

Solow, R. (1986) On the intergenerational allocation of natural resources. *Scandinavian Journal of Economics*, **88**(1), 141–9.

Squire, L. and van der Tak, H. (1975) *Economic Analysis of Projects*. Baltimore: Johns Hopkins University Press.

Stern, N. H. (1989) 'The economics of development: a survey'. *Economic Journal*, **99**.

Tarnoff, C. (1997) *Global Climate Change: The Role of US Foreign Assistance*. Congressional Research Service: Report for the Congress, 21 November.

Teitenberg, T. (1992) *Environmental and Natural Resource Economics*. New York: Harper Collins.

Temple, J. (1999) The new growth evidence. *Journal of Economics Literature*, **XXXVII**, 112–54.

Toth, F. (1999) Decision analysis for climate change. In M. Munasinghe, ed., *Climate Change and its Linkages with Development, Equity and Sustainability*. Geneva: Intergovernmental Panel on Climate Change.

UN (1993) *Agenda 21*. New York: United Nations.

UN (1996) *Indicators of Sustainable Development: Framework and Methodology*. New York: United Nations.

UNDP (1998) *Human Development Report*. New York: United Nations Development
Programme.

UNEP, IUCN, and WWF (1991) *Caring for the Earth*. Nairobi: United Nations
Environmental Programme.

WCED (1987) *Our Common Future*. Oxford: Oxford University Press.

Westra, L. (1994) *An Environmental Proposal for Ethics: The Principle of Integrity*. Lanham
MA: Rowman and Littlefield.

World Bank (1997) *Expanding the Measures of Wealth: Indicators of Environmentally
Sustainable Development*. Washington: World Bank.

World Bank (1998) *Environmental Assessment Operational Directive 4.01*. Washington:
World Bank.

4

Interactions between climate and development

In this chapter, we take the first steps towards applying some of the basic elements introduced in the previous chapter, in order to make development more sustainable in the context of climate change. In Chapter 3, we outlined the importance of two inter-related issues that need to be investigated, using the sustainomics framework: (a) the implications of climate change and climate change responses (including adaptation and mitigation) for sustainable development, and (b) the implications of development strategies and policies in both developing and developed countries, for climate change and climate change response options, and vulnerability to climate change impacts, including associated synergies and trade-offs.

4.1   Circular relationship between climate change and sustainable development

The full cycle of cause and effect between climate change and sustainable development is summarized in Figure 4.1, in which we outline an integrated assessment modelling framework (IPCC 2001b). Each socioeconomic development path in the bottom right-hand quadrant of the figure – driven by the forces of population, economy, technology, and governance – gives rise to different levels of greenhouse gas emissions. These emissions accumulate in the atmosphere, increasing the greenhouse gas concentrations and disturbing the natural balance between incident solar radiation and energy re-radiated from the Earth, as indicated in the Climate Domain box on the left-hand side. Such changes give rise to the enhanced greenhouse effect that increases radiative forcing of the climate system. The resultant changes in climate will persist well into the future, and impose stresses on the human and natural systems shown in the top right-hand quadrant

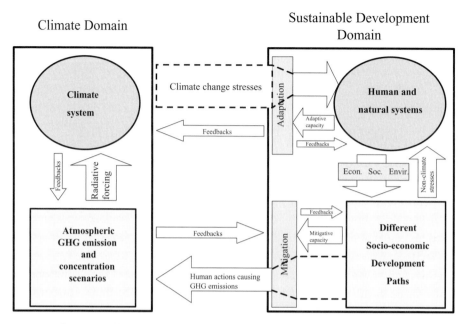

Figure 4.1. Climate change – an integrated framework considering socio-economic development. *Source:* IPCC (2001).

(Sustainable Development Domain). Such impacts will ultimately have effects on socioeconomic development paths, thus completing the cycle. The development paths also have direct effects on the natural systems, in the form of non-climate stresses, e.g. changes in land use leading to deforestation and land degradation.

The climate and sustainable development domains interact in a dynamic cycle, characterized by significant time delays. Both impacts and emissions, for example, are linked in complex ways to underlying socioeconomic and technological development paths. Development paths strongly affect the capacity to both adapt to, and mitigate climate change in, any region. The adaptive capacity of human and ecological systems support adaptation in these systems, which will reduce the severity of impacts. Analogously, superior mitigative capacity in human society could enhance prospects for greenhouse gas mitigation efforts, which help to reduce future greenhouse gas emissions. In this way, adaptation and mitigation strategies are connected dynamically with changes in the climate system and the prospects for ecosystem adaptation, food production, and long-term economic development. Finally, the economic, social, and environmental dimensions of sustainable development (described in Chapter 3) provide a useful framework for analysing the effects of climate change on future socioeconomic development scenarios. Feedbacks occur throughout the cycle, and changes in one part of the cycle influence other components in a dynamic manner, through multiple paths.

To summarize, climate change impacts are part of the larger question of how complex social, economic, and environmental subsystems interact and shape prospects for sustainable development. There are multiple links. Economic development affects ecosystem balance and, in turn, is affected by the state of the ecosystem. Poverty can be both a result and a cause of environmental degradation. Material- and energy-intensive lifestyles and continued high levels of consumption supported by non-renewable resources, as well as rapid population growth, are not likely to be consistent with sustainable development paths. Similarly, extreme socioeconomic inequality within communities and between nations may undermine the social cohesion that would promote sustainability and make policy responses more effective. At the same time, socioeconomic and technology policy decisions made for non-climate-related reasons have significant implications for climate policy and climate change impacts, as well as for other environmental issues. In addition, critical impact thresholds, and vulnerability to climate change impacts, are connected directly to environmental, social and economic conditions, and institutional capacity.

The First Assessment Report of the Intergovernmental Panel on Climate Change (IPCC 1990) focused mainly on the climate domain (left-hand half of Figure 4.1). The *Second Assessment Report* (IPCC 1995) strengthened and refined the analysis of climate aspects while expanding the coverage to include impacts on social and ecological systems (top right-hand quadrant of Figure 4.1). The *Third Assessment Report* (IPCC 2001) provided new information and evidence with regard to all quadrants of figure. In particular, a new contribution has been to fill-in the bottom right-hand quadrant of Figure 4.1 by exploring alternative development paths and their relationship to greenhouse gas emissions, and by undertaking preliminary work on the linkage between adaptation, mitigation, and development paths. However, the *Third Assessment Report* did not achieve a fully integrated assessment of climate change, because of the incomplete state of knowledge. In this book, we pay greater attention to the bottom right-hand quadrant, acknowledging that socioeconomic development paths determine not only the rate and magnitude of climate change, but also are impacted-upon by the consequences of climate change.

4.2   Principles for applying the sustainomics framework to climate change

The climate change problem fits quite readily within the broad conceptual framework of sustainomics described in Chapter 3. Decision-makers are beginning to show more interest in the assessment of how serious a threat climate change poses to the future basis for improving human welfare (Munasinghe 2000; Munasinghe & Swart 2000). In particular, increased greenhouse gas emissions and other unsustainable practices are likely to undermine the security of nations and communities through economic, social, and environmental impoverishment as well as inequitable distribution of adverse impacts, with undesirable consequences such as large numbers of 'environmental' refugees (Lonergan 1993; Ruitenbeek 1996; Westing 1992). Some of the potential linkages, and the sustainomics-related principles and concepts that could be applied in this context, are outlined below.

4.2.1   *Economic, social, and environmental risks and opportunities*

First, global warming poses a significant potential threat to the future economic wellbeing of large numbers of human beings. In its simplest form, the economic efficiency viewpoint will seek to maximize the net benefits (or outputs of goods and services) from the use of the global resource represented by the atmosphere. Broadly speaking, this implies that the stock of atmospheric assets, which provide a sink function for greenhouse gases, needs to be maintained at an optimum level. As indicated in the case study in Section 4.5.1, this target level is defined as the point at which the marginal greenhouse gas abatement costs are equal to the marginal avoided damages. The underlying principles are based on optimality and the economically efficient use of a scarce resource, i.e. the global atmosphere.

Second, climate change also could undermine social welfare and equity in an unprecedented manner. In particular, more attention needs to be paid to the vulnerability of social values and institutions, which are already stressed due to rapid technological changes (Adger 1999). Especially within developing countries, erosion of social capital is undermining the basic glue that binds communities together, e.g. the rules and arrangements that align individual behaviour with collective goals (Banuri *et al.* 1994). Existing international mechanisms and systems to deal with transnational and global problems are fragile, and unlikely to be able to cope with worsening climate change impacts.

Furthermore, both intra- and intergenerational equity are likely to be worsened (IPCC 1996a). Existing evidence demonstrates clearly that poorer nations and disadvantaged groups within nations are especially vulnerable to disasters (Banuri 1998; Clarke & Munasinghe 1995). Climate change is likely to result in inequities due to the uneven distribution of the costs of damage, as well as the necessary adaptation and mitigation efforts – such differential effects could occur both among and within countries. Some of the impacts of recent large-scale disasters such as El Niño might provide useful case study material. Inequitable distributions are not only ethically unappealing, but also may be unsustainable in the long run (Burton 1997). For example, a future scenario that restricts per capita carbon emissions in the South to 0.5 t per year while permitting a corresponding level in the North of over 3 t per year will not facilitate the co-operation of developing countries, and therefore is unlikely to be durable. More generally, inequity could undermine social cohesion and exacerbate conflicts over scarce resources.

Third, the environmental viewpoint draws attention to the fact that increasing anthropogenic emissions and accumulations of greenhouse gases might perturb significantly a critical global subsystem – the atmosphere (UNFCCC 1993). Changes in the global climate (e.g. mean temperature, precipitation, etc.) also could threaten the stability of a range of critical, interlinked, physical, ecological, and social systems and subsystems (IPCC 1996b). Environmental sustainability will depend on several factors, including:

1. Climate change intensity, e.g. magnitude and frequency of extreme climatic events.
2. System vulnerability, e.g. exposure to climate impacts and sensitivity to such impacts.
3. System resilience, i.e. ability to recover from impacts.

Damage may be unevenly distributed and sometimes irreversible. Although developed countries are responsible for much of historical greenhouse gas emissions, their strong economies leave them better positioned than developing countries to cope with changes in climate. Climate change will have economic consequences, because the damage it causes and the measures people take to adapt to a new climate regime will impose both monetarily quantifiable and non-quantifiable costs.

On the more positive side, climate change response also provides some opportunities for furthering sustainable development. The international acknowledgement of climate change as a risk for current and future generations has highlighted also the current vulnerability of populations and ecosystems to a broad range of existing environmental issues (of which climate change is only one). Reducing the vulnerability of socioeconomic and ecological systems to climate change also offers opportunities to: (a) reduce their vulnerability to the broader set of threats to sustainable development; (b) improve their resilience; (c) increase the efficiency of resource use; (d) reduce human pressures on the local, regional and global environment, and (e) enhance human wellbeing.

4.2.2   *Relevant principles for policy formulation*

When considering climate change response options, several principles and ideas that are used widely in environmental economics analysis would be useful; these include; (a) the polluter pays principle; (b) economic valuation; (c) internalization of externalities, and (d) property rights. In the polluter-pays principle it is argued that those responsible for damaging emissions should pay the corresponding costs. The economic rationale is that this provides an incentive for polluters to reduce their emissions to optimal (i.e. economically efficient) levels. Here, the idea of economic valuation becomes crucial. Quantification and economic valuation of potential damage from polluting emissions is an important prerequisite. In the case of a common property resource such as the atmosphere, greenhouse gas emitters can pollute freely without penalties. Such 'externalities' need to be internalized by imposing costs on polluters that reflect the damage caused. An externality occurs when the welfare of one party is affected by the activity of another party who does not take these repercussions into account in his/her decision-making, e.g. no compensating payments are made. The theoretical basis for this is well known since Pigou (1932) originally defined and treated externalities in rigorous fashion. In this context, the notion of property rights is relevant also to establish that the atmosphere is a valuable and scarce resource that cannot be used freely and indiscriminately.

An important social principle is that climate change should not be allowed to worsen existing inequities, although climate change policy cannot be expected to address all prevailing equity issues. Some special aspects include:

1. The establishment of an equitable and participative global framework for making and implementing collective decisions about climate change.
2. Reducing the potential for social disruption and conflicts arising from climate change impacts.
3. Protection of threatened cultures and preservation of cultural diversity.

From the social equity viewpoint, the polluter-pays principle mentioned above is based not only on economic efficiency, but also on fairness. An extension of this idea is the principle of recompensing victims, ideally by using the revenues collected from polluters. There is also the moral/equity issue concerning the extent of the polluters' obligation to compensate for past emissions, i.e. a form of environmental debt. As mentioned earlier, weighting the benefits and costs of climate change impacts according to the income levels of those who are affected, also has been suggested as one way of redressing inequitable outcomes. Kverndokk (1995) argued that conventional justice principles would favour the equitable allocation of future greenhouse gas emission rights on the basis of population. Equal per capita greenhouse gas emission rights (i.e. equal access to the global atmosphere) is consistent also with the UN Human Rights Declaration underlining the equality of all human beings.

Traditionally, economic analysis has addressed efficiency and distributional issues separately, i.e. the maximization of net benefits is distinct from who might receive such gains. Recent work has sought to interlink efficiency and equity more naturally. For example, environmental services could be considered public goods, and incorporated into appropriate markets as privately produced public goods (Chichilnisky & Heal 2000). Some social equity and economic efficiency interactions are discussed in Box 3.2.

Several concepts from contemporary environmental and social analysis are relevant for developing climate change response options, including the concepts of durability, optimality, safe limits, carrying capacity, irreversibility, non-linear responses, and the precautionary principle. Broadly speaking, durability and optimality are complementary and potentially convergent approaches (see earlier discussion). Under the durability criterion, an important goal would be to determine the safe limits for climate change within which the resilience of global ecological and social systems would not be threatened seriously. In turn, the accumulations of greenhouse gases in the atmosphere would have to be constrained to a point which prevented climate change from exceeding these safe margins. It is considered important to avoid irreversible damage to biogeophysical systems and prevent major disruption of socioeconomic systems. Some systems may respond to climate change in a non-linear fashion, with the potential for catastrophic collapse. Thus, the precautionary principle argues that lack of scientific certainty about climate change effects should not become a basis for inaction, especially where relatively low-cost steps to mitigate climate change could be undertaken as a form of insurance (UNFCCC 1992).

4.3   Sustainable development and adaptation

Sustainable development and adaptation to climate change are interlinked. The great majority of sustainable development strategies are not related to climate change, but they could make adaptation more successful. Similarly, many climate change adaptation policies certainly will help to make development more sustainable.

Adapting to climate change can be spontaneous or planned. Planning is needed to minimize the costs of negative impacts and maximize benefits of positive impacts. Adaptation efforts must be combined with mitigation, since controlling emissions is vital. Even an immediate and dramatic cut in greenhouse gas emissions will not fully prevent climate change impacts, as these gases respond with a time lag.

The most vulnerable (i.e. the extent to which climate change may damage or harm a system) ecological and socioeconomic systems are those with the greatest sensitivity (i.e. the degree to which a system will respond to a given change in climate) to climate change, the greatest exposure, and the least ability to adapt. Ecosystems already under stress are particularly vulnerable, e.g. fragmented systems. Social and economic systems tend to be more vulnerable in developing countries with weaker institutions and economies, e.g. areas with high population density, low-lying coastal areas, flood-prone areas, arid lands, and islands, etc.

Strategies for adapting to climate change and preventing losses are, for example: (a) barriers against sea-level rise; (b) reducing losses (e.g. redesigning crop mixes); (c) spreading or sharing losses (e.g. government disaster relief); (d) changing land use or activity (e.g. relocating agriculture from steep slopes), or (e) restoring a site, e.g. historical monument that has been vulnerable to flood damage. Successful strategies need advances in technology, management and law, finance and economics, public education, training and research, and institutional changes. Notably, the ability to incorporate climate change concerns into development plans can help ensure that new investments in infrastructures reflect likely future conditions. Although the crafting of adaptation policies is complicated by uncertainty, many adaptation policies will help promote sustainable development (e.g. improving natural resource management, or better social conditions), and as a result make sense to be implemented. These issues are elaborated in much more detail in Chapters 5–7.

4.4   Sustainable development and mitigation

Sustainable development and mitigation of climate change also are interlinked. The great majority of sustainable development strategies are not related to climate change, but they could make mitigation more successful. Similarly, many climate change mitigation policies will certainly help to make development more sustainable.

Like adaptation, mitigation has costs and benefits. Many variables need to be considered in the cost equation. Some include:

1.   Internationally agreed timetables and targets for emissions reductions.
2.   Global population and economic trends.
3.   Development of new technologies.
4.   Rate of capital replacement.
5.   Discount rates used for putting a current value on future benefits.
6.   Co-benefits of mitigation action for other environmental, social, and economic issues.
7.   Actions of industry and consumers in response to climate change related policy.

Policies to minimize the risk by reducing greenhouse gas emissions also will come with a price tag and vary widely due to the high degree of uncertainty. Although immediate action sometimes may seem more expensive than waiting, delays could lead to greater risks and therefore greater long-term costs. An early effort towards controlling emissions would increase the long-term flexibility of human responses to work towards stabilizing atmospheric concentrations of greenhouse gases. In Table 8.5 we give a more comprehensive overview of the pros and cons of advancing or delaying action.

Many cost-effective technologies and polices are available (e.g. hybrid-engine cars, wind turbines, advances in fuel cell technology, end-use energy efficiency in building, transport, manufacturing and industry to reduce emissions, etc.) that not only mitigate greenhouse gas emissions, but also address other development objectives by increasing resource efficiency and decreasing environmental pressure. Governments need to promote these solutions actively by addressing institutional and other barriers first. Economic incentives can be used to influence investors and consumers. For example, deposit funds could encourage people to trade-in their cars and appliances for more energy-efficient models; manufacturers could be rewarded for selling climate-friendly goods, or penalized for not doing so. In many cases, market prices do not include externalities such as pollution, etc. Prices could incorporate climate change concerns by introducing or removing taxes or subsidies. For example, a tax on oil, coal, or gas would discourage fossil fuel use, and help reduce carbon dioxide emissions. Tradable emission permits could also offer a cost-efficient, market-driven approach to controlling emissions.

Energy policy is a key to enhancing the cost effectiveness of decreasing emissions, e.g. optimal energy mix. Incentives for investing in cost-effective and energy-efficient technologies are essential. For example, improved building design, new chemicals for refrigeration and insulation, more efficient refrigerators and cooling/heating systems. Power plant emissions of both air pollutants and greenhouse gases can be reduced by switching to renewable sources, e.g. solar, hydro, wind, biomass, etc. Technology innovation, energy efficiency, and an emphasis on renewable energy sources will be essential to stabilize greenhouse gas concentrations in the next 50–100 years, and also will have more immediate benefits for relaxing environmental pressures. Governments need to remove barriers (e.g. cultural, institutional, legal, informational, financial or economic) that may slow the spread of low-emission technologies.

The transport sector is particularly important, since it is the most rapidly growing source of greenhouse gas emissions in most developed and developing countries. Its heavy reliance on fossil fuels makes controlling greenhouse gas emissions particularly difficult, while various air pollutants (e.g. particulate matter and ozone precursors) are released, especially in urban areas. New technologies can increase the efficiency of automobiles and reduce the emissions per kilometre travelled. Switching to less carbon-intensive fuels not only will reduce local and regional air pollution – which is of serious concern in most cities in the developing world as well as in industrialized countries – but also help reduce carbon emissions, e.g. biofuels, and fuel-cell-powered vehicles. Some policies to reduce emissions from transport could include:

1. Use of renewable energy technologies.
2. Improvements in maintenance and operating practices.
3. Policies to reduce traffic congestion.
4. Urban planners to encourage low-emissions transport, e.g. trams and trains, bicycles, and walking.
5. Impose user fees or encourage trade-in of older vehicles that do not meet national emissions standards.

Climate-friendly transport policies can help in promoting economic development while minimizing the local costs of traffic congestion, road accidents, and air pollution.

Deforestation and agriculture account for increases in levels of local and regional air pollution as well as greenhouse gas emissions. Forests need to be protected and better managed. Deforestation can be controlled by: (a) decreasing pressures of agriculture on forestry; (b) slowing down population growth; (c) involving local people in sustainable forest management; (d) harvesting commercial timber sustainably, and (e) addressing socioeconomic and political forces that spur migration into forest areas. Sustainable forest management can generate biomass as a renewable resource to substitute for fossil fuels. Improved management practices to increase agricultural productivity could enable agricultural soils to absorb more carbon, at the same time leading to higher organic matter content and enhanced productivity. Methane emissions from livestock could be reduced with new feed mixtures, since the emissions are a result of inefficient digestion of feed. Such options also increase the overall efficiency of livestock management. Methane emissions from wet rice cultivation can be reduced significantly through changes in irrigation practices and fertilizer use, which can be quite compatible with a more general safeguarding or improvement of the current productivity, while reducing environmental pressures. Nitrous oxides from agriculture can be minimized with new fertilizers and fertilization practices, some of which not only reduce nitrous oxide emissions, but more generally reduce nitrogen losses in the form of polluting nitrates, nitrogen oxides, and ammonia, thus enhancing fertilizer efficiency. The above options are discussed in Chapter 9.

Thus, it is possible to decrease emissions while generating environmental and economic benefits, e.g. decreased pollution, and increased forest cover, etc. The costs of climate change policies can be minimized through 'no regrets' strategies. Such strategies make environmental sense whether or not the world is moving towards rapid climate change. For example, removing market imperfections (e.g. fossil fuel subsidies), and generating double dividends, e.g. use revenue from taxes to finance reductions in existing distortionary taxes. Public participation (i.e. stakeholders, individuals, communities, and businesses) is an important aspect of effective policies. Education and training also are vital for effective policies, e.g. conserving energy, building codes to maximize sunlight, and use of solar power, etc. Equity aspects of policy need to be considered, i.e. cost efficiency and fairness.

Climate change response should adopt actions aimed at mitigation, adaptation, and research. Each country should select its own optimal policy in order to reduce the risk of climate change, while promoting sustainable development.

4.5   Climate change and sustainable development: interactions at the global level

Three simplified examples are presented below, to illustrate the application of some of the concepts discussed earlier.

4.5.1   *The interplay of optimality and durability in determining appropriate global greenhouse gas emission target levels*

Optimization- and durability-based approaches can facilitate the determination of target greenhouse gas emission levels (Munasinghe 1998a). Under an economic optimizing framework, the ideal solution would be first to estimate the long-run marginal abatement cost (MAC) and the marginal avoided damage (MAD) associated with different greenhouse gas emission profiles (see Figure 4.2(a)), in which the error bars on the curves indicate measurement uncertainties (IPCC 1996a). The optimal emission levels would be determined at the point at which future benefits (in terms of climate change damage avoided by reducing 1 unit of greenhouse gas emissions) are just equal to the corresponding costs (of mitigation measures required to reduce that unit of greenhouse gas emissions), i.e. MAC = MAD at point R_{OP}, where R_{OP} is optimum reduction.

Durable strategies become more relevant when we recognize that MAC and/or MAD might be quantified poorly and are uncertain. The example shown in Figure 4.2(b) assumes that MAC is better defined than MAD. First, MAC is determined using technoeconomic least cost analysis (an optimizing approach), while MAD is ignored (for the time being). Next, the target emissions are set on the basis of a safe but affordable minimum standard, at R_{AM}. This value is the upper limit on mitigation costs that will still avoid unacceptable socioeconomic disruption. The choice made here is more consistent with the durability approach.

Finally, Figure 4.2(c) indicates an even more uncertain world, in which neither MAC nor MAD is defined. In this case, the emission target is established on the basis of an absolute standard, R_{AS}, or safe limit. For example, such a safe standard

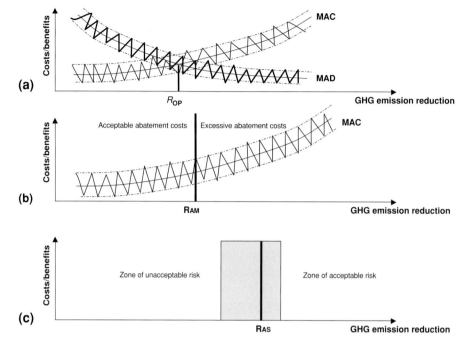

Figure 4.2. Determining mitigation targets: (a) cost–benefit optimum, (b) affordable safe minimum standard, (c) absolute standard. Adapted from IPCC (1996b).

could seek to avoid an unacceptably high risk of damage to ecological (and/or social) systems, without necessarily valuing either mitigation or damage costs in monetary terms. This last approach would be based mainly on the durability concept.

4.5.2   *Combining efficiency and equity to facilitate South–North co-operation for climate change mitigation*

Greenhouse gas mitigation efforts will require worldwide co-operation. Figure 4.3 clarifies the basic rationale for greater South–North resource transfers and technical co-operation, and also highlights how the sustainomics approach elucidates the complex interaction of economic efficiency and social equity (based simply on income distribution) in addressing the climate change problem. The vertical bars indicate the marginal abatement costs for two countries (X is a developing or southern country and Y is an industrialized or northern nation). In other words, the bars show the net additional costs per unit of greenhouse gas emissions reduced by mitigation schemes (over and above the costs of conventional technologies, and including all ancillary costs and benefits). The figure reflects the situation that relatively lower-cost options for greenhouse gas emissions reduction would

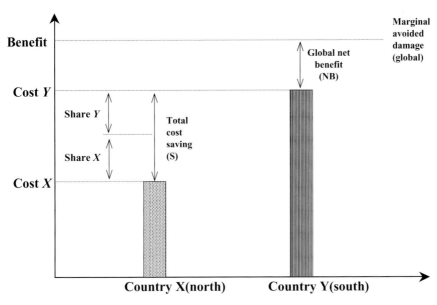

Figure 4.3. Rationale for south-north cooperation and interplay of efficiency and equity. *Source:* Munasinghe and Munasinghe (1993).

be in the developing country, whereas more costly alternatives would lie in the industrialized nation.

The global benefit of mitigation is shown by the upper horizontal line representing the MAD accruing to the entire global community due to greenhouse gas emission reductions. The marginal avoided damage realized by the developing country alone would be negligibly small, since abatement measures undertaken by any given nation will yield predominantly worldwide benefits well beyond the borders of that country. Clearly, if the developing country acted just in its own interests, it would be unwilling to incur any incremental costs of mitigation. In such a situation, only so-called 'win–win' or 'no regrets' options would be pursued, such as energy efficiency schemes for which cost–benefit analysis will show a net economic gain even without considering greenhouse gas abatement benefits, i.e. in which the value of conventional energy savings exceed project costs. Here, the global benefits of mitigation would be considered an 'externality' by the developing country.

From the perspective of the entire world community, all mitigation options should be pursued in all countries, up to the point at which additional costs of the marginal unit of emissions curtailed are equal to the corresponding benefits of avoided global warming damages. In this case, the avoided damages that are 'external' to the mitigating country, should be 'internalized' from the global viewpoint.

First, we explore the implications of this broad environmental rationale for resource transfers from the North to the South. In this context, consider a representative greenhouse gas mitigation project (e.g. reafforestation) in the developing country X, in which the additional costs of greenhouse gas emission reduction is less than the global avoided damage. From the perspective of the global community, it would be *economically efficient* for them to finance any additional costs (e.g. on a grant basis) in the developing country. They would thereby 'internalize' and capture the worldwide net mitigation benefits equivalent to global net benefit, *NB*, plus total cost saving, *S*.

Second, we make the case for a bilateral transfer of resources from an industrialized to a developing country. Consider the cost of a project (e.g. conversion of coal plants) which seeks to reduce greenhouse gas emissions in the industrialized country Y. This country could realize a cost saving if it were able to pursuade the developing country X to undertake the mitigation measure, while still achieving the same global emissions reduction benefit. The minimum compensation acceptable to country X would be its cost X. The maximum payment country Y would be willing to make, is *S* (cost Y − cost X). The foregoing could be the basis for bilateral co-operative schemes such as joint implementation and/or the Clean Development Mechanism under the Kyoto Protocol. To the extent that global net benefit, *NB*, and total cost saving, *S*, are significant, it would be both *equitable* and *efficient* for the industrial nation to give the poorer developing country more resources than the (minimum) break-even reimbursement cost X. For example, the cost saving could be shared in the proportions share Y and share X between the industrial and developing nations, respectively, where share X + share Y = S. In other words, the equity principle of sustainomics would favour such a sharing of potential cost savings between the two co-operating nations, based on the following underlying ethical arguments.

1. Both the historical and current levels of per capita greenhouse gas emissions from the industrial country are likely to be many times the corresponding contribution from the developing nation.
2. The per capita income and ability to pay of the industrial country would be many times greater than those of the developing country.

This also would provide a greater incentive for the developing country to participate in such a scheme. The same argument was proposed originally in the case of South–North co-operation to reduce ozone-depleting substances under the Montreal Protocol (Munasinghe & King 1992; Munasinghe & Munasinghe 1993).

4.5.3   *Equity and efficiency in emissions trading*

Equity and efficiency principles may be applied usefully in allocating the mitigation burden among various countries, to achieve a target level of desirable

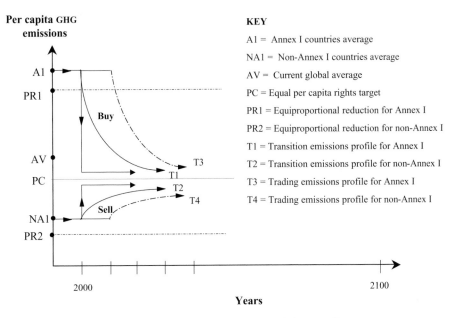

Figure 4.4. Combining efficiency and equity in emissions trading.

worldwide greenhouse gas emission in the future (see also Section 4.5.1). Suppose that there is international agreement to aim for stabilization of greenhouse gas concentrations, e.g. in the range of 550 ppmv in the next 200 years (see Chapter 2). This would fix the total global emissions profiles in the future.

Consider two contrasting rules for allocating rights to the fixed global level of future emissions year by year, among the different nations.

1. Equal per capita emission rights for all human beings, based on ethics and human right. The total national 'right to emit' would then be the product of the population and the basic per capita emissions quota, with all national quotas summing up to the desired global emissions target.
2. Equiproportional reductions of emissions, based on so-called 'grand-fathering'. In this case, all countries would reduce emissions by the same percentage amount relative to some pre-agreed baseline year, to achieve the desired global emissions target.

The dynamics of this allocation process are shown in Figure 4.4. The line PC indicates the constant level of per capita emissions, if the total global emissions target were allocated equally to all human beings during the decision-making time horizon.

The points AI and NA1 represent the average current per capita greenhouse gas emissions of the Annex I (industrialized), and non-Annex I (developing) countries, respectively. Although Figure 4.4 is not exactly to scale, AI is considerably larger and NAI is somewhat less than PC. Thus, the industrialized countries would need to cut back greenhouse gas emissions significantly if they were to meet the PC criterion – which would entail economic costs (depending on the severity of the curtailment in each country). On the other hand, the developing countries have some 'headroom' to increase their per capita emissions, as incomes and energy consumption grow.

The alternative allocation rule is based on equiproportional reductions of emissions. In this case, all countries would reduce emissions by the same percentage amount relative to some pre-agreed baseline year, in order to achieve the desired global emissions target. Assuming that the global average emission rate per capita per year is slightly higher than PC, this implies that all countries would need to curtail carbon emissions by a small amount (say about 10 per cent), to meet the PR criterion (as shown by the broken lines PR1 and PR2 in Figure 4.4). Clearly, such an outcome would be highly inequitable, since it would severely restrict growth prospects in the developing world – where per capita energy consumption is quite low, to begin with (Munasinghe 1995b).

Thus, both developed and developing countries would have reasons to oppose strict enforcement of the PC and PR approaches. Another related equity issue is whether past emissions should be considered also, or ignored in deciding the current and future quotas, since over 80 per cent of carbon accumulated up to 2000 has resulted from fossil fuel use in the industrialized world. Clearly, the industrialized countries have used up a significant share of the 'global carbon space' available to humanity while driving up atmospheric carbon dioxide concentrations from the pre-industrial norm of 280 ppmv to the current level of about 380 ppmv. Therefore, the developing countries argue that responsibility for past emissions should be considered when future rights are allocated.

Correspondingly, it would be in the interest of the industrialized countries to use a fixed base year population (in the year 2000, for example) as the multiplier of the per capita emissions right (e.g. PC in Figure 4.4) in determining total national emission quotas. This would effectively penalize those countries that had high population growth rates, since their allowed national quota (determined by the base year population) would have to be divided up among more people in the future.

In practice, it is possible that some intermediate requirement that falls between PC and PR might emerge eventually from the collective decision-making process. For example, PC may be set as a long-term goal. In the shorter run, pragmatic considerations suggest that both the industrialized and transition countries be given a period of time to adjust to the lower greenhouse gas emissions level, in order to avoid undue economic disruptions and hardship – especially to poorer groups within those countries (see transition emissions paths T1 and T2 in

Figure 4.4). Even if some industrialized nations might argue that the goal of PC emissions rights for all individuals is too idealistic or impractical, the directions of adjustment are clear. Net carbon dioxide emissions per capita in industrialized countries should trend downwards, while such emissions in developing countries will increase with time. This result will emerge even if the objective is a more equitable distribution of per capita emissions, rather than absolute equality of per capita emissions.

Another adjustment option could be the facilitation of an emissions trading system. For example, once national emissions quotas have been assigned, a particular developing country may find that it is unable fully to utilize its allocation in a given year. At the same time, an industrialized country might find it cheaper to buy such 'excess' emissions rights from the developing nation, rather than undertaking a much higher cost abatement programme to cut back emissions and meet its own target (see also Section 4.5.2). More generally, the emissions trading system would permit emissions quotas to be bought and sold freely on the international market, thereby establishing an efficient current price and eventually a futures market for greenhouse gas emissions. In Figure 4.4, new transition emission profiles T3 and T4 might emerge, creating potential space that permits buying and selling of carbon rights.

4.6   Climate change and sustainable development: interactions at the national and subnational levels

While much of the work on climate change issues has focused on the global or regional level, its eventual impact and ultimate responses will be relevant mainly at the national and subnational levels. As seen in the previous sections, climate change is likely to undermine the sustainability of future development. Correspondingly, the choice of development paths will have as great an (indirect) influence on climate change as mitigation and adaptation policies designed explicitly for climate change. Therefore, climate change strategy needs to be harmonized with national sustainable development policies.

4.6.1   *National economywide policies*

Scope of policies

The most powerful economic management tools currently in common use are economywide reforms that include structural adjustment packages. Economywide (or countrywide) policies consist of both sectoral and macroeconomic policies that have widespread effects throughout the economy. Sectoral measures mainly involve a variety of economic instruments, including pricing in key sectors (e.g. energy or agriculture) and broad sectorwide taxation or subsidy programmes, e.g. agricultural production subsidies, and industrial investment incentives. Macroeconomic measures are even more sweeping, ranging from exchange rate, interest rate, and wage policies, to trade liberalization, privatization, and similar programmes. Since space limitations preclude a comprehensive review of interactions between economywide policies and climate change, we examine briefly several examples that provide a flavour of the possibilities involved (Munasinghe 1997).

Range of impacts

On the positive side, liberalizing policies (e.g. the removal of price distortions and promotion of market incentives) have the potential to improve economic growth rates while increasing the value of output per unit of greenhouse gas emitted, i.e. so-called 'win–win' outcomes. For example, reforms that improve the efficiency of energy use could reduce economic waste and lower the intensity of greenhouse gas emissions. Similarly, improving property rights and strengthening incentives for better land management not only yield economic gains, but also reduce deforestation of open access lands, e.g. due to 'slash and burn' farming.

At the same time, growth inducing economywide policies could lead to increased greenhouse gas emissions, unless the macroreforms are complemented by additional environmental and social measures. Such negative impacts on climate change invariably are unintended and occur when some broad policy changes are undertaken while other hidden or neglected economic and institutional imperfections persist (Munasinghe 1998a). In general, the remedy does not require reversal of the original reforms, but rather the implementation of additional complementary measures (both economic and non-economic) that mitigate climate change. For example, export promotion measures and currency devaluation might increase the profitability of timber exports (see Box 4.1). This in turn, could further accelerate deforestation that was already under way due to low stumpage fees and open access to forest lands. Establishing property rights and increasing timber charges would reduce deforestation, without interrupting the macroeconomic benefits of trade liberalization. Similarly, market-oriented liberalization could lead to economic expansion and the growth of wasteful energy-intensive activities in a country in which subsidized energy prices persisted. Eliminating the energy price subsidies could help to reduce net greenhouse gas emissions while enhancing macroeconomic gains. Countrywide policies could also influence adaptation, negatively or positively. For example, national policies that encouraged population movement into low-lying coastal areas might increase their vulnerability to future impacts of sea-level rise. On the other hand, government actions to protect citizens from natural disasters (e.g. investing in safer

Box 4.1   Action impact matrix: a tool for policy analysis
and formulation

The sustainomics approach seeks to identify and analyse
economic–environmental–social interactions, and thereby formulate more
sustainable development policies. One tool that would facilitate the
implementation of such an approach is the action impact matrix (a simple
example is shown in Table 4.1, although an actual action impact matrix
would be very much larger and more detailed (Munasinghe & Cruz 1994).
Such a matrix helps to promote an integrated view, meshing development
decisions with priority economic, environmental, and social impacts. The far
left column lists examples of the main development interventions (both
policies and projects), while the top row indicates some typical sustainable
development issues. Thus, the elements or cells in the matrix help to:

1.   Identify explicitly the key linkages.
2.   Focus attention on methods of analysing the most important
     impacts.
3.   Suggest action priorities and remedies.

At the same time, the organization of the overall matrix facilitates the
tracing of impacts, as well as the coherent articulation of the links among a
range of development actions, both policies and projects.

A stepwise procedure, starting with readily available data, has been used
effectively to develop the action impact matrix in several country studies
that have been initiated recently, e.g. Ghana, Philippines, and Sri Lanka. This
process has helped to harmonize views among those involved (economists,
ecologists, sociologists, and others), thereby improving the prospects for
successful implementation (Munasinghe & Cruz 1994).

Screening and problem identification

One of the early objectives of the action impact matrix-based process
is to help in *screening and problem identification* – by preparing a preliminary
matrix that identifies broad relationships and provides a qualitative idea of
the magnitudes of the impacts. Thus, the preliminary action impact matrix
would be used to prioritize the most important links between policies and
their sustainability impacts. For example, in column 2 of Table 4.1, a
currency devaluation aimed at improving the trade balance, may make
timber exports more profitable and lead to deforestation of open-access
forests. Column 3 indicates severe land degradation and biodiversity loss.

Further down the same column, one appropriate remedy might involve complementary measures to strengthen property rights and restrict access to the forest areas.

A second example shown in column 3 involves increasing energy prices closer to marginal costs – to improve energy efficiency, while decreasing air pollution and greenhouse gas emissions. A complementary measure indicated in column 4 consists of adding pollution taxes to marginal energy costs, which will further reduce air pollution and greenhouse gas emissions. Increasing public sector accountability will reinforce favourable responses to these price incentives, by reducing the ability of inefficient firms to pass on cost increases to consumers or to transfer their losses to the government. In the same vein, a major hydroelectric project is shown lower down in Table 4.1 as having two adverse impacts (inundation of forested areas and village dwellings), as well as one positive impact (the replacement of fossil fuel power generation, thereby reducing air pollution and greenhouse gas emissions). A reafforestation project coupled with resettlement schemes may help address the negative impacts.

This matrix-based approach therefore encourages the systematic articulation and co-ordination of policies and projects to make development more sustainable. Based on readily available data, it would be possible to develop such an initial matrix for many countries and projects.

Analysis and remediation

This process may be developed further to assist in *analysis* and *remediation*. For example, more detailed analyses and modelling may be carried out for those matrix elements in the preliminary action impact matrix that had been identified already as representing high-priority linkages between economywide policies and economic, environmental, and social impacts. This, in turn, would lead to a more refined and updated action impact matrix that would help to quantify impacts and formulate additional policy measures to enhance positive linkages and mitigate negative ones.

The types of more detailed analyses that could help to determine the final matrix would depend on planning goals and available data and resources. They may range from fairly simple to rather sophisticated economic, ecological, and social models, in the sustainomics toolkit.

Table 4.1. *A simplified preliminary action impact matrix*[1]

		Impacts on key sustainable development issues			
A. Activity/Policy	Main objective	Land degradation biodiversity loss	Air pollution greenhouse gas emissions	Resettlement social effects	Others
A. Macroeconomic and sectoral policies	Macroeconomic and sectoral improvements	Positive impacts due to removal of distortions Negative impacts mainly due to remaining constraints			
A.1 Exchange rate	Improve trade balance and economic growth	(−H) (deforest open access areas)			
A.2 Energy pricing	Improve economic and energy use efficiency		(+M) (energy efficiency)		
Others					
B. Complementary measures and remedies[2]	Specific socioeconomic and environmental gains	Enhance positive impacts and mitigate negative impacts (above) of broader macroeconomic and sectoral policies			
B.1 Market based			(+M) (pollution tax)		
B.2 Non-market based		(+H) (property rights)	(+M) (public sector accountability)		
C. Investment projects	Improve effectiveness of investments	Investment decisions made more consistent with broader policy and institutional framework			
C.1 Project 1 (hydro-electric dam)		(−H) (inundate forests)	(+M) (displace fossil fuel use)	(−M) (displace people)	
C.2 Project 2 (re-afforestate and relocate)		(+H) (replant forests)		(+M) (relocate people)	

Source: Munasinghe and Cruz (1994).

Notes: [1] A few examples of typical policies and projects as well as key economic, environmental, and social issues are shown. Some illustrative but qualitative impact assessments are also indicated: thus + and − signify beneficial and harmful impacts, while H and M indicate high and moderate intensity. The action impact matrix process helps to focus on the highest priority socioeconomic and environmental issues.

[2] Commonly used market based measures include effluent charges, tradable emission permits, emission taxes or subsidies, bubbles and offsets (emission banking), stumpage fees, royalties, user fees, deposit-refund schemes, performance bonds, and taxes on products (such as fuel taxes). Nonmarket based measures comprise regulations and laws specifying environmental standard (e.g. ambient, emission, and technology standards) which permit or limit certain actions ('dos' and 'don'ts').

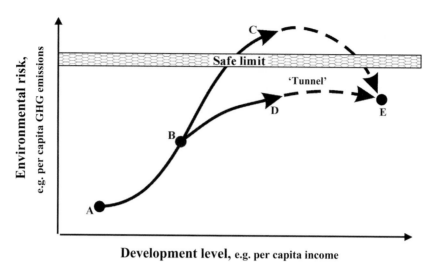

Figure 4.5. Environmental risk versus development level. *Source:* Munasinghe (1998a).

physical infrastructure or strengthening the social resilience of poorer communities) could help to reduce vulnerability to extreme weather events associated with future climate change (Clarke & Munasinghe 1995).

In this context, the sustainomics approach helps to identify and analyse economic–environmental–social interactions, and formulate effective sustainable development policies, by linking and articulating these activities explicitly. Implementation of such an approach would be facilitated by constructing a simple action impact matrix (Munasinghe 1997). Such a matrix could help promote an integrated view, by meshing development and climate related decisions with priority economic, environmental, and social issues (see Box 4.1).

4.6.2   *Restructuring growth*

Economic growth continues to be a widely pursued objective of most governments, and the sustainability of long-term growth is a key issue (Munasinghe, Sunkel & de Miguel 2001). In particular the reduction of the intensity of greenhouse gas emissions is an important step in mitigating climate change (Munasinghe & Swart 2000). Given that the majority of the world population lives under conditions of absolute poverty (e.g. over 3 billion subsist on less than US $1 per day), a climate change strategy that does not unduly constrain growth prospects in those areas would be more attractive. A sustainomics-based approach would seek to identify measures that modify the structure of development and growth (rather than restricting it), so that greenhouse gas emissions are mitigated and adaptation options enhanced.

The above approach is illustrated in Figure 4.5, which level shows how a country's greenhouse gas emissions might vary with its level of development. One would expect carbon emissions to rise more rapidly during the early stages of development (along AB), and begin to level off only when per capita incomes are higher (along BC). A typical developing country would be at a point such as B on the curve, and an industrialized nation might be at C. The key point is that if the developing countries were to follow the growth path of the industrialized world, then atmospheric concentrations of greenhouse gases would soon rise to dangerous levels. The risk of exceeding the safe limit (shaded area) could be avoided by adopting sustainable development strategies that would permit developing countries to progress along a path such as BD (and eventually DE), while also reducing greenhouse gas emissions in industrialized countries along CE.

As outlined earlier, growth-inducing economywide policies could combine with imperfections in the economy to cause environmental harm. Rather than halting economic growth, complementary policies may be used to remove such imperfections and thereby protect the environment. It would be fruitful to encourage a more proactive approach whereby the developing countries could learn from the past experiences of the industrialized world by adopting sustainable development strategies and climate change measures that would enable them to follow development paths such as BDE in Figure 4.5 (Munasinghe 1997). Thus, the emphasis is on identifying policies that will help delink carbon emissions and growth, with the curve in Figure 4.5 serving mainly as a useful metaphor or organizing framework for policy analysis.

This representation also illustrates the complementarity of the optimal and durable approaches discussed earlier. It has been shown that the higher path ABC in Figure 4.5 could be caused by economic imperfections that make private decisions deviate from socially optimal ones (Munasinghe 1998b). Thus, the adoption of corrective policies that reduce such divergences and thereby reduce greenhouse gas emissions per unit of output would facilitate movement along the lower path ABD. From the durability viewpoint, reducing the higher level of environmental damage at C would especially be desirable in order to avoid exceeding the safe limit or threshold representing dangerous accumulations of greenhouse gases (shaded area in Figure 4.5).

Several authors have estimated econometrically the relationship between greenhouse gas emissions and per capita income using cross-country data, and found curves with varying shapes and, sometimes, turning-points (Cole, Raynes & Bates 1997; Holtz-Eakin & Selden 1995; Sengupta 1996; Unruh & Moomaw 1998). One reported outcome is an inverted U-shape (called the environmental Kuznet's curve) e.g. like the curve ABCE in Figure 4.5. However, in most cases – and unlike the situation for other pollutants such as acidifying substances – such 'delinking' of income increases and environmental pressures has not yet been proven for greenhouse gases. This does not, however, exclude the possibility that in the future still higher incomes can be – or should be – associated with decreasing levels of greenhouse gas emissions. In order to achieve the ultimate objective of the climate convention, greenhouse gas concentrations have to be stabilized, which implies that emissions have to be reduced considerably, starting in the developed countries, but eventually also in the developing countries. This leads to the more socially and environmentally optimal path BDE, which could be viewed as a sustainable development 'tunnel' through the environmental Kuznet's curve (Munasinghe 1995a).

4.6.3   Subnational and project-level interactions

The procedures for conventional environmental and social impact assessment at the project/local level (which are now well accepted worldwide), may readily be adapted to assess the effects of microlevel activities on greenhouse gas emissions (World Bank 1998). The OECD (1994) has pioneered the 'Pressure–State–Response' framework to trace socioeconomic–environment linkages. This approach begins with the pressure (e.g. population growth and emissions), then seeks to determine the state of the environment (e.g. ambient pollutant concentration), and ends by identifying the policy response, e.g. pollution taxes. In Box 4.2, we illustrate how a multicriteria analysis-based analysis at the project level could provide balanced treatment of economic, social, and environmental considerations. The project evaluated involves the case of an improved woodburning stove.

Conventional economic valuation of environmental impacts is a key step in incorporating the results of project level environmental impact assessment into economic decision-making, e.g. cost–benefit analysis. At the macroeconomic level, recent work has focused on incorporating environmental considerations, e.g. depletion of natural resources and pollution damage into the system of national accounts (Atkinson *et al.* 1997; *UNSO* 1993). These efforts have yielded useful new indicators and measures (e.g. the system of environmentally adjusted environmental accounts, green gross national product (GNP), and genuine savings) that adjust conventional macroeconomic measures to allow for environmental effects. Costanza (2000) seeks to broaden the definition of valuation to include:

1.   Efficiency-based values (conventional economic willingness-to-pay).
2.   Fairness-based values (that capture community or social preferences).
3.   Sustainability-based values (that are related to contributions to system-wide and global functions).

At the same time, national policy-makers routinely make many key macro-level decisions that could have (often inadvertent) impacts on both climate change

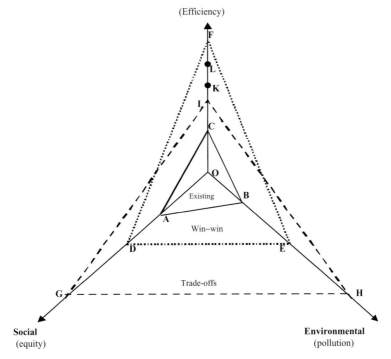

Figure 4.6. Analysing the sustainability of an improved woodburning stove using multicriteria analysis. *Source:* Munasinghe (1995b).

mitigation and adaptation which are far more significant than the effects of local economic activities. These pervasive and powerful measures are aimed at addressing economic development, environmental sustainability and social equity issues, which invariably have much higher priority in national agendas than climate change. For example, many macroeconomic policies seek to induce rapid growth, which in turn potentially could result in greater levels of greenhouse gas emissions or increase vulnerability to the future impacts of climate change. More attention needs to be paid to such economywide policies, the environmental and social linkages of which have not been explored adequately in the past.

Clearly, climate change strategies and policies consistent with other national development measures are more likely to be effective than isolated technological or policy options. In particular, the highest priority needs to be given to finding 'win–win policies' that not only make local and national development more sustainable, but also enhance climate change adaptation and mitigation efforts. Such policies could help build support for climate change strategies among the traditional decision-making community and, conversely, make climate change specialists more sensitive to sustainable development needs. They would

Box 4.2 Multicriteria analysis of a fuelwood stove project

Multicriteria analysis offers policy-makers an alternative when progress toward multiple objectives cannot be measured in terms of a single criterion, e.g. monetary values. Take the case of an efficient woodburning stove – an end-use option for sustainable energy development. While the economic value of such a stove is measurable, its contribution to social and environmental goals is not easily valued monetarily. As shown in Figure 4.6, outward movements along the axes trace improvements in three indicators: economic efficiency (net monetary benefits), social equity (improved health of poor energy users), and environmental pollution (reduced deforestation and greenhouse gas emissions).

We may assess the policy options as follows. First, triangle ABC represents the existing method of burning fuelwood (typically placing the cooking pot on three bricks). In this case, the indicators of economic efficiency, social equity, and overall environmental impact are all bad, because the stove uses fuelwood inefficiently, increases smoke inhalation (especially of women and children in poor households), and worsens greenhouse gas emissions and pressure on forest resources. Next, triangle DEF indicates a 'win–win' future option based on an improved woodburning stove, in which all three indices improve. The economic gains would include monetary savings from reduced fuelwood use and increased productivity from reductions in acute respiratory infections, lung disease, and cancer caused by pollutants in biomass smoke. Social gains would accrue from the fact that the rural poor benefit the most from this innovation, e.g. the lighter health and labour burden on women and children, and the reduced time spent on collecting fuelwood, thereby increasing time spent on other productive activities. The environment benefits occur because more efficient use of fuelwood will reduce both deforestation and greenhouse emissions resulting from inefficient combustion.

After realizing such 'win–win' gains, other available options would require trade-offs. In triangle GHI, further environmental and social gains are attainable only at the expense of sharply increasing costs. For example, shifting from fuelwood to liquid petroleum gas or kerosene as a fuel may increase economic costs, while yielding further environmental and social benefits. A policy-maker may not wish to make a further shift from DEF to GHI without knowing the relative weights that society places on the three indices – in sharp contrast to the move from ABC to DEF, which is unambiguously desirable. Such social preferences are often difficult to determine explicitly, but it is possible to narrow the options. Suppose a small economic cost, FL, yields the full social gain DG, while a large economic cost, LI, is required to realize the environmental benefit EH. Here, the social gain may better justify the economic sacrifice. Further, suppose that budgetary constraints limit costs to less than FK (where FL < FK < LI). Then, sufficient funds exist only to pay for the social benefits, and the environmental improvements will have to be delayed.

A recent study of power system planning in Sri Lanka has demonstrated the versatility and balance inherent in the multicriteria analysis approach (Meier & Munasinghe 1995). In this case, end-use energy efficiency, and demand side management measures, including fluorescent lighting, high efficiency electric motors, and pricing policy, provided 'win–win' options. They proved superior to all other alternatives (e.g. supply options such as hydroelectricity, and oil- and coal-fired generating plants), in relation to three key attributes: (a) a social indicator based on the effects of air quality on human health; (b) an environmental indicator based on an index of biodiversity loss, and (c) an economic indicator measured by monetary costs. Conversely, several prominent hydropower projects were excluded because they performed poorly in terms of both biodiversity loss and economic costs.

reduce the potential for conflict between two powerful current trends: the growth oriented, market based economic reform process, and protection of the global environment.

References

Adger, W. N. (1999) Social vulnerability to climate change and extremes in coastal Vietnam. *World Development*, **27**(2), 249–69.

Atkinson, G., Dubourg R., Hamilton, K., Munasinghe, M., Pearce, D. W. and Young, C. (1997) *Measuring Sustainable Development: Macroeconomics and the Environment.* Cheltenham: Edward Elgar.

Banuri, T. (1998) Human and environmental security. *Policy Matters*, **3**.

Banuri, T., Hyden, G., Juma, C. and Rivera, M. (1994) *Sustainable Human Development: From Concept To Operation: A Guide For The Practitioner.* New York: United Nations Development Programme.

Burton, I. (1997) Vulnerability and adaptive response in the context of climate and climate change. *Climatic Change*, **36** (1–2), 185–96.

Chichilnisky, G. and Heal, G., eds. (2000) *Environmental Markets: Equity and Efficiency.* New York: Columbia University Press.

Clarke, C. and Munasinghe, M. (1995) Economic aspects of disasters and sustainable development. In M. Munasinghe and C. Clarke, eds., *Disaster Prevention for*

Sustainable Development. Geneva and Washington: International Decade of Natural Disaster Reduction and World Bank.

Cole M. A., Rayner, A. J. and Bates, J. M. (1997) Environmental quality and economic growth. *Department of Economics Discussion Paper 96/20*. Nottingham: University of Nottingham, pp. 1–33.

Costanza, R. (2000) Ecological sustainability, indicators and climate change. In M. Munasinghe and R. Swart, eds., *Climate Change and its Linkages with Development, Equity and Sustainability*. Geneva: Intergovernmental Panel on Climate Change.

Holtz-Eakin, D. and Selden, T. M. (1995) Stoking the fires? CO_2 emissions and economic growth. *National Bureau of Economic Research Working Paper Series*, **4248**, pp. 1–38.

IPCC (1990) Houghton, H. T., Jenkins, G. J. and Ephraums, J. J., eds., *Climate Change: the IPCC Scientific Assessment*. Cambridge: Cambridge University Press.

IPCC (1995) *The Second Assessment Report*. Geneva: Intergovernmental Panel on Climate Change.

IPCC (1996a) (J. P. Bruce, H. Lee and E. F. Haites, eds.) *Climate Change 1995: Economic and Social Dimensions of Climate Change*. Contribution of Working Group III to The Second Assessment Report of the Intergovernmental Panel on Climate Change. Cambridge: Cambridge University Press.

IPCC (1996b) (R. T. Watson, M. C. Zinyowera, R. H. Moss and D. J. Dokken, eds.) *Climate Change 1995: Impacts, Adaptations, and Mitigation of Climate Change: Scientific–Technical Analyses*. Contribution of Working Group II to The Second Assessment Report the Intergovernmental Panel on Climate Change. Cambridge: Cambridge University Press.

IPCC (2001) *Climate Change 2001. Climate Change 2001:* Synthesis Report. Intergovernmental Panel on Climate Change *Third Assessment Report*. Cambridge: Cambridge University Press.

Kverndokk, S. (1995) Tradeable CO_2 emission permits: initial distribution as a justice problem. *Environmental Values*, **4**, 129–48.

Lonergan, S. L. (1993) Impoverishment, population and environmental degradation: the case for equity. *Environmental Conservation*, **20**(4), 328–34.

Meier, P. and Munasinghe, M. (1995) *Incorporating Environmental Concerns into Power Sector Decision-Making: A Case Study of Sri Lanka*. Washington: World Bank.

Munasinghe, M. (1995a) Making growth more sustainable. *Ecological Economics*, **15**, 121–4.

Munasinghe, M. (1995b) *Sustainable Energy Development*. Washington: The World Bank.

Munasinghe, M., ed. (1997) *Environmental Impacts of Macroeconomic and Sectoral Policies*. Solomons MD and Washington: International Society for Ecological Economics and World Bank.

Munasinghe, M. (1998a) Climate change decision-making: science, policy and economics. *International Journal of Environment and Pollution*, **10**(2), 188–239.

Munasinghe, M. (1998b) Countrywide policies and sustainable development: are the linkages perverse? In T. Teitenberg and H. Folmer, eds. *The International Yearbook of International and Resource Economics*. London: Edward Elgar.

Munasinghe, M. (2000) *Development, Equity and Sustainability in the Context of Climate Change.* Geneva: Intergovernmental Panel on Climate Change.

Munasinghe M. and Cruz, W. (1994) *Economywide Policies and the Environment.* Washington: World Bank.

Munasinghe, M. and King, K. (1992) Accelerating ozone layer protection in developing countries. *World Development*, **20**, 609–618.

Munasinghe, M. and Munasinghe, S. (1993) Enhancing North–South cooperation to reduce global warming. *Proceedings, IPCC Meeting on Climate Change, Montreal.* Geneva: Intergovernmental Panel on Climate Change.

Munasinghe, M. and eds., R., Swart (2000) *Climate Change and its Linkages with Development, Equity and Sustainability.* Geneva: Intergovernmental Panel on Climate Change.

Munasinghe, M., Sunkel O. and de Miguel, C. eds. (2001) *The Sustainability of Long Term Growth.* London: Edward Elgar.

OECD (1994) *Environmental Indicators.* Paris: Organization for Economic Co-operation and Development.

Pigou, A. C. (1932) *The Economics of Welfare.* London: Macmillan.

Ruitenbeek, H. J. (1996) Distribution of ecological entitlements: implications for economic security and population movement. *Ecological Economics*, **17**, 49–64.

Sengupta, R. (1996) *Economic Development and CO_2 Emissions.* Institute for Economic Development. Boston: University of Boston.

IPCC (2001)

UNSO (1993) *Integrated Environmental and Resource Accounting.* Series F, No. 61. New York: United Nations Statistical Office.

UNFCCC (1993) *Framework Convention on Climate Change: Agenda 21.* New York: United Nations Framework Convention on Climate Change.

Unruh, G. C. and Moomaw, W. R. (1998) An alternative analysis of apparent EKC-type transitions. *Ecological Economics*, **25**, 221–9.

Westing, A. (1992) Environmental refugees: a growing category of displaced persons. *Environmental Conservation*, **19**(3), 201–7.

World Bank (1998) *Environmental Assessment Operational Directive 4.01.* Washington: World Bank.

5

Adaptation to climate change: concepts and linkages with sustainable development

5.1   Introduction to adaptation

5.1.1   Rationale for adaptation

Global losses due to climate-related disasters have increased by a factor of 40 since the 1960s (IPCC 1998). While these disasters may be related to climate variability rather than change, the losses underline the vulnerability of societies to climatic events. As we have seen in Chapter 1, impacts from climate change have been observed already and, because of continued greenhouse gas emissions and the inertia of the climate system, the impacts are projected to continue and increase in the future (McCarthy *et al.* 2001). Even if international mitigation is successful and greenhouse gas concentrations are stabilized, the climate will continue to change for quite some time, and adaptation would be a key component of any climate change response strategy. Mitigation by stabilizing greenhouse gas concentrations would help to delay climatic impacts and reduce their severity, but can never fully substitute for adaptation.

Adaptation reduces the impact of climate stresses on human and natural systems, while mitigation lowers the potential greenhouse gas emissions. Development paths affect strongly the capacity to both adapt to, and mitigate, climate change in any region. In this way, adaptation and mitigation strategies are connected dynamically with changes in the climate system and the prospects for ecosystem adaptation, food production, and long-term economic development.

It is interesting to note that, notwithstanding the apparent inevitability of climate change, there has been a very strong bias towards mitigation both in the political debate (e.g. in the United Nations Framework Convention on Climate Change (UNFCCC)) and in scientific enquiry. The reasons for adaptation originally receiving less attention include 'political incorrectness' (talking about adaptation suggests that mitigation would be less important, adaptation suggests fatalism), and lack of precision (there are large uncertainties about what exactly to adapt to, including adaptation in international negotiations appearing to be more difficult than mitigation (see also Pielke 1998)). Furthermore, at the outset, impact analyses did not take into account any adaptation, suggesting that there is no other option than mitigation (Tol, Fankhauser & Smith 1998). One may even think of a situation in which a country may consider itself fully able to cope with the climatic changes locally and thus may not be interested in participating in international negotiations on a co-ordinated climate-response strategy (Toth *et al.* 2001). Furthermore, many believe that there is no need specifically to study adaptation, because it would be likely to happen anyway, without any significant costs, e.g. through natural selection or market forces (Kates 2000). Finally, there was a lot of initial optimism that mitigation would be quite possible, probably based on the positive experiences with the internationally co-ordinated abatement of ozone-depleting substances and acidification. Later on, however, climate change appeared to be a much harder problem to address. Pielke (1998) also notes that even if climate change could be mitigated successfully, adaptation would still be very relevant, since many current developments increase vulnerability to climatic events (development of marginal lands and lands at risk to extreme events, increased dependence on highly technical interdependent systems, increased water and food demands, etc.).[1]

Climate change impacts and adaptation are part of the larger question of how complex social, economic, and environmental subsystems interact and shape prospects for sustainable development. Economic development affects ecosystem balance and, in turn, is affected by the state of the ecosystem. Poverty can be both a result and a cause of environmental degradation. Material and energy-intensive lifestyles and continued high levels of consumption supported by non-renewable resources along with high population growth are likely to be unsustainable. Socioeconomic and technology policy decisions made for non-climate-related reasons have significant implications for climate policy and climate change impacts, as well as for other environmental issues. Critical impact thresholds and vulnerability to climate change impacts are connected directly to environmental, social, and economic conditions and institutional capacity.

Here, we discuss adaptation from two viewpoints. First, understanding adaptation is important to assess climate impacts. Article 2 of the UNFCCC (its 'ultimate objective') targets stabilization of greenhouse gas concentrations at a level, and within a timescale, that would allow ecosystems to adapt naturally to climate

---

[1] One may argue against this that economic growth with higher incomes and higher health and educational standards and lower importance of climate-sensitive sectors such as agriculture is likely to lead to lower vulnerability (Fankhauser, Smith & Tol 1999).

change while not threatening food production or hindering sustainable economic development. Thus, the ultimate objective of the UNFCCC is dependent directly on the adaptive capacity of natural, agricultural, and economic systems. Adaptation can, to some extent, alleviate climate change impacts and is one of the policy responses to climate change that has to be considered alongside mitigation. For example, the UNFCCC considers adaptation as part of a balanced response to climate change, calling — among other aspects of adaptation — for co-operation 'in preparing for adaptation to the impacts of climate change, developings and elaborating appropriate and integrated plans for coastal zone management, water resources and agriculture, and for the protection and rehabilitation of areas, particularly in Africa, affected by drought and desertification, as well as floods' (UNFCCC 1992, Article 4).[2] The Kyoto Protocol also recognizes the importance and urgency of adaptation, stipulating that a share of the proceeds from certified project activities under the Clean Development Mechanism should be used, among other things, to assist developing country parties that are particularly vulnerable to the adverse effects of climate change to meet the costs of adaptation (UNFCCC 1997). In the Marrakech Accords, a special adaptation fund and a fund for supporting the least developed countries in their climate-change response actions, including adaptation, were agreed.

In the next three chapters we discuss various aspects of adaptation. In this chapter, we first discuss concepts and methods, and place adaptation in the context of broader development issues. Then we elaborate the concept of adaptive capacity, and finally discuss potential costs and benefits of adaptation. In Chapter 6, we discuss more specifically the characteristics of adaptation options for different systems and sectors: (a) hydrology and water resources; (b) natural and managed ecosystems; (c) coastal zones and marine ecosystems; (d) energy, industry, and settlements; (e) financial resources and services, and (f) human health. In Chapter 7, adaptation options are viewed from a regional perspective, taking the regions as distinguished in Intergovernmental Panel on Climate Change (IPCC) assessments: Africa, Asia, Australia and New Zealand, Europe, Latin America, North America, polar regions, and small island states.

5.1.2   Concepts and definitions

Before discussing adaptation options and their costs and benefits, we need to define some key terms. Box 5.1 provides definitions of some basic concepts. To what extent systems can adapt to climate change is dependent on their

---

[2] A three-stage programme is being developed to implement the agreed actions, guided by decisions of the Conference of Parties. In Stage I, the emphasis is on planning, performing studies of vulnerability, and policy options, followed by Stage II, in which more concrete measures would be designed and capacity for their implementation be built, while, finally, Stage III measures would be implemented (UNDP 2001).

Box 5.1 Climate-change sensitivity, adaptive capacity, vulnerability, and resilience

Sensitivity is the degree to which a system is affected, either adversely or beneficially, by climate-related stimuli. Climate-related stimuli describe here all the elements of climate change, including mean climate characteristics, climate variability, and the frequency and magnitude of extremes. The effect may be direct (e.g. a change in crop yield in response to a change in the mean, range or variability of temperature) or indirect, e.g. damages caused by an increase in the frequency of coastal flooding due to sea level rise.

Adaptive capacity is the ability of a system to adjust to climate change (including climate variability and extremes), to moderate potential damages, to take advantage of opportunities, or to cope with the consequences.

Coping ability, the degree to which a system can grapple successfully with a stimulus, is an element of adaptive capacity.

Vulnerability is the degree to which a system is susceptible to, or unable to cope with, adverse effects of climate change, including climate variability and extremes. Vulnerability is a function of the character, magnitude, and rate of climate variation to which a system is exposed, its sensitivity, and its adaptive capacity.

Resilience is the amount of change a system can undergo without changing state, or the degree to which the system rebounds, recoups, or recovers from a stimulus.

Sources: McCarthy *et al.* (2001); Wheaton and MacIver (1998).

---

vulnerability. A system's vulnerability is not only dependent on its *sensitivity* to climatic changes, but also on its *capacity to adapt*. Simply put: vulnerability is determined by the difference between the projected impacts of climate change (without taking into account adaptation) and the adaptive capacity. A system can be sensitive to climate change, but not very vulnerable. For example, in a particular area, a crop may be very sensitive to changes in precipitation, but the level of development and flexibility of the local farming systems may make it relatively easy to shift to other crops. Conversely, a system may not be very sensitive, but vulnerable to climate change nevertheless. For example, climatic change may only lead to minor changes in the distribution of the prevalence of malaria parasites, but the poor condition of local public health and the public health systems in some of the least developed regions may yet make the population very vulnerable to such small changes. Understanding *social vulnerability* (e.g. in the area of food security, water scarcity, public health, and poverty issues) is important for planning societal adaptation. A system is called *resilient* with respect to a particular degree of

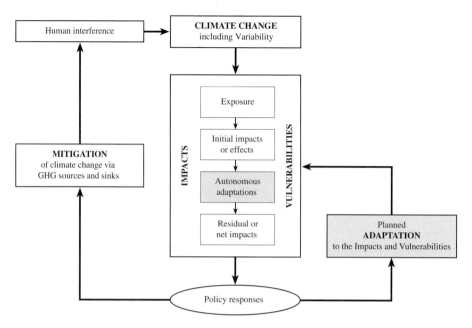

Figure 5.1. Places of adaptation in an integrated climate change response framework. *Source:* Smit *et al.* (2001).

climate change if its fundamental characteristics, determining its 'state', do not change with climate change. Increasing the resilience of a system decreases its vulnerability, and may be highly compatible with goals other than adaptation to climate change. For example, and as will be discussed in Chapter 7, in coastal zone management the emphasis used to be on technologically protecting coasts (e.g. by building seawalls, raising dikes), but increasing the resilience of the local communities (e.g. by enhancing their social capacity) in the coastal zones is an option that deserves increasing attention.

A key concept in adaptation is the ability of a community or natural system to adapt to climate change: *adaptive capacity*. One may want to distinguish between adaptive capacity and *coping capacity*. Banuri *et al.* (2001) suggest that, rather than seeking to anticipate and fix particular problems, the purpose of policy should be to develop coping capacity, stressing the relationship between risk, resilience, and governance. While the emphasis in adaptive capacity is on actively mitigating adverse consequences of climate change, the emphasis in coping capacity is on living with the consequences. Natural and social systems usually can cope with a range of environmental conditions, as long as these vary within certain bounds: the coping range. If environmental conditions change (e.g. the climate), the conditions may move out of this coping range (see Figure 1.17). By adaptation, the coping range can be moved, to have it envelop the environmental conditions again. Figure 5.1

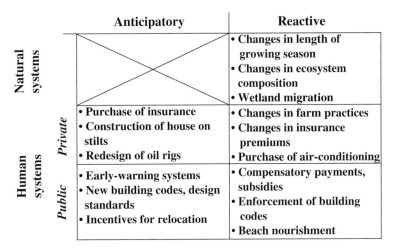

		Anticipatory	Reactive
Natural systems			• **Changes in length of growing season** • **Changes in ecosystem composition** • **Wetland migration**
Human systems	*Private*	• **Purchase of insurance** • **Construction of house on stilts** • **Redesign of oil rigs**	• **Changes in farm practices** • **Changes in insurance premiums** • **Purchase of air-conditioning**
	Public	• **Early-warning systems** • **New building codes, design standards** • **Incentives for relocation**	• **Compensatory payments, subsidies** • **Enforcement of building codes** • **Beach nourishment**

Figure 5.2. Types of adaptation and examples. Adapted from McCarthy *et al.* (2001).

indicates how adaptation can be placed in the pressure–state–impact–response cycle for climate change. Adaptation can happen: (a) *autonomously*, in an *unmanaged* fashion, or (b) as part of a climate-change response strategy, in a *planned*, and *managed*, way (see Figure 5.2 for types of adaptation and examples). 'Autonomous' is not well defined, because it depends on the view of the actor. Often, 'autonomous' implies that there is no government policy involved, but individuals and private firms adapt without such interference (Leary 1999). This does not imply that these individuals or firms do not plan or manage their adaptive actions, or that such 'autonomous' adaptation would be costless. Adaptation can be *reactive* or *anticipatory* (or *proactive*), depending on the timing, goal and motive (IPCC 1998). Reactive adaptation takes place after the impacts of climate change have occurred. Autonomous adaptation usually is reactive, e.g. a farmer may decide to change to other crop varieties after a number of years of poor harvests due to apparent changes in climate. Proactive adaptation is based on the expectation that climate will change rather than on its actual impacts, e.g. oil companies may take into account future rises in sea-level when designing and constructing offshore oil-platforms.

Adaptation is important because it can reduce adverse impacts and enhance beneficial impacts, especially for human systems (McCarthy *et al.* 2001). There is a wide variety of adaptation strategies, e.g. enhancing coastal protection against the risk of sea-level rise, promotion of drought-resistant varieties of agricultural crops, developing insurance schemes against damages from extreme weather events, and others that will be discussed in the next two chapters. Obviously, natural, unmanaged systems cannot design and implement climate change adaptation strategies on their own. However, there are limited options that can be implemented to facilitate, enable and accelerate adaptation by natural systems. An example is the establishment of corridors facilitating species migration.

5.1.3   *Adaptation in the context of development: issues and actors*

It is rather difficult to separate the analysis of adaptation options and its costs and benefits from broader development issues. Just as in the area of mitigation, adaptation decisions are unlikely to be taken for reasons of climate change alone. Most often, the risk of climate change and the desirability of adapting to projected changes is just one of the arguments for a particular investment being made or a policy being developed. Taking into account the economic, social, and ecological components of sustainable development, one may distinguish three objectives of adaptation: (a) it must be economically efficient; (b) advance social goals, and (c) be environmentally sustainable (Burton & Lim 2001). Adaptation measures are likely to be implemented only if they are consistent with, or integrated with, decisions or programmes that address non-climatic stresses (McCarthy *et al.* 2001). Alternatively, as Kane and Yohe (2000) put it even more strongly, the additional stress of climate change may be dwarfed by the stresses of other political, economic, social, or cultural forces. Measures enhancing adaptative capacity and advancing sustainable development can go hand in hand. Reducing the vulnerability to future changes in climate generally also reduces vulnerability to the negative effects of current climate variability. The latter may be a more urgent strategy to advance present development objectives, and decreased vulnerability to future climate change a mere secondary benefit.

Poor regions, characterized by low levels of technological advances, poor information availability, weak institutions, low educational levels, and unequal access to resources, are generally highly vulnerable, i.e. have low adaptive capacity (McCarthy *et al.* 2001). Kates (2000) answers the question: 'can the global poor adapt to global climate change?' with 'yes, but with great difficulty and much pain'. Poor countries and communities have demonstrated that they can cope with climate extremes and other natural hazards, but usually the toll in terms of human suffering and economic costs has been very high. Enhancing the ability to cope with future climate changes is likely to enhance the ability to cope with current climate variability and extreme weather events, and in this sense constitutes a 'no regrets' option because of its immediate benefits. It is because of these reasons that adaptation is a particularly important issue for developing countries.

Adaptation can imply behavioural changes, technological solutions, institutional adjustments, or a mix of these. Therefore, often a multitude of actors are involved: citizens, governments, the private sector, or combinations of these. The roles of these groups may not only be different in bringing about the changes and implementing the solutions and adjustments, they are also likely to change over time. They are linked through an intricate network cutting across geographic scales: optimal adaptation at the individual level may not be optimal on the regional level. Many studies do not adequately take into account the dynamics of the roles of the various actors. For example, the premise in many adaptation studies that existing institutions can cope efficiently and equitably with climate change, and that institutional gaps may be easily and costlessly filled, is likely to be incorrect (Kane & Yohe 2000). Thus, new institutional arrangements may have to be developed. Nevertheless, of course, understanding the past and present provides very useful knowledge about adaptive capacity. Lessons from past and current experiences to cope with climate variability and natural hazards (e.g. in agriculture, water resources, human settlements, human health, and coastal zones) can help in improving capabilities to anticipate or respond to climate change impacts (IPCC 1998). In Section 5.2, we elaborate upon approaches to enhance adaptive capacity. Most adaptive actions will be implemented by households and private agents, because it is in their interest to do so. Governments, however, also have a very important role to play: they can provide information, optimize legal and policy frameworks to facilitate adaptation, and protect public goods against the risks of climate change and climate variability (Leary 1999).

5.1.4   *Mitigation or adaptation?*

Mitigation and adaptation clearly are linked, because the more successful mitigation would be, the easier it would be to adapt to the residual climate change. Theoretically, costs and benefits of mitigation and adapation strategies should be compared, but fundamental differences in the nature of such strategies make this a difficult job (see also Section 5.3.4). At the global level, estimates of the costs of climate change mitigation and of climate change damage are in the same order of magnitude: at most some percentage points of gross domestic product (GDP). At the national or sectoral level, as discussed in Chapters 8 and 9, mitigation costs can be high, especially for the type of stringent climate targets that would be required to avoid adverse climate impacts in this century. Also, adaptation costs can be high locally. Countries that may have to face high adaptation costs may not be significant contributors to the problem, because they may have low greenhouse gas emissions. Thus, the balance of adaptation and mitigation activities will vary significantly between countries.

Mitigation activities only have benefits in terms of avoided climate change on long timescales; inertia in the climate system makes it likely that positive direct effects of mitigation (i.e. reduced climate impacts) may not be detectable for some time to come. Even very stringent climate policies will only have noticeable effects in terms of avoided climate change and associated limitation of adverse impacts in the second half of the century (see, for example, Dai *et al.* 2001). This view on the issue would change, however, if the ancillary benefits of mitigation policies (e.g. in terms of abatement of local air pollution) would be taken into account, which

Table 5.1. *Arguments for short-term adaptation or mitigation.*

Considerations in favour of short-term emphasis on adaptation	Considerations in favour of short-term emphasis on mitigation[a]
✓ Adaptation measures often have direct benefits reducing climate vulnerability, plus sometimes direct benefits in the context of wider socioeconomic development objectives	✓ Mitigation measures have ancillary benefits that often address urgent problems, like local air pollution and limited access to modern technologies
✓ Direct benefits of mitigation in terms of avoided climate change impacts will not be detectable for some decades, while benefits of adaptation are more immediate	✓ Delays in the climate system require early mitigation, especially for low stabilization goals, while effective adaptation measures can be put into place more quickly, when needed
✓ Climate impacts occur already today and are projected to increase	✓ Reduces risks of rapid nonlinear impacts, adaptation to which would be costly
✓ Adaptation measures are largely independent of international co-operation; their benefits generally accrue directly to those implementing the measures	✓ Eventual emissions reductions needed for stabilization are large, and international agreement on stepwise achievements of such reductions in a timely manner will take much time
✓ In developing countries: many regions are vulnerable today (agriculture in semi-arid areas, large low-lying coastal areas, drought-and flood-prone areas)	✓ In developed countries: because of their historical responsibility for the problem and the availability of resources, as agreed in the UNFCCC

⬇

Different local characteristics will require a different balance between adaptation and mitigation in different places and different times

[a] See Table 8.5 for a comparison of arguments for rapid versus delayed mitigation responses.

often is not the case. These ancillary benefits, which are discussed in more detail in Chapter 10, are usually more immediate. At the same time, many adaptation measures have short-term local benefits that are related directly to climate variability and change, and can have modest costs. For the long term, one may argue that the costs of climate change impacts and adaptation to these impacts may increase rapidly in the future, and even dramatic non-linear, irreversible, impacts may ensue in the longer term if no mitigation action is implemented. This also would affect the balance. Table 5.1 compares some of the arguments.

Mitigation and adaptation have in common that they both reduce climate change damage, but there are also important differences in terms of distribution of actions over time and place (see Toth *et al.* 2001). Much of the burden of climate change mitigation would be in the developed countries and benefit other generations in other regions, primarily developing regions. That is, if one would ignore the short-term, ancillary benefits of mitigation policies. Generally, climate and ancillary benefits of adaptation measures accrue in the shorter term, and to those actually implementing those measures locally. Free-riding (by those who do not contribute to the activities but benefit from their effects) is a greater problem in mitigation than in adaptation (Toth *et al.* 2001). Mitigation policies on the global scale are effective only if major emitters implement their commitments, while most adaptation policies are carried out by those for whom averted damage exceeds the expected costs (Jepma & Munasinghe 1998). Households and firms will find it in their self-interest to adapt (Leary 1999), because generally the benefits fall on themselves. The benefits of mitigation largely are external to these actors; hence, the weak social basis of mitigation.

In several impact studies, climate change impacts and their economic effects are considered to be mitigated by adaptation measures, which are implicitly assumed to have no costs. The most recent assessments suggest that the bias towards mitigation analysis is misguided. Adaptation does not necessarily happen all by itself and is probably not costless at all (Smit *et al.* 2001). As a result, adaptation has been receiving increasing attention by the scientific research community recently, and methods are being developed to analyse the role of adaptation in climate change impacts as a specific response strategy. However, to date the understanding of adaptation is still insufficient reliably to address adaptive capacities, or to evaluate the costs and benefits of different adaptation options in depth. In Section 5.3.4, quantitative methods to analyse mitigation and adaptation in an integrated fashion (e.g. cost–benefit analysis) are discussed.

One final point that we would like to make about the evaluation of adaptation and mitigation strategies is that often they cannot clearly be distinguished, especially in the longer term, when the distinction is blurred if we consider the differences from the viewpoints of adaptive and mitigative capacities. Many of the determinants of adaptive capacity (see Section 5.2) are similar to those determining the mitigative capacity of a society or community, and enhancing these factors thus would stimulate both adaptation and mitigation at the same time. This is consistent with the view that a more integrated, holistic, approach to climate change response, putting both adaptation and mitigation in the context of broader development objectives, is needed urgently in order to tackle the climate change problem effectively.

5.1.5 *Approaches and methods to planned adaptation strategies*

Methodological inquiries can look at adaptation in two ways: (a) adaptation as an important component of impact and vulnerability analysis, or

---

Box 5.2   Impacts and adaptation assessment: the IPCC approach

1. Define problem (including the study area, its sectors, etc.).
2. Select method of assessment most appropriate to the problems.
3. Test method/conduct sensitivity analysis.
4. Select and apply climate change scenarios.
5. Assess biophysical and socioeconomic impacts.
6. Assess autonomous adjustments.
7. Evaluate adaptation strategies:
   (a) define the objectives;
   (b) specify the climate impacts of importance;
   (c) identify the adaptation options;
   (d) examine the constraints;
   (e) quantify measures and formulate alternative strategies;
   (f) weigh objectives and evaluate trade-offs;
   (g) recommend adaptation measures.

Source: Carter *et al.* (1996).

(b) the costs and benefits of adaptation as specific climate change response strategy. Scientific research on regional, national or local risks posed by climate change has been biased towards impact analysis without systematic inclusion of adaptation. The methodology for impact and adaptation assessment suggested by the IPCC includes both, but emphasizes impact and vulnerability analysis (Carter *et al.* 1996). It was developed to help countries evaluate their vulnerability to climate change and the options to adapt to it, and was applied in many countries in country studies. Its seven steps are listed in Box 5.2.

In this chapter on adaptation, we will focus on step 5, 6 and 7 in Box 5.2, addressing autonomous and planned adaptation. First, objectives have to be defined. Usually, these are a combination of reducing vulnerability to climate variability and change, and pursuing broader (sustainable) development objectives, according to local priorities. Second, the types of climatic changes that are important in the local circumstances have to be specified, e.g. sea-level rise, increased summer temperatures, increased droughts, etc. Third, a comprehensive analysis has to be made of suitable adaptive response options. Carter *et al.* (1996) mention six possible options: (a) preventing loss; (b) tolerating loss; (c) spreading or sharing loss; (d) changing use or activity; (e) changing location, and (f) restoration. In all these areas, a combination of various ways of implementing adaptation options may be considered, including legal, financial, economic, technological, educational,

or training instruments. Fourth, the implementation of some of the options identified may be hindered by social, cultural, institutional, or other barriers. These
have to be identified and understood. Fifth, a quantitative assessment of the alternative strategies of combinations of the identified options needs to be performed,
using formal models, historical evidence, expert judgement, or other sources of
information. In the sixth step, the options are weighed against the objectives formulated in the first one, choosing locally relevant criteria. This would lead to one
or more preferred strategies, which, in the seventh step, would be recommended to
policy-makers. The IPCC *Technical Guidelines* were further elaborated at the system
and sector level by the United Nations Environment Programme (UNEP), (Feenstra
et al. 1998).

Application of the full IPCC method of Box 5.2 in various country studies has
focused on the first five steps (on impacts analysis), because most information
is available on these issues and because of limited resources, leading to neglect
of the last two steps (on concrete adaptation strategies). For adaptation, social
factors such as human behaviour, institutional capacity, and cultural issues, are
more important than biophysical impacts of climate change, but research and
assessment in the area of climate change has only gradually broadened from
the natural to the social sciences. This is reflected in the emphasis on natural factors in the 1996 IPCC methodology. Because of this weakness, because
of the long-term nature of climate scenarios and the associated impacts, and
because of the decoupling from more immediate development issues, results
of these studies have not resulted in concrete policy actions (Burton & Lim
2001).

Another weakness of the IPCC *Technical Guidelines* is that the emphasis in the
last two steps on adaptation is very much on the phase of identification, evaluation, and implementation of specific adaptation options. There is little attention to
the assessment of the broader socioeconomic context of adaptation, non-technical
aspects of adaptation, issues like raising awareness, public participation, and information gathering, and post-implementation monitoring and evaluation (Klein
et al. 1999).

In order to develop assessments of adaptation options more relevant to short-
term policies in developing countries, the United Nations Development Programme (UNDP) in collaboration with UNEP developed a new approach for usage
in the UNFCCC context (Burton & Lim 2001; see Box 5.3). While the IPCC framework is emphasizing climate modelling and long-term scenarios, the complementary adaptation policy framework emphasizes near- and medium-term adaptation. For example, adapting to extreme events or changed climate variability in
the short term is more policy-relevant than planning from adaptation to gradual changes in averages that are projected to occur over many decades. The two

Box 5.3   A new approach: the adaptation policy framework

The adaptation policy framework is underpinned by four *general principles*. It:

1. Pays greater attention to recent climate experience, impacts, and adaptation as part of the development of a baseline for adaptation analysis.
2. Ensures adaptation to climate variability and extreme events are explicitly included as a step towards reducing vulnerability to longer-term climate change.
3. Adopts a stronger focus on vulnerability in the present as well as the future in order to ground future policy in present-day experience.
4. Includes specific consideration of current development policies and proposed future activities and investments, paying particular attention to those activities that may tend to increase vulnerability to climate change or are maladaptive.

In effect, this requires:

5. Applying alternative ways of characterizing future climates in order to capture climate and weather variables more relevant to adaptation decisions.
6. Applying an analytical framework to socioeconomic scenarios to help strengthen the ability to assess vulnerability and the capacity to adapt.
7. Integrating adaptation strategies and measures with natural hazard reduction and disaster prevention programmes, and other relevant programmes.
8. Taking other atmospheric, environment, and natural resource issues into account.

A number of *specific initiatives* to be carried out for implementing the adaptation policy framework include:

9. Collecting and reporting data related to past adaptation and adaptation capacity.
10. Determining the vulnerabilities of greatest and most pressing concern.
11. Determining where adaptation has been, and can be, most effective.
12. Strengthening economic analysis.
13. Establishing priorities for adaptation.

14. Developing national strategies for adaptation, and integrating them into national economic and sustainable development planning.
15. Building capacity for adaptation.
16. Supporting outreach, extension, and educational programmes on adaptation.
17. Ensuring stakeholder and public participation.
18. Addressing regional and transboundary issues in adaptation.

Source: Burton & Lim (2001).

methods (short-term adaptation assessment and long-term impact assessment) provide complementary information and can be applied in parallel (see Figure 5.3). It is recommended that adaptation policy assessment occurs in two steps: based on the assessment of present vulnerability and subsequently on the assessment of future vulnerability. Thus, options can be identified that would address both present vulnerability to climate variability and future vulnerability to projected climatic changes.

5.2   Adaptive capacity

5.2.1   *Enhancing adaptive capacity*

Adaptation refers to the adjustment in human and natural systems in response to climate change stresses and their effects, which moderates damage, and helps to exploit opportunities for benefit, e.g. building sea-walls, developing drought- and salt-resistant crops. Adaptive capacity is the ability of a system to adjust to climate change.

Strengthening adaptive capacity is a key option, especially in the case of the most vulnerable and disadvantaged groups. Adaptive capacity itself will depend on the availability and distribution of: (a) economic, natural, social, and human resources; (b) institutional structure and access to decision-making processes; (c) information public awareness and perceptions, available technology and policy options, and (d) ability to spread risk. Adaptive capacity in the context of climate change can be enhanced in the following ways.

1. *Identifying stakeholders* and engaging them in the process.
2. *Assessing generic adaptive capacity*, i.e. the available resources and capabilities of the persons involved. A healthy population will have a greater ability to adapt to climate change. Likewise, with a better education and

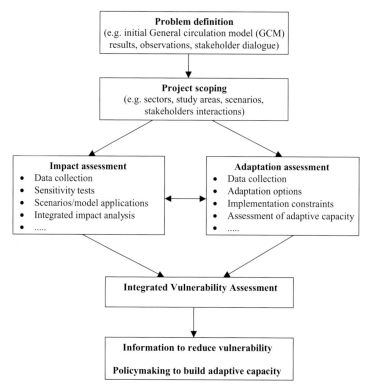

Figure 5.3. Complementary approaches for impacts and adaptation assessment. *Source:* Burton and Lim (2001).

training, the population is likely to be less vulnerable to changes in climate. Less poverty, improvements in economic status, and better availability of resources, are all positive aspects that would help communities adapt to climate change.

3. *Assessing specific adaptive capacity*, i.e. risk, geographical distribution, social, and institutional capabilities. A better distribution of income, diversification of income, high levels of stakeholder participation, and good institutional adaptive capacity, are some positive aspects that would increase the adaptive capacity of a society.

5.2.2   *Determinants of adaptive capacity*

In the previous section, the term *adaptive capacity* was introduced to describe the ability of natural or social systems to adapt to climate change. One advantage of the term is that it can be used in a way that recognizes the social, economic, institutional, and environmental context in which adaptation takes place. The concept thus allows us to consider adaptation options as part of the broader sustainable development policies. One could distinguish six important determinants of adaptive capacity (McCarthy *et al.* 2001):

1. *Economic resources.* Sometimes it is said that the best way to enhance adaptive capacity is by getting rich. Indeed, vulnerability is related directly to poverty, and increasing incomes and improving access of poor communities to financial means is likely to decrease vulnerability to climatic impacts. However, other, sometimes related, factors play a role too.

2. *Access to appropriate technology.* Many of the means to adapt to climate change involve the use of technologies, e.g. warning systems, protective structures, crop-breeding, and flood control structures. The access to these technologies, or the ability to develop them in a way suitable to local circumstances, is one of the key factors underpinning the adaptive capacity of a community.

3. *Availability of information and skills.* Focused planning for adaptation is possible only when information about local vulnerability and options to reduce that vulnerability is available, and skills at hand to use the information effectively. This implies appropriate education and involvement of all relevant actors.

4. *Infrastructure.* Adaptive capacity depends on the strength and weaknesses of the physical infrastructure of a region or community. For example, building developments in low-lying coastal areas or dependency on climate-sensitive energy sources decrease the adaptive capacity of communities.

5. *Adequate institutions.* Adaptation can be hindered by poorly developed institutions. Strong, well-organized institutions characterized by stable and effective policy arrangements can support adaptation activities more easily.

6. *Equitable access to resources.* Adaptive capacity not only is dependent on availability of resources, but also on the equitable access of individuals and communities to those resources. In this context, it is important to note that, rather than enhancing adaptive capacity of poor countries, efforts may focus on enhancing adaptive capacity of the poor in those countries.

5.2.3 *Enhancing adaptive capacity: links with development, equity and sustainability issues at different scales*

In general, since the capability to adapt depends very much on the state of development, promoting sustainable development would enhance adaptive capacity. More particularly, the IPCC suggest six requirements for enhancing the adaptive capacity of systems or countries, closely related to the above determinants of adaptive capacity (Smit *et al.* 2001):

1. A stable and prosperous economy.
2. A high degree of access to technology at various levels and in all sectors.
3. Clear roles and responsibilities for implementation of adaptation strategies.
4. Effective dissemination systems, and adaptation discussion and innovation forums.
5. Social institutions and arrangements for enhancing equitable access to resources.
6. Absence of factors compromising the present adaptive capacity.

Consequently, meeting these requirements would enhance adaptive capacity. It is important that sustainable development actions take climate change into account, because climate change impacts may seriously affect sustainable development aspirations. Nevertheless, such attention to climate issues is still remarkably scarce in ongoing development programmes. It may be evident that specific actions to meet the above requirements for enhancing adaptive capacity generally also help development. Smit *et al.* (2001) draw a list from the literature:

1. Improve access to resources.
2. Reduce poverty.
3. Lower unequal distribution of resources and wealth among groups.
4. Improve education and information.
5. Improve infrastructure.
6. Diminish intergenerational inequities.
7. Respect accumulated local experiences.
8. Moderate long-standing structural inequities.
9. Assure that responses are comprehensive and integrative, not just technical.
10. Ensure active participation by concerned parties.
11. Improve institutional capacity and efficiency.

More concrete recommendations would depend on local resources and other circumstances. Nevertheless, the list makes clear that most ways of enhancing adaptive capacity are, in fact, ways of promoting development and equity. Only in a few specific cases would enhancing adaptive capacity be an action targeting vulnerability to climate change alone.

Table 5.2. SRES *scenarios and vulnerability.*

SRES scenario	Expected climate impacts[a]	Projected adaptive capacity[b]	Vulnerability[c]
A1FI	+ +	+	+
A1B	+	+	0
A1T	+	+	0
A2	+ +	0	+ +
B1	0	+ +	−
B2	+	+	0

[a] + + high adverse impacts   [b] + + large adaptive capacity   [c] + + very vulnerable
   +   intermediate climate impacts   +   moderate adaptive capacity   +   vulnerable
   0   low climate impacts   0   low adaptive capacity   0   somewhat vulnerable
                                                     −   little vulnerable

5.2.4   *Future projections of adaptation and adaptive capacity*

Unfortunately, most scenarios of socioeconomic development and associated potential climate change do not pay attention to adaptation. Because *Emissions Scenarios: A Special Report of Working Group III of the Intergovernmental Panel on Climate Change* (SRES) (see Chapter 2) was published only in early 2000, and the first climate model results based on the SRES became available only in late 2000, the IPCC's contribution of Working Group II to the *Third Assessment Report* could not draw upon impact and adaptation analysis based on the SRES. This applies even more to the analysis of adaptation related to stabilization scenarios, which were addressed in the *Third Assessment Report*. While earlier IPCC scenario analysis did not offer good possibilities for considering vulnerability and adaptation, the SRES scenarios provide some background, a.o. because of the narrative components describing societal developments. Taking the simplified relationship: vulnerability equals climate impacts minus adaptive capacity, one can argue that B1 is the scenario with the lowest vulnerability (see Table 5.2). The reasons are: (a) the projected climate impacts are the lowest; (b) income is high and most equally distributed, and (c) there is much attention for social and environmental sustainability in this scenario. Scenario A2 is on the other end of the scale: vulnerability is highest because climate impacts are highest, income lowest and unevenly distributed, and there is relatively little interest in environmental and social sustainability. Both A1 and B2 are in between, but for different reasons. Climate impacts in both scenarios may be of comparable magnitude, but income levels are higher in A1, decreasing vulnerability.[3] In B2, vulnerability is lowered not so much by high incomes, but

---

[3] Vulnerability in A1FI is somewhat higher than in A1B and A1T because the projected climate impacts in this variant are higher.

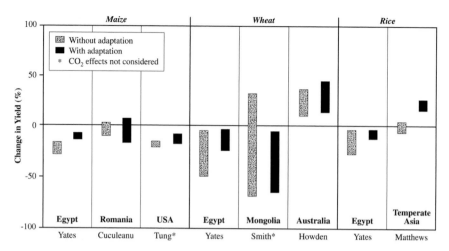

Figure 5.4. Changes in crop yields with and without adaptation. Endpoints of each range represent collective high and low percentage change values derived from all climate scenarios used in the study. *Source:* Gitay *et al.* (2001).

by an emphasis on social and environmental aspects of sustainable development at local scales.

The table underlines that differences between socioeconomic developments in various scenarios is exacerbated further by climate change. In a scenario such as A2, people generally are equipped poorly to adapt to climate change while the changes to which they have to adapt are the greatest. In a scenario like B1, people are equipped best to deal with the adverse consequences of climate change, but the magnitude of those consequences is lowest. In other words: pursuing a future focusing on sustainable development, especially when this is realized worldwide in a collaborative effort like in the B1 scenario, has multiple benefits even without explicit climate policies: climate change is limited as a side-effect, and the capacity to deal with the residual climatic changes is boosted.

Very few scenarios explicitly exploring long-term adaptive capacity have been developed at the sectoral level. An exception is scenario analysis that evaluates future *agricultural productivity*. A number of regional studies have addressed the effectiveness of adaptive strategies to cope with, or counteract, yield losses caused by climatic changes, or maybe even achieve gains (Gitay *et al.* 2001). Generally, such studies conclude that adaptation can reduce damage to crops significantly (see Figure 5.4). However, uncertainties are large. On the one hand, adaptation options considered are limited (adjustments in planting dates, fertilizer application rates, irrigation practices, cultivar selection) and may underestimate the total available array of opportunities. On the other hand, while earlier studies without adaptation basically considered 'dumb farmers' who would not adapt at all, most of these new studies assume 'clairvoyant farmers' who anticipate a particular climate change, and may thus overestimate adaptation actions. Analysis at the farm or farming system level supports the conclusion that adaptation can reduce significantly climate impacts in agriculture, and also confirms the hypothesis that this is more true in well-developed regions than in developing ones with poor access to resources and information. The authors of some studies have attempted to address the question of how much warming global agriculture can cope with before prices rise. This question is important in view of food security. Some models suggest that global temperature increases up to 2.5 °C may be accommodated by the world food production system without prices increasing. However, there is very low confidence in this finding, since studies do not all agree and uncertainties in them are very large.

Thus, if rapid diffusion of technologies and information is ensured, the agricultural sector may not be very vulnerable to small to moderate climatic changes, with the exception of those regions in developing countries that are vulnerable already to current climate variability. This is different *for sensitive natural systems*, of which the adaptive capacity is generally low and likely to be lowered still by future land-use and land-cover changes (Carter *et al.* 2001), as well as other human stresses, e.g. air pollution. Nevertheless, ecosystems can adapt to a certain degree, or 'move along' with climate change through migration. Many vegetation models are used to evaluate climate impacts project shifts in vegetation by implicitly taking into account such migration possibilities (the 'ecosystem movement paradigm'). That is, they project shifts of potential vegetation rather than actual shifts.

If climate change is slow enough, and if natural and human barriers do not prevent it, plant species can migrate and animals along with it. Here, the rate of change is crucial. Unfortunately, the maximum observed migration rates of tree species in paleoclimatic studies are slower than required for survival at current or projected rates of climate change. However, it is not clear if these observed rates reflect maximum intrinsic rates. Some evidence suggests that this is the case, while other research indicates that even higher rates of migration would be possible (see, for example, Malcolm and Markham 2000). It is likely that some species move faster than others, changing ecosystem composition. In general, current knowledge suggests that many habitats will change at a rate approximately 10 times faster than the rapid changes during the recent postglacial period. This applies not only to high latitudes and altitudes, but also to some subtropical and dry ecosystem types. Barriers to ecosystem migration will exacerbate climate-induced species loss, e.g. islands (Malcolm *et al.* 2002).

It is proposed in another viewpoint that new ecosystems are likely to differ from current ones, since rates of migration and adaptation will vary between species. There are likely to be in situ changes in species composition (the 'ecosystem modification paradigm') that are very difficult to project. These points complicate assessment of the conditions for allowing 'ecosystems to adapt naturally to climate change' and 'to ensure that food production is not threatened' in Article 2 of the UNFCCC.

5.3   Future costs and benefits of adaptation

5.3.1   Introduction

Finally, in this chapter we discuss possibilities to assess the costs and benefits of adaptation. The economic cost–benefit approach requires the valuation of mitigation, damage, and adaptation costs.[4] Cost–benefit analysis is used widely for climate change analysis, although there is debate as to whether it is applicable to the climate change problem at all, mainly because of two reasons as follows:

1. There are large uncertainties with reference to the magnitude, and sometimes even sign, of the costs and benefits of both mitigation actions and impacts; this is relevant particularly since some long-term impacts may be very large and possibly irreversible, but also costs of mitigation may be very high if pursued too aggressively.

2. Cost–benefit analysis focuses on economic efficiency, and while efficiency and equity can be addressed simultaneously, they are not necessarily compatible in all circumstances; cost–benefit analysis is not considered the ideal way to assess equity issues across regions and generations.

In view of these problems, other approaches can be considered. Pearce et al. (1996) consider two alternative approaches to evaluate climate change responses: (a) the sustainability approach, and (b) the consensus approach. The sustainability perspective gives the highest priority of avoiding intolerable damages by climate change, regardless of mitigation costs. Thus, this framework is not particularly suitable for assessing the costs of adaptation. The consensus approach acknowledges that neither maximizing net benefits (in the cost–benefit approach) without attention to environmental and social concerns, nor environmental sustainability not accounting for human welfare effects, can be the sole objective. The consensus approach takes a precautionary approach to protect the environment unless the costs of doing so are very high. It is similar to what has been called the 'safe minimum standard' approach. All methods in some sense weigh costs and benefits,

---

[4] Beyond cost–benefit analysis, various assessment approaches exist for making development more sustainable, including multicriteria analysis and others. For a general introduction, see Chapter 3, and for specific methods applicable to mitigation assessment, see Chapter 8.

albeit not always in unified, monetary terms, and sometimes mitigation costs or expected damage are considered 'unacceptable'.

Thus, even if cost–benefit analysis may not be the sole method to assess climate change response strategies, it can at least provide valuable partial information for policy development, i.e. about economic efficiency. Munasinghe *et al.* (1996) set out several different techniques to assess project decisions:

1. Traditional project level cost–benefit analysis, comparing (monetized) costs and benefits of alternative project alternatives.
2. Cost-effectiveness cost–benefit analysis, determining least-cost options to meet a desired level of benefits (a variant of the approach mentioned above).
3. Multicriteria analysis, including multiple objectives of which economic efficiency is only one criterion.
4. Decision analysis, evaluating decision-making under uncertainty, a.o. allowing for considering irreversibility.

Adaptation costs can be evaluated in different ways at different levels of scale. For some purposes (e.g. comparison of adaptation options at the project level) one may, in a limited sense, focus on the evaluation of the net investment costs involved (see Section 5.3.2). For a more generic evaluation of adaptation strategies, or a comparison with costs and benefits of mitigation options, a more encompassing approach considering opportunity costs would be more appropriate, taking into account the welfare implications of no-action scenarios versus mitigation scenarios (Jepma *et al.* 1996, and see Section 5.3.3).

5.3.2   *Project-level investment costs*

Analysis of investment costs is relevant for evaluating adaptation costs at the project level for the short term at local scales. Determining the investment costs of adaptation projects involves addressing a broad range of questions (see also Boxes 5.2 and 5.3):

1. Adaptation to what?
2. How to adapt?
3. When to adapt?
4. Who pays?
5. What would have happened without adaptation?
6. What are the costs and benefits unrelated to climate change response?

The answer to the first question *adaptation to what?*, depends on local circumstances: one may have to adapt to changing temperatures, sea-level rise, or changes in precipitation and water availability. Options for *how to adapt?*, can be very broad-ranging, such as protection (e.g. sea-walls against sea-level rise); accommodation (e.g. adjust lifestyles to a changing climate); retreat or migration (e.g. to higher elevations because of sea-level rise), or adjustment, e.g. shifting to other crop varieties. Clearly, adaptation is not limited to concrete projects, but can involve a broad variety of actions, e.g. behavioural adjustments of households, firms or institutions, management changes, legal measures, information campaigns, etc. Some of these actions can be costed relatively easily, others pose difficulties.

The question *when to adapt?*, is important from an economic point of view. Here, the issue of no-regrets options is significant: those measures that would have benefits even without climate change would have no net additional costs from a climate change perspective, and there may be no clear reason to postpone the implementation of such options.[5] For adaptation options with a positive cost, the discount factor used is an important consideration (see Box 5.3). Moreover, there is the question of uncertainty and the weighting of future information. If climatic change would not materialize in the projected way, the investment costs may be wasted. If there would be a chance that increasing knowledge may reduce uncertainties about future climate change, it would be rational to postpone adaptation investments. However, there are also large uncertainties about the possibility that uncertainties about crucial factors (e.g. local or regional climatic changes) would or would not be resolved adequately in the future.

A next question is: *who pays?* 'Autonomous' adaptation is generally considered costless. However, this assumption may be questioned, since no adaptation would take place without human choices, which usually entail some form of costs. Leary (1999) defines 'autonomous' adaptation as adaptation by households and private actors acting autonomously, excluding responses that require collective actions involving public decision-making. A full cost–benefit analysis would take costs for both autonomous and planned adaptation into account as they are incurred by the various stakeholders.

Often measures will be taken or projects initiated which have adaptation only as one of the multiple objectives, or projects will be adapted to some extent to accommodate adaptation objectives. Then, a key question is: what would the costs of the measure or project have been without taking into account adaptation?, i.e. *what would have happened without adaptation?* Which part of the project costs should be attributed to climate change adaptation would vary per option. Many investments, including climate change adaptation, may lead to only small cost

---

[5] But one may even consider postponing or foregoing these no-regrets adaptation options in order to invest in climate mitigation measures that would decrease future adaptation costs (Jepma *et al.* 1996). The long timescales involved in decreasing impacts through mitigating greenhouse gas emissions make this trade-off a theoretical one.

increments, but for other projects, climate change adaptation, may be the prime mover. Ideally, this question also addresses impacts or damage avoided by the project. Valuing impacts is a difficult job, as it can include different categories of values: use values (direct use values, indirect use values or option values), and non-use values (see Box 3.4). As one goes from left to right in Figure 3.3, valuation becomes more difficult and controversial, and is included less often in analyses. A related question is: *which costs and benefits does the project imply for issues other than climate change?* Generally, the costs and benefits of a project are dependent very much on the boundaries selected for the analysis (what is taken into account and what is not).

Finally, the criteria used for doing the analyses and comparing project alternatives determine the ranking of projects. A multicriteria analysis that takes into account other factors than monetary costs is likely to lead to a different outcome than an analysis that takes into account only direct monetary costs.

5.3.3   *Welfare costs of adaptation strategies*

A full cost–benefit analysis of the implications of adaptation measures for social welfare at a more aggregate level than project appraisal would take into account the benefits of adaptation in terms of avoided climate change damage, and the costs of the remaining unmitigated damage, as well as indirect costs and benefits. From an economic perspective, adaptation measures may be considered justified if the additional costs of adaptation do not exceed the benefits from reduced damages, plus the net indirect or ancillary benefits, if any. However, while such a comprehensive analysis makes a lot of sense from an economic perspective, the analyst performing the analysis will find various problems on the way. Jepma and Munasinghe (1998) mention three:

1. *Assessment of non-market damages.* The monetary evaluation of various damage categories is highly controversial. For example, the question of if and how the loss of threatened ecosystems or human injuries or loss of human lives can or should be assessed in monetary terms, remains an issue of intense debate amongst experts. Some argue that these benefits can be measured in the context of an economic cost–benefit framework, e.g. through measurements of willingness to pay or willingness to accept compensation for losses. Others oppose the monetary valuation of some non-market damages because they consider that some damages should be avoided regardless of the costs (the 'strong sustainability' perspective). The choice of method depends on the social, political, economic, and cultural context.

2. *Intergenerational equity and discounting.* The costs of adaptation measures are incurred immediately, while many of the benefits in terms of avoided climate impacts accrue to future generations. So, how should we take the benefits to these future generations into account when making investment decisions now? Usually, the projected future damages are discounted to arrive at the present value of those damages. If this should be done and, if so, how, is again a matter of profound debate. The discount factor establishes how a certain benefit would be valued now if it would occur some years into the future. As such, it reflects an individual's time preference: does one want to have the benefits now rather than in the future? The higher one chooses the discount factor, the lower the present value of the projected damages (see also Box 5.4).

3. *Valuing ancillary costs and benefits.* As discussed above, many adaptation measures will have ancillary benefits for the present generation, which may have to do with decreasing vulnerability to current climate variability, or enhancing general socioeconomic development. Often, these benefits may even be more important than adaptation to future climate change, or at least as important. Therefore, we prefer the term 'co-benefits'. Taking these co-benefits into account and allocating the portion of the costs associated with climate change adaptation is a major hurdle that adaptation analysts have to take.

In evaluating adaptation strategies, one may distinguish between costs and benefits of autonomous adaptation by private agents and collective adaptation co-ordinated by public bodies, and take into account uncertainty with respect to climate change. Welfare implications – and, hence, ranking of adaptation options – would depend on the question of whether climate impacts materialize or not, and on the evaluation of autonomous adaptation versus adaptation by private agents and governments collectively. If climate change would not materialize, adaptation policies might still be useful because they are likely to enhance the capacity to deal with current climate variability. Evaluating adaptation options for a scenario with no climate change and a climate change scenario provides information about the robustness of the options.

Table 5.3 gives an overview of measures of social welfare to compare and rank adaptation strategies elements and enable a comprehensive cost–benefit analysis. 'Current adaptation policies' acknowledge that public (governments) and private (firms, households, farmers) actors continuously take climate considerations into account when making decisions, and assume that the infrastructure, and institutional rules and programmes that shape how climate affects society, do not change (Leary 1999). Examples include crop insurance, disaster relief policies, water

Box 5.4 Discounting and adaptation

Discounting allows economic effects occurring at different times to be compared. In climate change – where impacts are projected on the longer term but mitigation and adaptation expenditures have to be made on short timescales – this tool is indispensable for cost–benefit analysis. However, IF future impacts should be discounted, and if so at which rate, remains a hotly debated issue. Because of the long times involved in climate change, the choice of discount rate affects the net present value of damage and mitigation costs – and, hence, policy recommendations – in a huge way. Nevertheless, after the oil crisis in the 1970s, a consensus approach was developed (Lind *et al.* 1982), triggered to enable the assessment of alternative investments in energy systems that would bring along high costs for the current generation but benefits primarily for the next. The subsequent problem of climate change, among others, implied time horizons of multiple generations, and in the 1980s a debate re-emerged about the appropriateness of discounting and the choice of discount rate. Two approaches have been proposed (Arrow *et al.* 1996) to select a suitable discount rate: the prescriptive approach (how should impacts be valued from a normative perspective?) and the descriptive approach (which preferences do people actually have now?).

The prescriptive approach generally takes the point of view that risks to future generations – of both serious, adverse, climate change impacts and the need for rapid future abatement of emissions – should be minimized. Hence, this approach is consistent with a low discount rate (the social rate of time preference, usually well below 3 per cent), favouring higher current mitigation expenditures. The descriptive approach pays more attention to time preferences in the real world today, taking into account that mitigation expenditures replace other investments, leading to significantly higher discount rates (usually well above 3 per cent), justifying lower expenditures on mitigation (Arrow *et al.* 1996). The latter could be related to the individual rate of time preference (the trade-off between present and future benefits and costs) or the opportunity cost of capital, based on the marginal productivity of capital. These would be equal if there were efficient markets and no taxes, but in practice the range of individual time preference rates is very broad (Markandya 1999). Furthermore, the preferred discount rate in developing countries could be different from that in developed countries. Applying a descriptive approach, the opportunity costs of capital in developing countries is higher than in developed countries, suggesting discount rates of 10–12 per cent in developing countries against 4–6 per cent in developed countries (Markandya 1999). Economists generally believe that discount rates should be lower for issues with long time horizons such as climate change, and therefore discount rates could decrease with the time horizon. According to Newell and Pizer (2002), taking into account the uncertainty about future discount rates leads to higher valuation of future benefits, attaching higher value to short-term mitigation and adaptation measures. Analyses in support of policy development could take into account a range of discount rates to show the implications of the different approaches.

What does the choice of discount rate imply for adaptation? For high rates, the present value of future damage is low with or without adaptation, favouring neither mitigation nor adaptation. For low rates, the present value of future damage would be high, favouring not only mitigation, but also investments in planned adaptation to reduce future damages. Questions arise as to whether the discounted costs of short-term adaptation investments (e.g. coastal protection) can be compared with the discounted long-term damage avoided by these investments; one option may be to use discount rates that vary over time.

Sources: Arrow *et al.* (1996); Lind *et al.* (1992); Markandya (1999); Portney and Weyant (2001); Toth *et al.* (2002).

storage and delivery policies and infrastructure. 'Additional adaptation policies' refer to policies that go beyond this, by purposefully enhancing the capacity to deal with climate variability and change.

5.3.4   *Integrated assessment of adaptation and mitigation*

As noted in Section 5.1, there is an intimate link between adaptation and mitigation. If climate change is mitigated, there is less need for adaptation. The optimal blend is hard to find because of: (a) the divergent characteristics of the two strategies; (b) a variety a possible approaches; (c) differences between who pays and who reaps the benefits, and (d) a multitude of uncertainties. An attempt to compare the costs of mitigation against (avoidable) damage of climate change in order to evaluate the desirability of mitigation action is usual in integrated assessment. Sometimes, but not always, adaptation is taken into account when analysing the damages.

Since Jepma and Munasinghe (1998) argued that insufficient information was available to determine the ideal blend of mitigation and adaptation, more

Table 5.3. *Framework for welfare implications of adaptation policies and uncertainty about climate change (adapted and expanded from Leary 1999).*

| | Welfare implications of mitigation strategies | | | |
| | Climate variability | | Climate change | |
Scenario	Costs	Benefits	Costs	Benefits
Current adaptation policies	Welfare change through autonomous adjustments to current climate variability	Reduced damage due to enhanced resilience to current climate variability plus ancillary benefits of autonomous adjustments	Welfare change through residual climate impacts[a] and by autonomous adjustments	Reduced climate change damage[a] plus net ancillary benefits of autonomous adjustments
Additional adaptation policies	Net cost of public policy plus autonomous adjustments to current climate variability	Reduced damage due to resilience to current climate variability plus ancillary benefits of autonomous adjustments and public policy	Welfare change through residual climate impacts and autonomous adjustments plus net costs of public policy	Reduced climate change damage[a] plus net ancillary benefits of autonomous adaptation and public policy

[a] Benefits of damage avoided by mitigation is not included here, only damage avoided by adaptation. For example, a coastal protection project may protect a coast only partially, or an agricultural measure may reduce vulnerability to climate change, but not completely.

information has become available, but it could be questioned that the situation changed fundamentally. Taking into account not only the many gaps in knowledge, but also the subjective and controversial nature of the assumptions needed to perform a comparison of mitigation and adaptation costs, one may question the feasibility of ever achieving a credible comparison of mitigation and adaptation costs and benefits (see Box 5.5). This does not negate the importance of an inventory of what we do know for both strategies and appreciation of the potential for synergies between them, as well as the potential trade-offs.

Mitigation and adaptation combined determine the risks of adverse climate impacts and the costs to reduce those risks. There are several reasons why costs of mitigation and adaptation strategies are difficult to compare. In Section 5.3.3 we discussed some factors that make the valuation of impacts very difficult, including the valuation of non-market impacts, the selection of the appropriate discount factor, the valuation of ancillary costs and benefits, choice of a reference scenario, and the applied model approach. Another reason for comparisons between costs and benefits of adaptation and mitigation strategies being challenging is that adaptation costs often are determined in the context of a fixed climate change scenario (e.g. double carbon dioxide impacts), independent of mitigation actions. Furthermore, policy-making is fragmented: mitigation policy is developed in the context of climate change, while adaptation policies are often developed in the context of natural hazards (Kane & Shogren 2000). It is often suggested that at an aggregate level, mitigation and adaptation compete for scarce resources. A euro spent on mitigation cannot be spent on adaptation, and vice versa. Mitigation and adaptation can be complementary to each other, but also be substitutes, with possible undesirable trade-offs. One example of a link in the agricultural sector would be investments in intensification of agriculture (more energy, fertilizer input) to counteract potential adverse impacts of climate change, but at the same time exacerbating climate change through increasing greenhouse gas emissions. In general, it is very difficult to compare the euro spent on mitigation with the euro spent on adaptation, because they most likely will be spent by different actors. It is most likely that strategies that incorporate both mitigation and adaptation are more efficient than strategies only pursuing one of them (Toth *et al.* 2001). However, also from the scientific perspective, such integrated evaluations are seldom made, not least because of methodological and data problems.[6] Possible approaches include an economic efficiency approach, or a precautionary approach.

From an *economic efficiency* perspective, a framework similar to adaptation could be used for mitigation (see Table 5.4), and net marginal costs and net[7] marginal benefits of both strategies could be compared and overall welfare loss or macroeconomic costs minimized in an optimization framework (Toth *et al.* 2001). In

---

[6] Adaptation research and assessments of costs and benefits of adaptation are hindered by scarcity of data. Data are needed in a wide range of areas, including information on the climate in the past, present, and future, and on the biological, environmental, and social systems affected by climate and climate change, and their interactions (Basher 1999). Currently, consistent data on past and present data are easily available through long-standing national and international climate observation systems. Also access to future climate data in the form of output of climate models has been facilitated for climate impact analyst through the IPCC Data Distribution Center (http://ipcc-ddc.cru.uea.ac.uk/). Data on other environmental and socioeconomic indicators are less easily available, and generally coverage is poorest in developing countries. Particularly absent is information on issues of particular importance for impact and adaptation assessment, e.g. vulnerability, critical thresholds, resilience, coping ranges, and adaptation potential (Basher 1999).

[7] Taking into account costs and ancillary benefits for mitigation, and costs and ancillary benefits for adaptation, as well as residual climate change impacts.

Table 5.4. *Framework for welfare implications of mitigation policies and uncertainty about climate change (adapted and expanded from Leary 1999).*

| | Welfare implications of mitigation strategies | | | |
| | Climate variability | | Climate change | |
Scenario	Costs	Benefits	Costs	Benefits
Current mitigation policies	Net costs of autonomous adjustments, e.g. in energy sector	Ancillary benefits of autonomous mitigation	Net costs of autonomous adjustments, e.g. in energy sector	Reduced climate change damage[a] plus net ancillary benefits of autonomous adjustments
Additional mitigation policies	Net costs of public policies and autonomous adjustments	Ancillary benefits of autonomous mitigation plus public policies	Net costs of public policies and autonomous adjustments	Reduced climate change damage[a] plus net ancillary benefits of autonomous adaptation and public policy

[a] Benefits of damage avoided by mitigation is not included here, only damage avoided by adaptation.

Chapters 8, 9, and 10, we discuss mitigation options and their costs in detail. But, as Toth *et al.* (2001) note, sector- and country-specific damage functions and adaptation options and their costs are largely unknown, especially in developing countries. Also mitigation costs are very uncertain, as discussed in Box 5.3 and Chapters 8, 9, and 10. This is why most integrated assessment models and studies do not directly balance mitigation and adaptation costs, although some models address adaptation indirectly, e.g. because changes in prices, changing consumption, and changing production structures drive market adjustments. This approach emphasizes the differences between mitigation and adaptation while, as we noted above, many measures enhance both mitigative and adaptive capacity.

Alternatively, from a *precautionary perspective*, precautionary investments should be made in both adaptation and mitigation to hedge against the uncertainties involved in climate change (Toth *et al.* 2001). A risk-averse attitude based on the precautionary principle is characteristic of many countries, and is also embedded in the UNFCCC, which recommends precautionary measures be taken

Box 5.5   Are estimates of costs and benefits of adaptation more uncertain than those of mitigation?

It is interesting to note that often adversaries of mitigation action claim that climate change and its impacts are much too uncertain to warrant mitigation, while at the same time they seem to be certain that mitigation action would be too expensive. Proponents of mitigation action emphasize the high costs related to climate impacts and adaptation, while sometimes noting that mitigation costs are uncertain and probably low. Are there differences between what we know about the costs of mitigation, adaptation and climate impacts?

As to adaptation and (avoided) impacts, the uncertainties are of various kinds. First, the climatic changes to which one should adapt are very uncertain, notably the regional patterns of change, their magnitude, and the rate of climate change. The uncertainties are compounded further by the fact that many damages can be associated with extreme climatic events, or possible future non-linear changes or 'surprises' in the climate system. Second, climate impacts and adaptation potential are very much dependent on the level of development of the region involved, and that (future) level of development can be considered only through scenarios of socioeconomic development. Third, not all damage can be valued easily or in a non-controversial way, e.g. loss of biodiversity, or loss of human lives. Nevertheless, apart from such non-market damages, valuing market damages can be difficult, e.g. if market distortions affect prices. Fourth, when costing adaptation activities, these often have benefits other than reducing vulnerability to climate change, thus generating uncertainty with respect to the allocation of costs to climate change adaptation.

In addition, mitigation costing is plagued by a wide variety of uncertainties (Moss 1999). Some are similar to those mentioned above for adaptation. For example, mitigation costs are very dependent on the baseline, and the assumed socioeconomic development scenario used in the analysis. Moreover, allocating costs to mitigation is difficult, when one considers that many mitigative measures have benefits other than reducing greenhouse gas emissions, which may be as important. Similar to the valuation of impacts and adaptation, also for mitigation, factors that can scarcely be valued in monetary terms play a role, such as possible political advantages of solutions (e.g. decreased dependency on fuel imports from unstable regions), social or institutional changes, etc. Another key uncertainty in mitigation costs assessment is the 'autonomous' technological change, change that would take place anyway, without mitigative policies. Finally, we could mention the choice of policy instruments that would be used to implement mitigation actions, e.g. assumptions with reference to international co-ordination of policies, effectiveness of various types of instruments (market-based, regulatory, voluntary agreements, research and development enhancement) or the use of revenues from applying certain instruments cause uncertainties.

These significant uncertainties make it difficult to interpret results of cost–benefit analyses of climate change. However, they should not disqualify the analyses. Rather, if the uncertainties involved in the methods applied, and the assumptions selected, are mentioned explicitly and discussed, this can help in qualifying the results and making them more useful.

to anticipate, prevent or minimize the causes of climate change and mitigate its adverse effects. Thus, according to the UNFCCC, lack of knowledge should not be an argument for postponing actions, and mitigation as well as adaptation measures should be taken and implemented before uncertainties are resolved. It is very likely that not all uncertainties ever will be resolved, and decision-making can be expected to continue to take place under uncertainties. What does this approach imply for adaptation? Interpreted very strictly, adaptation measures would have to be taken to minimize adverse effects of climate change regardless of the costs, and regardless of the mitigation efforts that may lessen those effects as well. Applying the precautionary approach less rigidly, the safe minimum standard approach could be followed, implementing mitigation and adaptation options unless costs are considered to be unacceptable. This would not imply a fully fledged optimizing cost–benefit analysis searching for the ideal, most efficient, blend of adaptation and mitigation, but an iterative analysis of the costs and benefits of both strategies, and evaluation of the costs (are they acceptable?) and benefits (do they lower the risks sufficiently?) of both strategies. This approach would also be better suited for analyses of options that would enhance adaptive and mitigative capacity simultaneously, contributing to wider sustainable development objectives.

References

Arrow, K. J., Cline, W. R., Maler, K-G., Munasinghe, M., Squitieri, R. and Stiglitz, J. E. (1996) Intertemporal equity, discounting, and economic efficiency. In J. P. Bruce, Hoesung Lee and E. F. Haites, eds., *Climate Change 1995: Economic and Social Dimensions of Climate Change*. Cambridge: Cambridge University Press.

Banuri, T., Weyant, J., Akumu, G., Najam, A., Pinguelli Rosa, L., Rayner, S., Sachs, W., Sharma, R. and Yohe, G. (2001) Setting the stage: climate change and sustainable development. In B. Metz, O. Davidson, R. Swart and J. Pan, eds., *Climate Change 2001: Mitigation*. Cambridge: Cambridge University Press.

Basher, R. E. (1999) Data requirements for developing adaptations to climate variability and change. In *Mitigation and Adaptation Strategies for Global Change*, Vol. 4. pp. 227–37.

Bein, P., Burton, I., Chiotti, Q., Demeritt, D., Dore, M., Rothman D. and Vandierendonck, M. (1999) Costing Climate Change in Canada: Impacts and Adaptation. Presented at IPCC Meeting on *Costing Issues for Mitigation and Adaptation to Climate Change*. Tokyo: Global, Industrial and Social Progress Research Institute.

Burton, I. and Lim, B. (2001) *An Adaptation Policy Framework: Capacity Building for Stage II Adaptation, A UNDP-GEF Project*. New York: United Nations Development Program/Global Environment Facility.

Carter, T. R., Parry, M., Nishioka, S. N. and Harasawa, H. (1996) IPCC technical guidelines for assessing climate change impacts and adaptations. In R. T. Watson, M. C. Zinyowera, R. H. Moss and D. F. J. Dokken, eds., *Climate Change 1995: Impacts, Adaptations and Mitigation of Climate Change: Scientific–Technical Analyses*. Cambridge: Cambridge University Press.

Carter, T. R., La Rovere, E. L., Jones, R. N., Leemans, R., Mearns, L. O., Nakicenovic, N., Pittock, A. B., Semenov, S. M. and Skea, J. (2001) Developing and applying scenarios. In J. McCarthy, O. F. Canziani, N. A. Leary, D. J. Dokken and K. S. White, eds., *Climate Change 2001: impacts, adaptation, and vulnerability*. Cambridge: Cambridge University Press.

Dai, A., Wigley, T. M. L., Meehl, G. A. and Washington, W. M. (2001) Effects of stabilising atmospheric CO_2 on global climate in the next two centuries. *Geophysical Research Letters* (in press).

Fankhauser, S., Smith, J. B. and Tol, R. S. J. (1999). Weathering climate change. Some simple rules to guide adaptation investments. *Ecological Economics*, **30**, 67–78.

Feenstra, J. F., Burton, I., Smith J. B. and Tol, R. S. J. eds. (1998). Handbook on Methods for Climate Change Impacts Assessment and Adaptation Strategies, Version 2. Nairobi/Amsterdam: United Nations Environmental Programme and Institute for Environmental Studies, Vrije Universiteit.

Gitay, H., Brown, S., Easterling, W., Jallow, B., Antle, J., Apps, M., Beamish, R., Chaoin, T., Cramer, W., Frangi, J., Laine, J., Erda, L., Magnuson, J., Noble, I., Price, J., Prowse, T., Root, T., Schulze, E., Sirotenko, O., Sohngen, B. and Soussana, J. (2001) Ecosystems and their goods and services. In J. McCarthy, O. F. Canziani, N. A. Leary, D. J. Dokken and K. S. White, eds., *Climate Change 2001: Impacts, Adaptation, and Vulnerability*. Cambridge: Cambridge University Press.

IPCC (1998) *Summary Report to IPCC*. San Jose, Costa Rica: Workshop on Adaptation to Climate Variability and Change.

Jepma, C. J., Asaduzzaman, M., Mintzer, I., Maya, R. S. and Al-Moneef, M. (1996) A generic assessment of response options. In J. P. Bruce, Hoesung Lee and E. F. Haites, eds., *Climate Change 1995: Economic and Social Dimensions of Climate Change*. Cambridge: Cambridge University Press.

Jepma, C. J. and Munasinghe, M. (1998) *Climate Change Policy: Facts, Issues, and Analysis.* Cambridge: Cambridge University Press.

Kane, S. and Shogren, J. F. (2000) Linking adaptation and mitigation in climate change policy. *Climatic Change*, **45**, 75–102.

Kane, S. and Yohe, G. (2000) Societal adaptation to climate variability and change: an introduction. *Climatic Change*, **45**(1), 1–4.

Kates, R. (2000). *Cautionary tales: adaptation and the global poor. Climatic Change*, **45**(1), 5–17.

Klein, R. J. T., Nicholls, R. J. and Mimura, N. (1999) Coastal adaptation to cliomate change: can the IPCC *Technical Guidelines* be applied? *Mitigation and Adaptation Strategies for Global Change*, **4**(3–4), 239–52.

Leary, N. (1999) A framework for benefit–cost analysis of adaptation to climate change and climate variability. In *Mitigation and Adaptation Strategies for Global Change*, Vol. 4. pp. 307–18.

Lind R. C. (1982) A Primer on the major issues relating to the discount rate for evaluating national energy options. In R. C. Lind, ed. *Discounting for Time and Risk in Energy Policy*. Washington: Resources for the Future, pp. 257–71.

Malcolm, J. R. and Markham, A. (2000) *Global Warming and Terrestrial Biodiversity Decline*. Geneva: World Wildlife Fund for the Conservation of Nature.

Malcolm, J. R., Liu, C., Miller, L. B., Alnutt, T. and Hanssen, L. (2002) *Habitats: Global Warming and Species Loss in Globally Significant Terrestrial Ecosystems*. Switzerland: World Wildlife Fund for the Conservation of Nature.

McCarthy, J., Canziani, O. F., Leary, N. A., Dokken, D. J. and K. S. White, eds. (2001) *Climate Change 2001: Impacts, Adaptation, and Vulnerability*. Cambridge: Cambridge University Press.

Markandya, A. (1999) The Treatment of Discounting in the assessment for climate change. *Costing Issues for Mitigation and Adaptation to Climate Change*. Tokyo: Global, Industrial and Social Progress Research Institute.

Moss, R. (1999) Cost estimation and uncertainty. *Costing Issues for Mitigation and Adaptation to Climate Change*. Tokyo: Global, Industrial and Social Progress Research Institute.

Munasinghe, M., Meier, P., Hoel, M., Hong, S. W. and Aaheim, A. (1996) Applicability of techniques of cost–benefit analysis to climate change. In J. P. Bruce, Hoesung Lee and E. F. Haites, eds., *Climate Change 1995: Economic and Social Dimensions of Climate Change*. Cambridge: Cambridge University Press.

Newell, R. and Pizer, W. (2002) *Discounting the Benefits of Climate Change Mitigation: How Much do Uncertain Rates Increase Valuations?* Washington: Pew Center.

Pearce, D. W., Cline, W. R., Achanta, A. N., Fankhauser, S., Pachauri, R. K., Tol, R. S. J. and Vellinga, P. (1996) The social costs of climate change: greenhouse damage and the benefits of control. In J. P. Bruce, Hoesung Lee and E. F. Haites, eds., *Climate Change 1995: Economic and Social Dimensions of Climate Change*. Cambridge: Cambridge University Press.

R. A. Pielke (1998) Rethinking the role of adaptation in climate policy. *Global Environmental Change*, **8**(2), 159–70.

Portney, P. R. and J. P. Weyant, eds. (2001) *Discounting and Intergenerational Equity*. Washington: Resources for the Future.

Smit, B., Pilifosova, O., Burton, I., Challenger, B., Huq, S., Klein, R. J. T. and Yohe, G. (2001) Adaptation to climate change in the context of sustainability, development and equity. In J. McCarthy, O. F. Canziani, N. A. Leary, D. J. Dokken and K. S. White, eds., *Climate Change 2001: Impacts, Adaptation, and Vulnerability*. Cambridge: Cambridge University Press.

Tol, R. S. J., Fankhauser, S. and Smith, J. B. (1998) The scope for adaptation to climate change: what can we learn from the impact literature? *Global Environmental Change*, **8**(2), 109–23.

Toth, F. L., Mwandosya, M., Carraro, C., Christensen, J., Edmonds, J., Flannery, B., Gay-Garica, C., Hoesung Lee, Meyer-Abich, K., Nikitina, E., Rahman, A., Richels, R., Ruqiu, Y., Villavicencio, A., Wake, Y. and Weyant, J. (2001) Decision-making Frameworks. In B. Metz, O. Davidson, R. Swart and J. Pan. *Climate Change 2001: Mitigation*. Cambridge: Cambridge University Press.

UNFCCC (1992) *United Nations Framework Convention on Climate Change*. Bonn: United Nations Framework Convention on Climate Change.

UNFCCC (1997) *The Kyoto Protocol to the United Nations Framework Convention on Climate Change*. Bonn: United Nations Framework Convention on Climate Change.

Wheaton, E. and MacIver, D. (1998) *Working Paper on Adaptation to Climate Variability and Change*. In San Jose, Costa Rica: Workshop on Adaptation to Climate Variability and Change.

6

Vulnerability, impacts, and adaptation by sectors and systems

Natural and human systems are exposed to variations in climate. These include changes in average range and variability of temperature and precipitation as well as the frequency and severity of weather events. Some indirect effects of climate change include sea-level rise, changes in soil moisture and land and water conditions, changes in frequency of fire and pest infestation, and the distribution of disease vectors and hosts. The potential for a system to sustain adverse impacts is determined by its capacity to adapt to the changes.

6.1 Hydrology and water resources

6.1.1 Impacts of climate change on hydrology and water resources

Water stress is becoming an apparent problem in many parts of the world. The amount of water available per person is decreasing, whereas the ratio of volume of water withdrawn to volume of water potentially available is increasing (Arnell 1999; 2000), (see Figure 6.1).

This water stress is caused by many socioeconomic and natural changes unrelated to climate change, but is likely to be exacerbated by climate change. Approximately one-third (1.7 billion) of the population of the world presently live in countries that are water-stressed, i.e. using more than 20 per cent of their renewable water supply. This number is expected to increase to 5 billion by 2025.

Water, once revered for it's life-giving properties, has become a commodity. Throughout the world, human use of water has already led to dried-up and polluted rivers, lakes, and groundwater resources. Potable water is becoming increasingly scarce. By the year 2025, it is predicted that water abstractions will increase by 50 per cent in developing countries and 18 per cent in developed countries.

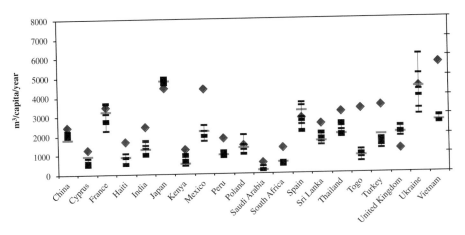

Figure 6.1. National water resources per capita (m³/yr) in 1990 and 2050 under several climate change scenario, for some countries. Diamonds represent 1990, long bars 2050 with no climate change, and short bars 2050 under different climate change scenarios. Source: Arnell *et al.* (2001).

Effects on natural ecosystems will be dramatic. The inevitable result of further human abstraction of water on this scale will be the degradation or complete destruction of key terrestrial freshwater and coastal ecosystems that are vital to life itself. Humankind needs to recognize that social wellbeing, economic stability, and the natural environment, are interdependent. Some strategies for achieving sustainable water could be to conserve and restore global freshwater resources. An ecosystem-based approach should be adopted within river basins, in addition to empowering the people to establish participatory, equitable and responsible water use. Political will and good governance is also essential to facilitate wise water use and prevent conflicts. Programmes should be initiated to raise awareness and strengthen capacity to change human behaviour to reduce water consumption and waste, promote recycling, and protect the ecosystems. Sharing of knowledge and technology is also essential to make water resources management more sustainable.

Streamflow

Projected climate change could further decrease streamflow and groundwater recharge in water-stressed countries (e.g. central Asia, southern Africa, and the area around the Mediterranean), but may increase these factors in others. Studies have shown an increase in streamflow in high latitudes. However, there is still a lack of confidence as to whether this is a result of climate change or other factors, e.g. response of river flows to natural hydrological variability and changes in land use and land cover by humans. The effect of climate change on streamflow and groundwater recharge is dependent largely on projected changes in precipitation.

The peak streamflow is expected to move from spring to winter (Bergstrom & Carlsson 1993). Higher temperatures mean that a greater proportion of the winter precipitation falls as rain rather than snow, and is thus not stored on the land surface until it melts in spring. In the colder areas, precipitation would still fall as snow, hence the areas that would be most affected by this change in temperature would be the 'marginal' zones, including central and eastern Europe and the southern Rocky Mountain chain, where a small temperature rise would reduce snowfall substantially.

Water quality

The quality of water is expected to be degraded due to higher water temperatures as increased temperatures alter the rate of operation of biogeochemical processes, and also lower the dissolved oxygen concentration in the water (Murdoch, Baron & Miller 2000). An increase in streamflow would help dilute the chemical concentrations, whereas a lower streamflow would increase concentrations (Frisk *et al.* 1997; Kallio *et al.* 1997). In lakes, changes in mixing may offset or exaggerate the effects of increased temperature.

Flood magnitude

There is likely to be an increase in flood magnitude and a decrease in low flows in many regions. Increase in flood magnitude and frequency is a result of a projected increase in the frequency of heavy precipitation events (Mirza *et al.* 1998). A projected increase in evaporation may lead to lower low flows, even where precipitation increases or shows little change.

Demand for water

Increased temperatures would lead to increased evaporation, hence causing an increase in irrigation water demand. Industrial and municipal demand will not be affected substantially by climate change (Shiklomanov 1998; Shiklomanov *et al.* 2000).

Hydrological variability

Factors such as system characteristics, changing pressures on the system, management systems, and which adaptations to climate change are implemented, would also influence water resources. Unmanaged systems are most vulnerable to climate change as they have no structure in place to buffer the effects of hydrological variability.

6.1.2   Adaptation to climate change impacts on hydrology and water resources

As discussed in Chapter 5, adaptation can be anticipatory or reactive. Anticipatory strategies include the incorporation of uncertainty into water resources planning and management, and improved ability to forecast stream-flow ahead of time. Such measures would significantly enhance water management and its use to cope with changing hydrological variability. Reactive adaptation includes short-term operational changes, e.g. temporary exploitation of new sources. Adaptation in the water sector can focus on protecting or enhancing the supply of water, but also on decreasing demand and increasing water-use efficiency (see Table 6.1).

In the past few years, emphasis has been placed on demand-side strategies, including the promotion of ways of managing and pricing water resources in order for them to be more effective. In practice, it is not easy to change water management practices. Institutional capacity, wealth, management philosophy, planning, scale, and organizational arrangements would help to decide which adaptive measures should be taken. There is a clear need to introduce sustainable water management practices into specific institutional settings – which is necessary even in the absence of climate change – to improve the effectiveness of water management. Practically all the above-mentioned adaptation measures also make sense in the absence of climate change, as they decrease the vulnerability to current variability in water supply.

6.2   Managed and natural ecosystems

6.2.1   Impacts on agriculture

The response of crop yield to climate change depends on the species, cultivar, soil conditions, treatment of carbon dioxide direct effects, and other locational factors. Studies have shown that a temperature increase of a few degrees would lead to an increase in temperate crop yields (Bowes *et al.* 1996; Casella, Soussana & Loiseau 1996). However, with larger amounts of warming, these yield responses become negative. In the tropics, where some crops are near their maximum temperature tolerance, yields tend to decrease with minimal changes in temperature. Higher minimum temperatures will be beneficial to some crops, especially in temperate regions, and detrimental to others, especially in low latitudes. High maximum temperatures generally will be detrimental to many crops. In addition, extreme events are likely to affect crop yields.

The impact of climate change on agriculture is expected to result in small changes in global income, with positive changes in more developed regions and smaller or negative changes in developing regions (Antle 1996; Reilly 1996; Smith

Table 6.1. *Supply and demand side adaptation options in water management.*

Supply side strategies	Demand side strategies
✓ Increasing capacity, e.g. building reservoirs or structural flood defences	✓ Changing institutional practices
✓ Catchment source control to reduce peak discharge for flood management	✓ Managing demand, e.g. water-efficient irrigation, water pricing initiatives
✓ Locks or dredging to manage water levels for navigation	✓ Introducing incentives to use less water
✓ Extracting more water from rivers or groundwater	✓ Legally enforcing of water use standards
✓ Interbasin transfer	✓ Increasing use of grey water
✓ Changing operating rules for existing structures and systems	✓ Reducing leakages
✓ Desalinization	✓ Developing non-water-based sanitation systems
✓ Seasonal forecasting	✓ Increasing irrigation use efficiency
✓ Increasing irrigation source capacity	✓ Increasing drought toleration
✓ Using low-grade water for industry and power station cooling	✓ Changing crop patterns
✓ Enhancing treatment works to minimize pollution	✓ Increasing water use efficiency and water recycling in industries and power station cooling
	✓ Increasing efficiency of hydro-turbines and encourage energy efficiency
	✓ Altering ship size and frequency of navigation
	✓ Reducing volume of effluents needing treatment, e.g. charging discharges
	✓ Managing catchments to reduce pollution runoff
	✓ Improving flood warning and dissemination
	✓ Curbing floodplain development
	✓ Introduce non-structural flood management methods, e.g land-use controls

et al. 1996). Increases in mean annual temperature of 2.5 °C would increase world food prices due to the inadequate supply of food to meet demand.

6.2.2   *Adaptation in agriculture*

Degradation of soil and water is likely to be intensified by adverse changes in temperature and precipitation (Pinstrup-Andersen & Pandya-Lorch 1998). Land use and management are shown to have a greater effect on soil conditions than the direct effect of climate change; thus, adaptation has the potential significantly to mitigate these impacts. Autonomous agronomic adaptation (e.g. altering planting dates and cultivar selections) have reduced temperate crop loss (Rosenzweig & Inglesias 1998; Parry *et al.* 1999). In the tropics, autonomous agronomic adaptation has allowed crops to be affected less adversely by climate change than without adaptation (see Figure 5.4). Other direct adaptation options, although more expensive, include changing land use allocations and fertilization rates, and developing and using irrigation infrastructure and drought-resistant crops. Moreover, in the case of livestock management, adaptation options exist, e.g. enhancing shade and introducing sprinkler systems.

In general, the transfer of appropriate technologies to boost agricultural production in a sustainable fashion can reduce vulnerability, but there could be costs associated with learning about, and gaining experience with, different crops, or greater need for irrigation that is not readily available. Investing in local institutional capacity should accompany transfers of new technology (Gitay *et al.* 2001).

6.2.3   *Natural terrestrial and freshwater ecosystems*

Today, there are relatively few ecosystems in the world that are not, to some extent, influenced by human society. The influence of climate change on these 'natural' systems will be determined by land and water management, adaptation, and interactions with other pressures. Adaptive capacity is greater in intensively managed lands and waters that produce marketed goods (see Sections 6.2.1 and 6.2.2), than in less intensively managed lands that produce non-marketed goods. Nevertheless, in the latter systems, some adaptation options are available (see Table 6.2). Several of these options are specifically to help natural ecosystems adapt to climate change.

Protecting vulnerable ecosystems (which are already under external stresses), can enhance their resilience to climate change. Establishing more connections among protected systems (rather than relying on fragmented areas) is often useful for existing species movements, but could also facilitate future redistribution or migration in case the climate shifts.

Table 6.2. *Adaptation in natural ecosystems, including low-intensity managed lands.*

System	Impact issues	Adaptation options
Biodiversity	Populations of many species are threatened by climate change (Stattersfield *et al.* 1998; UNEP 2000), decreasing habitat sizes, land use changes that fragment habitats (Wilson 1992). Loss of wildlife would affect low-income societies that depend on wildlife for subsistence living, and also the services provided by wildlife in ecosystems	Establishment of refuges, parks, and reserves with corridors to allow migration Use of captive breeding and translocation
Aquatic systems	Human responses to warmer climate, changing precipitation scenarios, and human population increases are likely to place greater demands on freshwater to meet the water needs for drinking, industry, and irrigation. Climate change will cause fish species to migrate polewards (Minns & Moore 1995). There will be a loss in habitat for cold-water fish and a gain in habitat for warm-water fish. There will also be an increase in extinctions and invasions of exotics (Dettmers & Stein 1996), and the exacerbation of existing pollution problems such as euthrophication (Horne & Goldman 1994), toxics (Magnuson *et al.* 1997; Schindler 1997), acid rain (Yan *et al.* 1996) and UVB radiation	Poleward transportation of fish across watershed boundaries to cooler waters Introducing better-adapted species
Wetlands	Most wetland processes are dependent on catchment-levels hydrology which is being altered by land use changes; thus, it is difficult to adapt to climate change	Small-scale restoration for key habitats is possible if sufficient water is available Where wetlands are used for agriculture, the impacts could be mitigated by choice of cropping method, including alternative crops and depth of drainage

(cont.)

Table 6.2. (*cont.*)

System	Impact issues	Adaptation options
Rangelands	It is difficult to separate the human impacts on rangelands from climate change impacts. However, adaptation and management strategies need to be implemented to prevent degradation	Landscape management Selection of plants and livestock Multiple cropping systems and agroforestry Community participation, development of public policy
Wildlife	Opportunities for adapting to expected changes in high-latitude and alpine ecosystems are limited as these systems are limited in moving poleward or upward. Careful management of wildlife resources could minimize climatic impacts on indigenous peoples depending on these resources. High levels of endemism in alpine floras and their inability to migrate upward causes these species to be very vulnerable	Establishing parks and reserves Introducing captive breeding and translocation Replacing lost ecosystem services
Forests and woodlands	Increasing carbon dioxide levels would increase net primary productivity, whereas increasing temperature may have either positive or negative effects. Largest and earliest climate change induced impacts are likely to occur in boreal forests due to changes in weather and nutrient cycling. In arid and semi-arid areas, climate change is expected to decease soil moisture and productivity. Terrestrial ecosystems appear to be storing increasing amounts of carbon. Studies show that terrestrial uptake may be caused by changes in uses and management of land rather than by increased carbon dioxide or climate change. If markets exist, prices will mediate adaptation through land and product management	Salvaging dead and dying timber Replanting new species that are better suited to the new climate Planting genetically modified species Intensifying or decreasing management Land and product management (e.g. substituting species in the production process for solid wood and pulp wood products, shifting harvests from one region to another, developing new technologies and products such as wood products manufactured with adhesives etc.), increasing global timber supply Maintaining fuelwood supplies and tree cover in developing countries through traditional management, agroforestry, small woodlot management and woodrows

6.3   Coastal zones and marine ecosystems

6.3.1   Impacts on coastal zones and marine ecosystems

Global climate change is expected to increase sea-surface temperature and sea-level (Levitus *et al.* 2000), while decreasing sea-ice cover (Rothrock, Yu & Maykut 1999) and changing salinity levels, wave climate (Young 1999) and ocean circulation (Cane *et al.* 1997). Changes in upwelling rates would have major impacts on coastal fish production. Fluctuations in fish populations are regarded as biological responses to medium-term climate fluctuations (Ware 1995), in addition to overfishing and other anthropogenic factors. Survival of marine mammals and seabirds is also affected by inter-annual and longer-term variability in oceanographic and atmospheric processes (Springer 1998).

There is increasing recognition of the role of climate–ocean systems in the management of fish stocks. This has led to new adaptive strategies based on determination of acceptable removable percentages of fish and stock resilience. Creating sustainable fisheries also depends on understanding synergies between climate-related impacts on fisheries and factors such as harvest pressure and habitat conditions. Expansion of marine aquaculture may partly compensate for reductions in ocean fish catch. Decreases in dissolved oxygen levels due to increased seawater temperatures and increased organic matter create conditions for the spread of disease in wild and aquaculture fisheries, as well as outbreaks of algal blooms on coastal areas (Anderson, Cambella & Hallegraeff 1998). Pollution and habitat destruction that accompany aquaculture may limit the survival success of wild stocks.

Climate change and sea-level rise would increase frequency and levels of sea-flooding, accelerate coastal erosion (Bird 1993) and seawater intrusion into freshwater sources. Island nations and highly diverse and productive coastal ecosystems are likely to suffer serious impacts. Low-latitude tropical and subtropical coastlines, especially in areas in which there is significant human population pressure, are very susceptible to climate change impacts. Inundation, salinization of potable groundwater and coastal erosion, will be accelerated with global sea-level rise.

Coral reefs are among the richest and, at the same time, most threatened ecosystems in the world. They are sensitive particularly to increased water temperatures, but their health is threatened also by marine water pollution, fishing practices, exploitation as building materials, and intensive tourism activities (McLean *et al.* 2001). Decreasing such stresses will help not only to conserve these valuable ecosystems today, but also increase their resilience against the impacts of future climate change.

High-latitude coastlines are subject to accelerated sea-level rise, more energetic wave climate (Solomon, Forbes & Kierstead 1994), a reduced sea-ice cover (Dallimore, Wolfe & Solomon 1996), and increased groundwater temperatures (Forbes, Shaw & Taylor 1997). This is likely to have severe impacts on settlements and infrastructure, and can necessitate coastal retreat.

6.3.2   *Adaptation to coastal and marine climate change impacts*

Three coastal adaptation strategies can be developed as follows.

1. *Protect*: protect land from sea so that existing land uses can continue, by constructing hard structures (e.g. sea-walls, groynes) as well as using soft measures, e.g. beach nourishment.
2. *Accommodate*: continue to occupy the land but make some adjustments, e.g. elevating buildings on piles, growing flood or salt tolerant crops.
3. *Retreat*: abandon the endangered coastal area.

Over the last few years, adaptation strategies have shifted away from hard protection structures to soft protection measures. Options include managed retreat and enhanced resilience of biophysical and socioeconomic systems, including the use of flood insurance to spread financial risk. Additional options include the collection of information and awareness-raising, planning and design, implementation, monitoring, and evaluation.

Maintenance or improvement of coastal defence systems by institution-building, avoiding investments in flood-prone zones, managed retreat and flood insurance, improving natural coastal protection systems (e.g. coral reefs, mangroves and other coastal wetlands), and disaster-preparedness and prevention programmes, are examples of strategies that would not only decrease vulnerability to current weather variability, but also increase preparedness to sea-level rise and other potential impacts of climate change in coastal lands (McLean *et al.* 2001).

Improvements in sustainable development and management could be based on integrated assessments of coastal zones and marine ecosystems and the better understanding of their interaction with human development and multiyear climate variability (Figure 6.2). Adaptation options for coastal and marine management are most effective when they are incorporated with policies in other areas, e.g. disaster mitigation, watershed management, and land use plans.

Recent experiences with fishery conflict have shown that adaptation can be difficult when a resource is exploited by multiple competing users who possess incomplete information about the resource. Improved fishery management includes better collection and sharing of information, modified fishing industry practices, protection of spawning areas and habitats. All these measures will

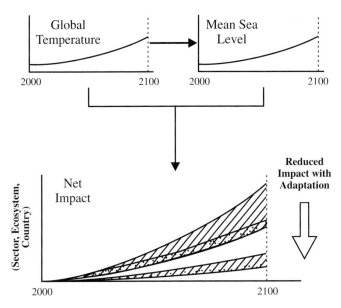

Figure 6.2. The role of adaptation in reducing potential impacts in the coastal zone from global temperature increase and sea level to the year 2100. The bottom panel is a schematic that shows the increasing costs or loss to an economic sector, ecosystem, or country. The area shown by cross-hatch indicates the range of possible impacts and how net impact can be reduced with adaptation. Stipple within the cross-hatched areas indicate the importance of a sector, ecosystem or country resilience as a component of net impact. Source: McLean *et al.* (2001)

contribute to harvests that remain within sustainable production levels. Climate change is expected to change further the distribution and abundance of marine animals. Therefore, establishing a sound and sustainable fishery industry now will reduce its vulnerability to future climate change impacts (McLean *et al.* 2001).

6.4 Energy, industry, and settlements

6.4.1 *Impacts on energy, industry, and settlements*

Global warming is expected to increase energy demand for space cooling (EIA 1998) and decrease energy use for space warming. The projected net effect on annual energy consumption is scenario- and location-specific. As for industry, a poleward intensification of agricultural, forestry, and mining activities has been observed, resulting in increased population and intensified settlement patterns (Cohen 1997). Manufacturing industry dependent on climate-sensitive natural resources (e.g. agriculture), would be affected severely. Very little is known about the impacts of climate change on industry and most information is highly speculative.

Human settlements may be affected by climate change in the following three ways.

1. Changes in productive capacity (e.g. agriculture and fisheries) or market demand for goods and services, e.g. demand from people living nearby and from tourism.
2. Physical infrastructure (including energy transmission and distribution systems), buildings, infrastructure (including transport) and industries, e.g. agroindustry, tourism, construction.
3. Direct effect on population through extreme weather, changes in health or migration.

The most widespread impacts of climate change include flooding, landslides, mudslides, and avalanches, driven by increases in rainfall intensity and sea-level rise. Settlements may not be designed adequately to support these impacts, and may result in other environmental impacts such as reduced air and water quality (Hardoy, Mitlin & Satterthwaite 2000; McGranahan & Satterthwaite 2000). Windstorms, water shortages, and fire are also experienced in some regions.

6.4.2   Adaptation in the energy and industry sectors, and human settlements

To be successful, adaptation options in settlements must be consistent with economic development. They must be environmentally and socially sustainable over time, and equitable (Table 6.3).

The local capacity to limit environmental health hazards and support human wellbeing in a settlement, also enhances the capacity to adapt to climate change, unless adaptation implies expensive infrastructure investment. Adaptation to a warmer climate will require the tuning of settlements to a changing environment, not just to warmer temperatures. Urban experts state that successful environmental adaptation cannot occur without locally based, technically and institutionally competent, and politically supported, leadership that has good access to national level resources.

Possible adaptation options involve taking into account climate change in the planning of settlements and their infrastructure, placement of industrial facilities, and reducing the effects of climate and extreme weather events that are of low probability and high consequences. Some tools for better environmental planning and management include market-based tools for pollution control, demand management and waste reduction, mixed-use zoning and transport planning, environmental impact assessments, capacity studies, strategic environmental plans, environmental audit procedures, and state-of-the-environment reports. Most of these options also address broader development objectives.

Table 6.3. *Adaptation in energy, industry and settlements.*

Planning and design	✓ Increase economic diversification
	✓ Enhance oasis development
	✓ Develop wind breaks
	✓ Develop irrigation and water supply systems
	✓ Improve sanitation, water supply, power distribution systems, solid waste collection
	✓ Enhance fire protection
	✓ Take climate change into account in zoning and land use planning
	✓ Adapt building codes to reduce resource use and temperature control
	✓ Reduce flood risks, e.g. flood barriers, managed retreat, hazard mapping
	✓ Take advantage of replacement schedules for buildings and infrastructure
Management	✓ Increase environmental and health education
	✓ Improve landscape management
	✓ Preserve and maintain environmental quality
	✓ Develop warning systems, evacuation plans, insurance, relief
	✓ Enforce building codes
	✓ Increase disaster preparedness
	✓ Introduce market mechanisms for efficient management of water supply, e.g. fix leaks
	✓ Introduce water wholesaling and delivery
	✓ Enforce pollution controls efficient public transport systems
Institutional changes	✓ Improve institutional capacity in environmental management
	✓ Establish partnerships between all responsible parties (government, private sector, nongovernmental organizations)
	✓ Adapt property rights to allow informal settlements to buy, rent, or build housing on safe sites
	✓ Improve access to technology

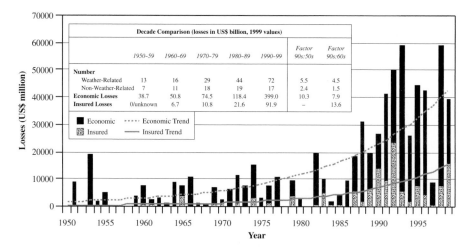

Figure 6.3. The costs of catastrophic weather events have exhibited an upward trend in recent decades. Yearly economic losses from large events increased 10.3 fold from US$4 billion per year in the 1950s to US$40 billion in the 1990s. Source: Vellinga *et al.* (2001)

6.5   Financial resources and services

6.5.1   Impacts on the financial and services sectors

The effect of climate change on the financial services sector is likely to increase through changes in spatial distribution, frequency, and intensity of extreme weather events. The financial services sector, that offers insurance and disaster relief, banking, and asset management services, can be considered as a unique indicator of potential impacts of climate change as it is very sensitive to weather-related damages as signs of potential climate change and its 'integrated' effects on other sectors. This sector is important in adaptation (e.g. through support of building codes and some land use planning), and providing financial services to represent risk-spreading mechanisms through which the cost of weather-related events are distributed among other sectors and throughout society. The costs of weather events have risen rapidly (Anderson 2000), despite significant efforts at fortifying infrastructure and enhancing disaster preparedness (Figure 6.3)

The observed upward trend in disaster losses is linked to socioeconomic factors such as population growth, increased wealth, and urbanization in vulnerable areas (Kunkel, Pielke & Changnon 1999), and is linked partly also to climatic factors such as changes in precipitation, flooding, and drought events (Pielke & Downton 2000). Increased weather-related losses could stress insurance companies to a point of impaired profitability, causing them to increase costs, leading to consumers withdrawing their coverage and demanding publicly funded compensation and relief (White & Etkin 1997; Kunreuther 1998).

6.5.2 Adaptation in the financial sector

Adaptation to climate change presents complex challenges, as well as opportunities, for the financial sector. There are various options to decrease the vulnerability of the financial sector to climate change. Examples include increasing firm size, diversification and integration of insurance with other financial services, as well as improved tools to transfer risk. Regulatory involvement in pricing, tax treatment of reserves, and the ability of firms to withdraw from at-risk markets would influence the resilience of the sector.

The effects of climate change are expected to be greatest in developing countries in terms of loss of life, effects on investment, and effects on the economy (see Figure 6.4). Weather disasters set back development, particularly when funds are redirected from development projects to disaster recovery projects. More extensive access to insurance, disaster preparedness and recovery resources would increase the ability of developing countries to adapt to climate change. Widespread introduction of microfinancing schemes and development banking could be an effective mechanism to help developing countries and communities adapt.

There is a need for: (a) better analysis of economic losses in order to determine the causes of disasters; (b) assessment of financial resources involved in dealing with climate change and adaptation; (c) evaluation of alternative methods to generate such resources; (d) deeper investigation of a sector's vulnerability, and (e) resilience to a range of extreme weather event scenarios. Also, more research is needed into how the sector could innovate to meet the potential increase in demand for adaptation funding in developed and developing countries in order to spread and reduce risks of climate change.

6.6   Human health

Global climate change will have diverse effects on human health, both positive and negative. Frequencies in extreme heat and cold, floods and droughts, and the profile of local air pollution and aeroallergens would affect the population health directly. Other impacts would occur from the impact of climate change on ecological and social systems, e.g. changes in infectious disease occurrence, local food production and undernutrition, and health consequences of population displacement and economic disruption. Population health status is multifactorial, as it depends on socioeconomic, demographic, and environmental issues, which are also changing.

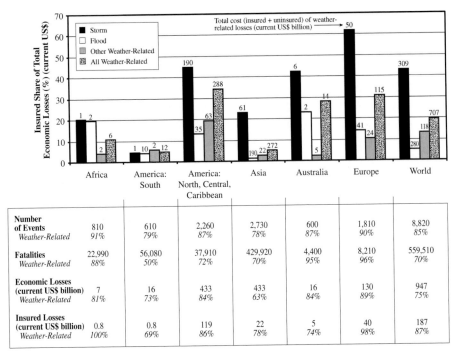

Figure 6.4. Regional insurance coverage for weather- and non-weather-related natural disasters from 1985–1999. The role of insurance in paying weather-related losses varies by event type and region, generally dominated by windstorms. Source: Vellinga *et al.* (2001).

Munich Re, 1999b: Topics 2000 – Natural Catastrophes, The Current Position (published statistics updated by Munich Re to reflect adjustments for 1999 year-end loss accounting). Munich Reinsurance Group, Geoscience Research Group, Munich, Germany, 126 pp.

In IPCC 2001, Climate Change 2001: Impacts Adaptation, and Vulnerability. Contribution of Working Group I to the Third Assessment Report of the Intergovernmental Panel on Climate Change [J. J. McCarthy, O. F. Canziani, N. A. Leary, D. J. Dokken, K. S. White (eds.)] Chapter 8.

In general, adaptation options for health impacts include social, institutional, technological, and behavioural methods. There is a basic need, especially in developing countries, for public health infrastructure to be strengthened and maintained. Adequate financial and public health resources to encompass training programmes, research to develop and implement more effective surveillance and emergency response systems, and sustainable prevention and control programmes are required. There is a need for continued research and further understanding of associations between extreme events and vector-borne diseases. Research into medical prevention and control (e.g. vaccines, methods to deal with drug-resistant strains of infectious agents, and mosquito control) is essential. More generally, research is required in order to evaluate adaptation measures, assess their environmental and health implications, and set priorities on adaptation strategies.

The ability of affected communities to adapt to health risks also depends on social, environmental, political, and economic circumstances. Autonomous adaptation can happen by a natural or spontaneous response to climate change by affected individuals. Purposeful adaptation is composed of planned responses to projected climate change, by governmental or other institutional organizations. Purposeful adaptation can also occur via deliberate modification of personal, family, and community lifestyles. Improved education can make people better aware of health risks and improve the general health status, again with a positive side-effect of enhancing resilience to possible adverse health impacts of climate change. Anticipatory adaptations are planned responses that take place in advance of climate change. Below, some specific health-related impacts and adaptation options are summarized.

Heat waves

Increased heat waves and intensity would increase the risk of death and serious illness, mainly in older people and the urban poor (Ando, Uchiyama & Ono 1998; Ando *et al.* 1998). The greatest increases in thermal stress are forecast for mid to high latitudes, especially in populations with maladapted architecture and limited air-conditioning. There is some evidence to show that in temperate countries, reduced winter deaths would outnumber increased summer deaths (Langford & Bentham 1995).

Legislative adaptation measures for thermal stress include building guidelines. Technical adaptation could include housing, public buildings, urban planning to reduce heat-island effects and increased air-conditioning. Early warming systems could also be adopted, along with changing clothing and having siestas.

Extreme weather events

Increased frequency of storms, floods (Menne *et al.* 1999), droughts (McMichael *et al.* 1996), and cyclones (Noji 1997) would affect human health adversely. Natural hazards can cause injury and loss of life, loss of shelter, population displacement, contamination of water supplies, loss of food production (causing hunger and malnutrition), increased risk of infectious disease epidemics, and damage to infrastructure for provision of health services (IFRC 1998).

Planning laws, building guidelines, forced migration, and economic incentives for buildings could be encouraged as legislative adaptation. Urban planning and storm shelters also could be provided in addition to adopting early warning systems.

Air quality

Climate change can affect air quality in urban areas. Increase in temperature increases formation of ground-level ozone, which has adverse effects on respiratory health (Patz *et al.* 2000).

Emission controls and traffic restrictions are important steps to reduce health impacts from poor air quality in general. Thus, such measures also will decrease the vulnerability of the population to the possible additional effects of climate change. Improved public transport, use of catalytic converters, car-pooling, and other ways of reducing local air pollution can be encouraged, in addition to the adoption of pollution-warning systems.

Vector-borne infectious diseases

Higher temperatures, changes in precipitation and climate variability would affect seasonality of vector-borne infectious diseases. These diseases are transmitted by blood-nourished organisms that depend on climate and other ecological factors for survival (see Table 6.4). A change in climatic conditions will increase the incidence of various types of water-and food-borne infectious diseases (Gubler 1998). Some technical adaptation mechanisms include vector control, vaccinations, impregnated mosquito nets, sustainable surveillance, prevention and control programmes. Health education and safe water storage practices should be encouraged. All these measures reduce the vulnerability of the population to health hazards, even in the absence of climate change.

Water-borne diseases

In order to protect the population better from water-borne diseases – for which the risk could be enhanced by climate change – watershed protection laws and water quality regulations can be initiated. Generic molecular screening of pathogens, improved water treatment (e.g. filters), and improved sanitation reduce risk. People could be trained to boil polluted water, wash hands, and adopt other hygienic behaviours, as well as be encouraged to use pit latrines.

Biotoxin poisoning

Climate change may cause changes in the marine environment that would alter the risk of biotoxin poisoning from the consumption of fish and shellfish

Table 6.4. (IPCC 2001b, Chapter 9, Table 9.1)

Main vector-borne diseases: population's at risk and burden of disease (WHO data)

Disease	Vector	Population at risk	Number of people currently infected or new cases per year	Disability adjusted life years lost[a]	Present distribution
Malaria	Mosquito	2400 million (40 per cent world population)	272 925 000	39 300 000	Tropics/subtropics
Schistosomiasis	Water snail	500–600 million	120 million	1 700 000	Tropics/subtropics
Lymphatic filariasis	Mosquito	1000 million	120 million	4 700 000	Tropics/subtropics
African trypanosomiasis (sleeping sickness)	Tsetse fly	55 million	300 000–500 000 cases yr^{-1}	1 200 000	Tropical Africa
Leishmaniasis	Sandfly	350 million	1.5–2 million new cases per year	1 700 000	Asia/Africa/ southern Europe/Americas
Onchocerciasis (river blindness)	Blackfly	120 million	18 million	1 100 000	Africa/Latin America/Yemen
American trypanosomiasis (Chagas's disease)	Triatomine bug	100 million	16–18 million	600 000	Central and South America
Dengue	Mosquito	3 000 million	Tens of millions cases per year	1 800 000[b]	All tropical countries
Yellow fever	Mosquito	468 million in Africa	200 000 cases per year	Not available	Tropical South America and Africa
Japanese encephalitis	Mosquito	300 million	50 000 cases per year	500 000	Asia

[a] Disability adjusted life year (DALY) = a measurement of population health deficit that combines chronic illness or disability and premature death (see Murray 1994; Murray & Lopez 1996). Numbers are rounded to nearest 100 000.

[b] Data from Gubler and Metzer (1999).

Sources: McMichael *et al.* (2001).

(WHO 1984). Biotoxins associated with warmer water could extend their range to high latitudes (Tester 1994).

Food supply

Changes in food supply could affect nutrition and health of the poor. The impacts of reduced food yields are greatest in developing countries (FAO 1999). Undernourishment increases the susceptibility of humans to infectious diseases.

References

Anderson, D. R. (2000) Catastrophe insurance and compensation: remembering basic principles. *CPCU Journal*, **53**(2), 76–89.

Anderson, D. M., Cembella, A. D. and Hallegraeff, G. M. eds., (1998) Physiological ecology of harmful algal blooms. In *Proceedings of NATO Advanced Study Institute, Bermuda, 1996*. Berlin: Springer-Verlag.

Ando, M., Uchiyama, I. and Ono, M. (1998) Impacts on human health. In S. Nihioka, and H. Harasawa, eds., *Global Warming: The Potential Impact on Japan*. Tokyo: Springer-Verlag, pp. 203–13.

Ando, M., Kobayashi, I. N., Kawahara, I., Asanuma, S. and Liang, C. K. (1998) Impacts of heat stress on hyperthermic disorders and heat stroke. *Global Environmental Research*, **2**, 111–20.

Antle, J. M. (1996) Methodological issues in assessing potential impacts of climate change on agriculture. *Agricultural and Forest Meteorology*, **80**, 67–85.

Arnell, N. W. (1999) Climate change and global water resources. *Global Environmental Change*, **9**, S31–49.

Arnell, N. W. (2000) *Impact of climate change on global water resources*: Vol. 2. Report to Department of the Environment, Transport and the Regions. Southampton: University of Southampton.

Bergstrom, S. and Carlsson, B. (1993) *Hydrology of the Baltic Basin*. Swedish Metereological and Hydrological Institute Reports. *Hydrology*, **7**, 21.

Bird, E. C. F. (1993) *Submerging Coasts: The Effects of Rising Sea Level on Coastal Environment*. Chichester: John Wiley and Sons.

Bowes, G., Vu, J. C. V., Hussain, M. W., Pennanen, A. H. and Allen, L. H. Jr. (1996) An overview on how rubisco and carbohydrate metabolism may be regulated at elevated atmospheric ($CO2$) and temperature. *Agricultural and Food Science in Finland*, **5**, 261–70.

Cane, M. A., Clement, A. C., Kaplan, A., Kushir, Y., Pozdayakov, R., Seager, S. and Zebaik, E. (1997) Twentieth century sea surface temperature trends. *Science*, **275**, 957–60

Casella, E., Soussana, J. F. and Loiseau, P. (1996) Long-term effects of CO_2 enrichment and temperature increase on a temperate grass sward, I: productivity and water use. *Plant and Soil*, **182**, 83–99.

Cohen, S. J. (1997) *Mackenzie Basin Impact Study Final Report*. Downsview ON: Environment Canada.

Dallimore, S. R., Wolfe, S. and Solomon, S. M. (1996) Influence of ground ice and permafrost on coastal evolution, Richards Island, Beaufort Sea Coast, NWT. *Canadian Journal of Earth Sciences*, **33**, 664–75.

Dettmers, J. M. and Stein, R. A. (1996) Quantifying linkages among gizzard shad, 'zooplankton, and phytoplankton in reservoirs. *Transactions of the American Fisheries Society*, **125**, 27–41.

EIA (1998) *International Energy Outlook*. DOE/EIA-0484(98). Washington: US Department of Energy. Energy Information Administration.

FAO (1999) *The State of Food Insecurity in the World 1999*. Rome: Food and Agriculture Organization of the United Nations.

Forbes, D. L., Shaw, J. and Taylor, R. B. (1997) Climate change impacts in the coastal zone of Atlantic Canada. In J. Abraham, T. Canavan and R. Shaw, eds., *Climate Variability and Climate Change in Atlantic Canada*. Ottawa: Environment Canada. 6, 51–66.

Frisk, T., Bilaletdin, A., Kallio, K. and Saura, M. (1997) Modeling the effects of climate change on lake euthrophication. *Boreal Environment Research*, **2**, 53–67.

Gitay, H., Brown, S., Easterling, W., Jallow, B., Antle, J., Apps, M., Beamish, R., Chapin, T., Cramer, W., Frangi, J., Laine, J., Lin, E., Magnuson, J., Noble, I., Price, J., Prowse, T., Root, T., Schulze, E., Sirotenko, O., Sohngen, B. and Soussana, J. (2001) Ecosystems and their goods and services. In *Climate Change 2001: Impacts, Vulnerability and Adaptations*, J. J. McCarthy, O. F. Canziani, N. A. Leary, D. J. Dokken and K. White, (eds.), Cambridge: Cambridge University Press.

Gubler, D. J. (1998) Climate change: implications for human health. *Health and Environment Digest*, **12**, 54–55.

Gubler, D. J. and Meltzer, M. (1999) The impact of dengue/dengue hemorrharic fever in the developing world. In K. Maramorosch, F. A. Murphy and A. J. Shatkin, eds., *Advances in Virus Research*, Vol. 53. San Diego: Academic Press, pp. 35–70.

Hardoy, J. E., Mitlin, D. and Satterthwaite, D. (2000) *Environmental Problems in an Urbanising World: Local Solutions for City Problems in Africa, Asia and Latin America*. London: Earthscan Publications.

Horne A. J. and Goldman, C. R. (1994) *Limnology* (2nd edn.). New York: McGraw-Hill.

IFRC (1998) *World Disaster Report 1998*. New York: International Federation of Red Cross and Red Crescent Societies, and Oxford: Oxford University Press.

Kallio, K., Rekolainen, S., Ekholm, P., Granlund, K., Laine, Y., Johnsson, H., and Hoffman, M. (1997) Impacts of climatic change on agricultural nutrient losses in Finland. *Boreal Environment Research*, **2**, 33–52.

Kunkel, K. E., Pielke, Jr. R. A., and Changnon, S. A. (1999) Temporal fluctuations in weather and climate extremes that cause economic and human health impacts: a review. *Bulletin of the American Meteorological Society*, **80**(6), 1077–98.

Kunreuther, H. (1998) Insurability conditions and the supply of coverage. In H. Kunreuther, and R. Roth, eds., *Paying the Price: The Status and Role of Insurance Against Natural Disasters in the United States*. Washington: Joseph Henry Press, pp. 17–50.

Langford, I. H. and Bentham, G. (1995) The potential effects of climate change on winter mortality in England and Wales. *International Journal of Biometeorology*, **38**, 141–7.

Levitus, S., Antonov, J. I., Boyer, T. P. and Stephens, C. (2000) Warming of the world ocean. *Science*, **287**, 2225–9.

Magnuson, J. J., Webster, K. E., Assel, R. A., Bowser, C. J., Dillon, P. J., Eaton, J. G., Evans, H. E., Fee, E. J., Hall, R. I., Mortsch, L. R., Schindler, D. W. and Quinn, F. H. (1997) Potential effects of climate change on aquatic systems: Laurentian Great Lakes and Precambrian shield region. *Hydrological Processes*, **11**, 825–71.

McLean, R. F., Tsyban, A., Burkett, V., Burkett, V., Codignotto, J. O., Forbes, D. L., Mimura, N., Beamish, R. J. and Ittekkot, V. (2001) Coastal zones and marine ecosystems. In J. J. McCarthy, O. F. Canziani, N. A. Leary, D. J. Dokken and K. White, eds., *Climate Change 2001: Impacts, Vulnerability and Adaptations*. Cambridge: Cambridge University Press.

McGranahan G. and Satterthwaite, D. (2000) Environmental health or sustainability? Reconciling the brown and green agendas in urban development. In C. Pugh, ed., *Sustainable Cities in Developing Countries*. London: Earthscan Publications, pp. 73–90.

McMichael, A. J., Haines, A., Sloof, R. and Kovats, S. eds., (1996) *Climate Change and Human Health*. (WHO/EHG/96.7). Geneva: World Health Organization.

Menne, B., Pond, K., Noji, E. K. and Bertollini, R. (1999) *Floods and Public Health Consequences, Prevention and Control Measures*. (UNECE/MP.WAT/SEM.2/1999/22). Rome: World Health Organization European Centre for Environment and Health.

Minns, C. K. and Moore, J. E. (1995) Factors limiting the distributions of Ontario's freshwater fishes: the role of climate and other variables, and the potential impacts of climate change. In R. J. Beamish, ed., *Climate Change and Northern Fish Populations*. Canada: Fish Aquatic Sciences, pp. 137–160.

Mirza, M. Q., Warrick, R. A., Ericksen, N. J. and Kenny, G. J. (1998) Trends and persistence in precipitation in the Ganges, Brahmaputra and Meghna Basins in South Asia. *Hydrological Sciences Journal*, **43**, 845–58.

Murdoch, P. S., Baron, J. S. and Miller, T. L. (2000) Potential effects of climate change on surface water quality in North America. *Journal of the American Water Resources Association*, **36**, 347–66.

Murray, C. J. L. (1994) Quantifying the burden of disease: the technical basis for disability-adjusted life years. *Bulletin of WHO*, **72**, 429–45.

Murray C. J. L. and Lopez, A. D., eds. (1996) *The Global Burden of Disease: Global Burden of Disease and Injury Series*, Vol. 1. Harvard School of Public Health. Boston: Harvard University Press.

Noji, E., ed., (1997) *The Public Health Consequences of Disasters*. Oxford and New York: Oxford University Press.

Parry, M., Fischer, C., Livermore, M., Rosenzweig, C. and Iglesias, A. (1999) Climate change and world food security: a new assessment. *Global Environmental Change*, **9**, S51–67.

Patz, J. A., McGeehin, M. A., Bernard, S. M., Ebi, K. L., Epstein, P. R., Grambsch, A., Gubler, D. J. and Reiter, P. (2000) The potential health impacts of climate variability and change for the United States: executive summary of the report of the health sector of the US National Assessment. *Environmental Health Perspectives*, **108**, 367–76.

Pielke R. A. Jr. and Downton, M. W. (2000) Precipitation and damaging floods: trends in the United States, 1932–1997. *Journal of Climate*, **13**(20), 3625–37.

Pinstrup-Andersen, P. and Pandya-Lorch, R. (1998) Food security and sustainable use of natural resources: a 2020 vision. *Ecological Economics*, **26**, 1–10.

Reilly, J. (1996) Agriculture in a changing climate: impacts and adaptation. In R. T. Watson, M. C. Zinyowera and R. H. Moss, eds., *Climate Change 1995: Impacts, Adaptations and Mitigation of Climate Change: Scientific Technical Analyses. Contribution of Working Group II to the Second Assessment Report of the Intergovernmental Panel on Climate Change*. Cambridge and New York: Cambridge University Press, pp. 429–67.

Rosenzweig, C. and Iglesias, A. (1998) The use of crop models for international climate change impact assessment. In G. Y. Tusji, G. Hoogrnboom and P. K. Thorton, eds., *Understanding Options for Agriculture Production*. Dordrecht: Kluwer, pp. 267–92.

Rothrock, D. A., Yu, Y. and Maykut, G. A. (1999) Thinning of the Artic sea ice cover. *Geophysical Research Letters*, **26**, 3469–72

Schindler, D. W. (1997) Widespread effects of climatic warming on freshwater ecosystems in North America. *Hydrological Processes,* **11**, 825–71.

Shiklomanov, A. I., Lammers, R. B., Peterson, B. J. and Vorosmarty, C. (2000) The dynamics of river water flow in the Artic Ocean. In E. L. Lewis, E. P. Jones, P. Lemke, T. D. Prowse and P. Wadhams, eds, *The Freshwater Budget of the Artic Ocean*. Dordrecht: Kluwer.

Shiklomanov, A. I. (1998) *Assessment of water resources and water availability in the World. Background Report for the Comprehensive Assessment of the Freshwater Resources of the World.* Stockholm: Stockholm Environment Institute.

Smith, J. B., Huq, S. Lenhart, S. Mata, L. J., Nemesova, I. and Toure, S. (1996) *Vulnerability and Adaptation to Climate Change: Interim Results from the U.S. Country Studies Program.* Dordrecht and Boston: Kluwer.

Solomon, S. M., Forbes, D. L. and Kierstead, B. (1994) *Coastal Impacts of Climate Change: Beaufort Sea Erosion Study.* Downsview ON: Canadian Climate Centre.

Springer, A. M. (1998) Is it all cc? Why marine bird and mammal populations fluctuate in the North Pacific. In G. Holloway, P. Muller and D. Henderson, eds., *Biotic Impacts of Extratropical Climate Variability in the Pacific*. Honolulu: National Oceanic and Atmospheric Administration and the University of Hawaii, pp. 109–20.

Stattersfield, A. J., Crosby, M. J., Long, A. J. and Wege, D. C. (1998) *Endemic Bird Areas of the World: Priorities for Biodiversity Conservation. Birdlife Conservation Series No. 7.* Cambridge: Birdlife International.

Tester, P. A. (1994) Harmful marine phytoplankton and shellfish toxicity potential consequences of climate change. *Annals of New York Academy of Sciences*, **740**, 69–76.

UNEP (2000) *Global Environmental Outlook 2000*. Nairobi: United Nations Environment Program.

Vellinga, P., Mills, E., Berz, G., Bouwer, L., Huq, S., Kozak, L. A., Palutikof, J., Schantenbächer, B. and Soler, G. (2001) Insurance and other financial services. In

J. J. McCarthy, F. Canziani, N. A. Leary, D. J. Dokken and K. S. White, eds., *Climate Change 2001: Impacts, Adaptation, and Vulnerability*. Cambridge and New York: Cambridge University Press.

Ware, D. M. (1995) A century and a half of change in the climate of the North East Pacific. *Fisheries Oceanography*, **4**, 267–77.

White, R. and Etkin, D. (1997) Climate change, extreme events and the Canadian insurance industry. *Natural Hazards*, **16**, 135–63.

WHO (1984) *Environmental Health Criteria 37: Aquatic (Marine and Freshwater) Biotoxins*. Geneva: World Health Organization.

Wilson, E. O. (1992) *The Diversity of Life*. New York: Norton.

Yan, N. D., Keller, W., Scully, N. M., Lean, D. R. S. and Dillon, P. J. (1996) Increased UV-B penetration in lakes owing to drought-induced acification. *Nature*, **381**, 141–3.

Young, I. (1999) Seasonal variability of the global ocean wind and wave climate. *International Journal of Climatology*, **19**, 931–50.

7

Vulnerability, impacts, and adaptation by geographic region

From the regional viewpoint, developing countries lying within the tropical areas are the most sensitive and vulnerable to climate change impacts. Many ecosystems in poorer countries are already under stress, and climate change impacts will further exacerbate the situation. Social and economic systems are also more vulnerable, because income levels are lower (including limited funds, human resources, and skills), and political, institutional, and technological support systems are weaker than in the industrialized world. An important implication is that adaptive capacity must be strengthened significantly, especially in the poorest and most vulnerable regions and countries. In this chapter, we will show that strengthening adaptive capacity generally is fully consistent with achieving broader development objectives.

The status of the major regions of the world, with reference to vulnerability, impacts, and adaptation, is reviewed briefly in this chapter. Necessarily, there is some overlap with the generic, sectoral approach of the previous chapter, but we make an attempt in this chapter to focus on adaptation options that may be of specific priority for the various regions.

7.1   Africa

7.1.1   Vulnerability

Because of the endemic poverty and the reliance of the rapidly growing population on natural resources and agriculture for the provision of basic needs and economic production (UNEP 2002), Africa is particularly vulnerable to environmental changes. Land degradation, deforestation, habitat degradation, water stress and scarcity, coastal area erosion and degradation, floods and droughts, and armed conflict, are some of the key environmental problems in Africa that

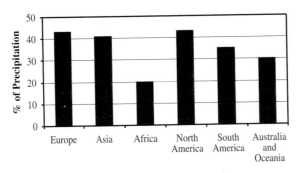

Figure 7.1. Comparative hydrology in world regions – total runoff as percentage of precipitation. *Source:* Desanker *et al.* (1997).

hamper development. According to UNEP (2002), tackling these problems is not an option, but a necessity. Many of these environmental pressures are likely to be exacerbated by climate change.

In view of that, Africa is highly vulnerable to climate change, particularly in relation to water resources (see Figure 7.1) (Reibsame *et al.* 1995), food production (WRI 1998), human health (WHO 1998), desertification (UNEP 1997), and coastal zones (Nicholls, Hoozemans and Marchand 1999). The diversity of African climates, high rainfall variability, and a sparse observational network, make predictions of future climate change difficult.

The overall capacity for Africa to adapt to climate change is generally very low due to the diversity of constraints facing many nations, notably the pervasively poor economic situation. Current technologies and approaches, especially in agriculture and water, are inadequate to meet projected demands under increased climate variability. Uncertainty of future conditions means that there is low confidence in projected costs of climate change. Individual states need to begin developing methodologies for estimating such costs based on their specific circumstances.

7.1.2   Water resources

Water availability varies considerably within countries, influenced by physical characteristics and seasonal patterns of rainfall. Increased evaporation from high temperatures is not compensated for enough by precipitation. Per capita water availability has decreased considerably since 1990 (see Figure 7.2). In 2000, about 300 million Africans risk living in a water-scarce environment (Sharma *et al.* 1996). The greatest impact will be felt by the poor, who have the most limited access to water resources. Vulnerability of water resources affects water supply for household use, agriculture and industry.

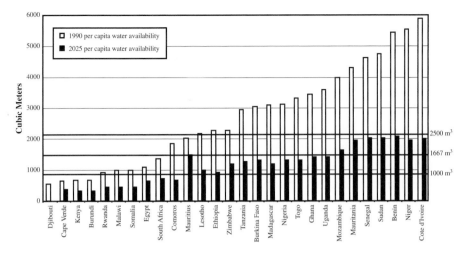

Figure 7.2. Water scarcity and people in Africa. *Source:* Desanker *et al.* (2001).

Some adaptation options include:

1.   Early warning systems to enable timely remedial measures.
2.   Shared basin management through international agreements, to ensure equity in access to, and accountability for, water supply and water quality management.
3.   Water use strategies, (especially pricing and demand management) in industry, settlements, and agriculture.
4.   Improved crop watering technology to be less wasteful.
5.   Increase water storage facilities, e.g. damming, synchronization of reservoir operations.
6.   Water recycling in industry.
7.   Supply restrictions.
8.   Monitoring to improve data reliability.
9.   National action plans that include long-term changes and 'no regrets' strategies.
10.   Greater regional co-operation in science, resource management, and development.
11.   Research into energy usage and alternate renewable energy at household and industrial levels that would be less vulnerable to climatic changes, as currently Africa is dependent on hydropower for its energy supply.

In order to minimize sensitivity to climate change, African economies should be more diversified, and agricultural technology should optimize water usage through efficient irrigation and crop development.

7.1.3   Agriculture and food security

Africa is clearly among the regions with the lowest food security and lowest ability to adapt to future changes. Food security is likely to worsen due to increased extremes, and temporal and spatial shifts in climate. Food production in most of Africa has not kept pace with increasing population over the past three decades (WRI 1998), i.e. food consumption exceeded domestic production by 50 per cent in the 1990s. Agriculture is not only a vital food source for Africa, but also the prevailing way of life. Fluctuations in annual food production resulting from climate change are increasing the reliance on food aid. Migration in search of better lands and opportunities places increasing pressure on the environment, including social amenities. If food insecurity prevents private investment in agricultural economies, resources for adapting to climate change may not keep pace with impacts. However, Africa has vast resources – human and natural – that can be tapped to make rapid gains in food security and thus reduce the risks of adverse climate change.

Changes in temperature, moisture levels, ultraviolet radiation, carbon dioxide levels, and pest and diseases, are likely to impact food production. Crop water balances will be affected by changes in precipitation, increased evapotranspiration, and increased water use efficiency resulting from elevated carbon dioxide levels. Production of staple food for millions of people (wheat, corn, rice, beans, potatoes) is likely to decrease due to global warming (Pimentel 1993).

Domestic livestock plays an important role in many African cultures and is concentrated mainly in the arid and semi-arid zones, and feeds on grasslands and savannas. Crop residues are an important supplement during the dry seasons. With increased temperatures, meat and milk production declines, as animals remain in the shade instead of foraging. Adaptation would require substitution with species such as the oryx, which are able to withstand higher temperatures. Drought and increased temperatures also adversely affect fish stocks in rivers and lakes. Some adaptation options include:

1. Seasonal forecasting, e.g. seasonal maize water stress forecast anticipates it 6 months prior to harvest based on water stress and historical global sea-surface temperature and sea-level pressure records.
2. Early warning systems to determine which geographic areas are unlikely to meet their subsistence needs before the next agricultural season.
3. Technology transfer and carbon sequestration, in a way that takes advantage of Africa's resources and human potential.
4. Windbreaks, mulching, ridging, and rock bunds.
5. Better soil and water conservation practices, better irrigation practices.
6. More disease- and drought-tolerant crop varieties.

7. Agroforestry programmes, e.g. in the Senegal river basin.

8. Improved pest and weed control.

9. Supplemented livestock diets to increase meat production.

10. Lower livestock numbers per hectare would reduce carbon dioxide and methane emissions.

11. Limited amount of heat resistance can be introduced through breeding programmes.

12. Access to international markets, to diversify economies and increase food security.

13. Remove distortions created by import barriers.

14. Link environmental degradation to economic policy.

15. Empower people to develop adaptive strategies towards sustainable livelihoods.

7.1.4   *Natural resources and biodiversity*

Irreversible losses of biodiversity could be accelerated through intensity, frequency, and extent of vegetation fires, habitat modification, and land use changes. Alteration of spatial and temporal patterns in temperature, rainfall, solar radiation, and winds will exacerbate desertification (Tucker, Dregne & Newcomb 1991).

Forests cover one-sixth of the continent's land area (FAO 1999). These areas provide ecosystem services of carbon sequestration, storing and transpiring water required for precipitation, maintaining soil fertility, and forming habitats for a diverse array of plant and animal species. They also provide firewood, structural timber, traditional medicines, staple foods, and drought emergency foods. A large part of the African population is dependent on forest resources. Dry woodlands and savannas will be increasingly subject to drying in the next century, as well as increasing land use intensity – including conversion to agriculture (Desanker *et al.* 1997). Species are likely to respond to changing climate and disturbance regimes individualistically, with substantial time-lags and periods of reorganization.

Some promising adaptation strategies to declining tree resources include:

1. Allow for natural regeneration of local species.
2. Energy-efficient cooking stoves.
3. Sustainable forest management.
4. Community-based natural resource management.

Africa occupies one-fifth of the global land surface of the Earth, and contains about one-fifth of its flora and fauna (Seigfried 1989). Biodiversity hotspots are threatened by a shift in rainfall seasonality, increases in temperature, and changes in atmospheric carbon dioxide. This biodiversity forms an important resource for the African people (e.g. food, fibre, fuel, shelter, medicinal, wildlife trade) and the economically important tourism industry. Land use conversion effects on biodiversity in affected areas will overshadow climate change effects for some time to come. Without adaptive and mitigating strategies, the impact of climate change will reduce the effectiveness of the reserve network, by altering ecosystem characteristics within it and causing species emigrations or extinctions.

Adaptation requires a risk-sharing approach to be taken between countries, which could include transboundry nature reserves.

7.1.5   *Human health*

Changes in temperature (see Figure 7.3) and rainfall will affect human health negatively. Geographic location (e.g. proximity to water), socioeconomic status (quality of housing) and knowledge about preventative measures, would exacerbate health risks. Following the 1997–8 El Niño event, malaria, Rift Valley fever and cholera outbreaks were recorded in East Africa. The meningitis belt has also expanded to the east. A weak infrastructure, land use change, and drug resistance by pathogens, have aggravated the spread of diseases. Increased malaria in Rwanda could be explained by changes in rainfall and temperature (Loevinsohn 1994). Cholera is a water- and food-borne disease, and increased incidence is linked to rises in sea-surface temperature (Colwell 1996). Meningitis infections usually start in the middle of the dry season and end a few months after the rains start (Greenwood 1984). Transmission could be affected by warming and reduced precipitation. Increased precipitation could increase the risk of infections of Rift Valley fever. Biomass burning and emissions from badly maintained vehicles along with increasing temperatures could increase the risk of respiratory diseases, and eye and skin infections (Boko 1992).

Adaptation to climate change would have to increase the understanding of how climate affects the transmission of these diseases. Some effective adaptation options include:

1. Early and effective preparedness.
2. Safe drinking-water technologies.
3. Insecticide-treated fabrics (e.g. mosquito nets and curtains) to decrease malaria infections (Lengeler 1998).
4. Remote sensing for forecasting risks of malaria and cholera (Hay, Snow & Rogers 1998).
5. Boiling and filtering water for drinking.

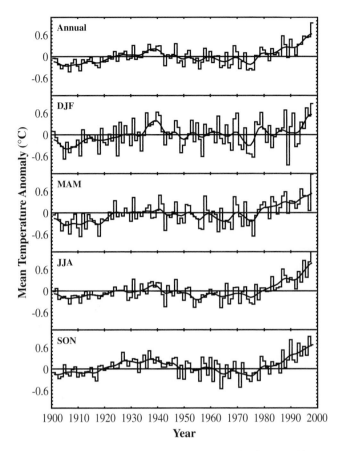

Figure 7.3. Mean surface air temperature anomalies for the African continent, 1910–1998, expressed with respect to 1961–1990 average, annual and four seasons (DJF, MAM, JJA, SON). Smooth curves result from applying a 10-year Gaussian filter. *Source:* Desanker *et al.* (2001).

7.1.6   *Settlements and infrastructure*

Increased frequency of flooding, heat waves, dust storms, hurricanes, and other extreme events cause degradation to infrastructure, leading to serious deterioration of systems that deliver social, health, and economic services. Sea-level rise, salt-water intrusion and flooding will have serious effects on African economies. Extreme events associated with climate change are likely to exacerbate management problems relating to pollution, sanitation, waste disposal, water supply, public health, infrastructure and technologies of production (IPCC 1966). High concentrations of human settlements are prevalent within 100 km of coastal zones (Singh *et al.* 1999), in areas of high economic potential, in river and lake basins, in close proximity to major transportation routes, and in places that enjoy pleasant climatic regimes.

Sea-level rise can be devastating to coastal zones and ports. Heavy precipitation could have severe negative impacts on road networks and air transport. Increased temperatures and air pollution lead to increased incidence of respiratory diseases.

Some adaptation options could include:

1. Regional integration among coastal zone states.
2. Coastal defence systems.
3. Design of infrastructure facilities (e.g. roads and telecommunications) to withstand extreme events.
4. Design buildings with natural ventilation, etc. to cope with increased temperatures.
5. Clean-air policies and stringent air-quality standards.
6. Flood control management technology.
7. Better understanding of hydrology of river basins.
8. Identify vulnerable areas and plan coping mechanisms.
9. Early warning systems against floods.
10. Innovation in building designs, e.g. to minimize urban flooding.
11. Regional co-operation in sharing hydro-electric potential.
12. More intensive use of renewable energy such as solar, wind, biomass, and biogas.

7.2   Asia

7.2.1   Vulnerability

Asia holds more than 60 per cent of the world's population, causing its natural resources to come under stress, with the resilience to climate change in most sectors being poor. Asia has the largest land mass in the world, with a very great diversity of landscapes, climates, and populations. Therefore, different sub-regions in Asia have very different vulnerabilities to climate change and need a diverse portfolio of adaptive options.

Environmental problems are similar to Africa, and include deforestation, water scarcity and pollution, land degradation, and coastal erosion. Even more than in Africa, Asian environmental problems are also concentrated in the large number of megacities, which suffer from waste management problems, unsustainable water supplies, and air pollution. Many of these problems, again, are likely to be worsened by climate change. Already, temperatures are showing clear increases (see Figure 7.4).

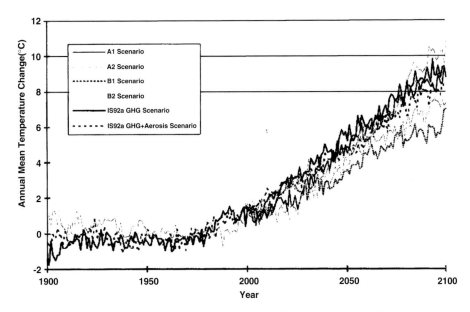

Figure 7.4. Future trends in area-averaged annual mean temperature increase over land regions of Asia as simulated by CCSR/NIES AOGCM for IS92a and SRES emission scenarios. *Source:* In *IPCC* 2001, *Climate Change 2001: Impacts Adaptation, and Vulnerability. Contribution of Working Group I to the Third Assessment Report of the Intergovernmental Panel on Climate Change* [J. J. McCarthy, O. F. Canziani, N. A. Leary, D. J. Dokken, K. S. White (eds.)] Chapter 11.

Food security appears to be of primary concern as crop production (Naylor, Falcon & Zavaleta 1997), (see Table 7.1) and aquaculture are threatened by thermal and water stresses, sea-level rise, increased flooding, and strong winds. Climate change would exacerbate current threats to biodiversity resulting from land use/cover change and population pressure. Many species – especially those in coastal areas and mountainous habitats – could experience large population declines. Competition between species in alpine areas may lead to new species taking over due to increased temperatures. Frequency of forest fires is expected to increase in Boreal Asia (Valendik 1996). Climate change is likely to affect primary productivity, species composition and migration, and the occurrence of pests and diseases (Melillo *et al*. 1996). Pronounced warming in high latitudes of the Northern Hemisphere could lead to the thinning or disappearance of permafrost (Nelson & Anisimov 1993). A change in precipitation is likely to affect the quality of water in rivers and lakes (Fukushima *et al*. 2000). Freshwater availability and quality is expected to be highly vulnerable to climate change (Arnell 1999a). Runoff is expected to increase in high latitudes and near the equator and decrease in mid latitudes (IPCC 1998). Developing countries in Asia are very vulnerable to extreme

Table 7.1. *Food security in Asia: changes in areas under rice cultivation in select Asian countries, 1979–99.* Source: *Lal* et al. *(2001).*

Country	Period	Total rice cultivation area (10 000 ha)	Change in rice cultivation area (10 000 ha)	Rate of change in rice cultivation area (ha yr^{-1})
Bangladesh	1979	10 160	310	14 762
	1999	10 470		
Cambodia	1979	774	1 187	56 524
	1999	1 961		
China	1979	34 560	−2 840	−135 238
	1999	31 720		
India	1979	39 414	3 586	170 762
	1999	43 000		
Indonesia	1979	8 804	2 820	134 286
	1999	11 624		
Malaysia	1979	738	−93	−4 429
	1999	645		
Myanmar	1979	4 442	1 016	48 381
	1999	5 458		
Nepal	1979	1 254	260	12 381
	1999	1 514		
Sri Lanka	1979	790	39	1 857
	1999	829		
Pakistan	1979	2 035	365	17 381
	1999	2 400		
Philippines	1979	3 637	341	16 238
	1999	3 978		
Thailand	1979	8 654	1 346	64 095
	1999	3 978		
Vietnam	1979	5 485	2 163	103 000
	1999	7 648		

climate events such as typhoons, cyclones (Walsh & Pittock 1998), droughts, and floods. Large deltas and low-lying coastal areas would be inundated by sea-level rise (Huang 1999; Walker 1998), (see Table 7.2). Increased atmospheric carbon dioxide concentration and a consequent rise in sea-surface temperature are likely to have serious damaging effects on reef accretion and biodiversity. Anomalies in water temperatures and changing ocean currents have led to low commercial fish catches in recent years (Yoshino 1998). Warmer and wetter conditions would

Table 7.2. *Sea-level rise in Asia: potential land loss and population exposed in Asian countries for selected magnitudes of sea-level rise and under no adaptation measures.* Source: *Lal* et al. *(2001).*

Country	Sea-level rise (cm)	Potential land loss (km²)	(%)	Population exposed (millions)	(%)
Bangladesh	45	15 668	10.9	5.5	5.0
	100	29 846	20.7	14.8	13.5
India	100	5 763	0.4	7.1	0.8
Indonesia	60	34 000	1.9	2.0	1.1
Japan	50	1 412	0.4	2.9	2.3
Malaysia	100	7 000	2.1	>0.05	>0.3
Pakistan	20	1 700	0.2	n.a.	n.a.
Vietnam	100	40 000	12.1	17.1	23.1

n.a. not applicable

increase the potential for heat-related (Ando 1998) and infectious (Colwell 1996) diseases.

7.2.2   Adaptation

Adaptation strategies will have to be differentiated across regions of Asia. Adaptation to climate change depends on the affordability of measures, access to technology, and biophysical constraints, e.g. land and water resource availability, soil characteristics, genetic diversity for crop breeding, and topography. Coping strategies would have to be adopted for land resources, water resources, and food productivity. In developing countries in Asia, options such as population growth control, poverty alleviation, and capacity building on food production, health care delivery, and water resource management, hold great potential in creating resilient social systems that are capable of withstanding the negative impacts of climate change. Adaptation measures should be designed to anticipate potential effects of climate change. It is difficult to generalize appropriate adaptation options across the diverse lands in Asia, but we attempt to list some possible priority options in various areas.

In the case of extreme weather events, a wide range of precautionary measures at regional and national levels, including awareness and acceptance of risk factors among communities, will help avert or reduce the impacts of disasters on economic and social systems. Early warning systems against floods and droughts will help avert major disasters.

In *agriculture*, adaptation options include: (a) choosing suitable agricultural crops; (b) soil conservation; (c) shifts in seed sowing dates; (d) optimum use of fertilizers and development of agrotechnologies; (e) shifts to crop types not vulnerable to evapotranspiration; (f) breeding heat-resistant crops by utilizing genetic resources that may be better adapted to drier and warmer climates; (g) maintaining seed banks; (h) encouraging agriculture in greenhouses; (i) preventing soil degradation; (j) genetically altering crop architecture and physiology in order to adapt to warmer environmental conditions and exploiting the potentially beneficial effects of carbon dioxide enhancement on crop growth and water use efficiency; (k) promoting use of high-yielding varieties and technological application, and (l) adopting pricing and marketing policies in order to reduce impacts on poor farmers.

The increasing numbers of people and growing economic production require increasing amounts of *water*, which is currently withdrawn in an unsustainable way. Protection and enhancement of water supplies, increased water efficiency and managed demand is crucial in order to ensure future availability of sufficient and safe water supplies for consumption and irrigation purposes. Options to achieve this, and at the same time reduce vulnerability to climate change impacts, include: (a) increasing the capacity of recycled water use and autonomous water use systems; (b) reducing water intake for industry in the dry season; (c) bottom-deepening along navigation canals to facilitate transport of goods through rivers; (d) runoff management and irrigation technology (i.e. river runoff control by reservoirs, water transfers and land conservation practices); and (e) rainwater harvesting, and other water conservation practices.

Because of the dwindling natural lands in Asia, the threat of *biodiversity* loss is particularly severe, mainly from current human pressures. At the same time, it is likely that biodiversity will suffer under increasing pressures from climatic changes. Possible adaptation options include: (a) the establishment of large and connected protected areas; (b) reducing habitat fragmentation and promoting development of migration corridors and buffer zones; (c) introducing integrated ecosystem planning and management; (d) encouraging mixed-use strategies; (e) using climate change resistant species to rehabilitate and restore affected ecosystems (although introduction of foreign species has to be done with caution), and (f) reducing deforestation and conserve natural habitats.

Adaptation options for *coastal resources*, include: (a) implementing coastal zone management plans; (b) modifying infrastructure to accommodate sea-level rise; (c) protecting lakes and reservoirs; (d) preparing resettlement plans to combat sea-level rise; (e) improving emergency preparedness for weather extremes (e.g. typhoons and storm surges); (f) protecting marine resources and wetlands, and (g) allowing for migration. Development of aquaculture, effective conservation, and sustainable management of marine and inland fisheries is needed so that living aquatic resources can continue to meet regional and national nutritional needs.

The growing population in Asia is moving increasingly into areas that are vulnerable to the impacts of climate change. Adaptation options in *infrastructure* include: (a) changing building structures in order to accommodate increasing populations and reduce the risk of floods and droughts; (b) building heat-resistant infrastructures and taking measures to reduce air and water pollution; (c) protecting the coast against floods and sea-level rise, and (d) planning for social and institutional adaptations in regions sensitive to sea-level rise, and the development and promotion of risk management.

Climate change impacts on *human health*, call for a need for adaptation options to be taken. These include: (a) developing technological/engineering solutions to prevent vector-borne diseases; (b) improving health care systems, including surveillance, monitoring and information dissemination; (c) improving public education and literacy rates; (d) improving the infrastructure for waste disposal, and (e) improving sanitation facilities.

In general, there is a need to continue monitoring variability in key climatic elements, and improvements need to be made to weather forecasting. Land use planning and reforms need also to be improved. Climate adaptation activities need to be co-ordinated among countries in the region. Public awareness and involvement in planning, adaptation, and mitigation strategies are essential, as traditional knowledge can be used for planning for the future.

7.3   Australia and New Zealand

7.3.1   *Vulnerability*

Australia and New Zealand, because of their affluence, have a larger capacity to adapt to climate change than most developing nations. However, Australia's ecosystems are vulnerable significantly to climate change, due to its varied climates and ecosystems, including deserts, rainforests, coral reefs, and alpine areas (IPCC 1998). New Zealand – a smaller and more mountainous country with a generally more temperate, maritime climate – may be less vulnerable to climate change than Australia (IPCC 1998). Climate change will add to the existing stresses (e.g. rapid population and infrastructure growth in vulnerable coastal areas, inappropriate use of water resources, and complex institutional arrangements), and hinder achievement of sustainable land use and conservation of terrestrial and aquatic biodiversity.

Competition for water is very high in parts of Australia between agriculture, power generation, urban areas, and environmental flows. Agriculture is vulnerable to reductions in rainfall and increasing temperatures. A major fraction of exports from Australia and New Zealand are agriculture and forestry products, which are sensitive to climate change, water availability, carbon dioxide levels and pests and diseases. Fisheries in the area are influenced by nutrient upwellings in the surrounding waters. The El Niño Southern Oscillation (ENSO) influences the recruitment of some fish species and the incidence of toxic algal blooms. Increasing temperature would threaten species currently growing near the upper limit of their temperature range. Fragmentation of habitats increases the vulnerability of species. Increased sea-level, atmospheric carbon dioxide, temperature, pollution, and damage from tropical cyclones, puts coral reef ecosystems at high risk. Increases in tropical cyclone intensity and possible changes in their location and frequency would have a major impact (Pittock *et al.* 1999). Increased frequency of high rainfall intensity would increase flood damage to settlements and infrastructure. Rising snowlines due to increased temperatures threaten the alpine ecosystems and the ski industry. Climate change will enhance the spread of some disease vectors due to floods, droughts, and increases in temperature, despite existing biosecurity and health services.

7.3.2   Adaptation

Adaptation to climate change as a means of maximizing gains and minimizing losses is important. Natural ecosystems in Australia and New Zealand have only a limited capacity to adapt, and many managed systems will face limits on adaptation imposed by costs, acceptability, and other factors. Including indigenous people into the planning process would be beneficial in deriving climate adaptation polices. Table 7.3 provides some suggestions for adaptation options in Australia and New Zealand.

1. *Water.* Improving water use efficiency, effective trading mechanisms for water, water pricing, more appropriate land use policies, reducing deforestation which increases runoff and the risk of flood-related injury and contamination of water.
2. *Agriculture.* Provision of climate information and seasonal forecasts, improved crop cultivars.
3. *Biodiversity.* Tree farming could alleviate dry land salinity and be a gain in the carbon-trading scheme.
4. *Settlements.* Revised engineering standards and zoning infrastructure development, air-conditioning and other measures to reduce exposure to heat.
5. *Health.* Improved biosecurity and health services, screens and doors to reduce exposure to disease vectors, quarantine and eradication of disease vectors.

Table 7.3. *Some impacts and adaptation options in Australia and New Zealand.*

Sector/issue	Impacts	Adaptation options
Hydrology and water supply	Water resources are already stressed in some areas, and are therefore highly vulnerable (Schreider *et al.* 1996). Increased salinization in some streams and coastal aquifers is seen	✓ Improved water-use efficiency and effective trading mechanisms for water ✓ Planning, allocation and pricing of water ✓ Use of alternative water supplies ✓ In the case of saltwater intrusions, retreat would be an option
Terrestrial ecosystems	A warming of 1 °C would threaten the survival of species that are currently growing near the upper limit of their temperature range (Whitehead, Leathwick & Hobbs 1992). Biodiversity loss, increased risk of fires and weed invasion are some effects of climate change on these systems (Noble *et al.* 1996)	✓ Changes in land use practice ✓ Landscape management ✓ Fire protection
Agriculture, grazing and forestry	Agricultural activities are vulnerable particularly to reductions in rainfall in southwest and inland Australia. There is an increased risk of forest fire due to drought. Spread of pests and diseases is apparent (Sutherst *et al.* 1996). Increase in carbon dioxide initially increases productivity, but is likely to be offset by further climate change later	✓ Provision of climate information and seasonal forecasts to land users to help them manage climate variability ✓ Improved crop cultivars ✓ More appropriate land use policies ✓ Management and policy changes ✓ Fire prevention ✓ Marketing and planning ✓ Niche and fuel crops ✓ Carbon trading ✓ Exclusion and spraying for insects ✓ Change farm practices and change industry ✓ Relocation

(cont.)

Table 7.3. (*cont.*)

Aquatic and coastal ecosystems	These systems are influenced by the extent and location of nutrient upwellings governed by prevailing winds and boundary currents. The ENSO influences the recruitment of some fish species (Harris *et al.* 1987) and the incidence of toxic algal blooms. Aquatic systems are also subject to high levels of eutrophication. Coral bleaching is of serious concern (Hoegh-Guldberg, Berkelmans & Oliver 1997). Salinization of some coastal freshwater wetlands is prevalent	✓ Physical intervention is needed to prevent the salinization of freshwater wetlands ✓ Change in water allocations ✓ Reduce nutrient flows ✓ Seed coral ✓ Monitoring and management of fisheries
Settlements and industry	Increased rainfall frequency and higher intensity would increase flood damage to settlements and infrastructure (IFRCRCS 1999; Pittock *et al.* 1999)	✓ Zoning for infrastructure development ✓ Disaster planning ✓ Engineering standards for buildings and infrastructure
Health	Climate change is likely to enhance the spread of diseases, heat-related illnesses (Hales *et al.* 2000), and photochemical air pollution	✓ Improved biosecurity and health services ✓ Quarantine ✓ Eradication or control of diseases ✓ Emission controls to prevent photochemical air pollution effects

7.4   Europe

Climate change impacts differ substantially among the subregions in Europe. More marginal and less wealthy areas will be less able to adapt. Climate change is likely to widen water resource differences between northern and southern Europe (Arnell 1999b). Risk of water shortage is projected to increase, particularly in southern Europe. Natural ecosystems will change, in view of increasing temperatures and atmospheric carbon dioxide concentration. Faunal and floral shifts, loss of important habitats, and loss in diversity of nature reserves are

Table 7.4. *Estimates of flood exposure and incidence for the coasts of Europe in 1990 and the 2080s. Estimates of flood incidence are highly sensitive to assumed protection standards and should be interpreted in indicative terms only. The former Soviet Union is excluded.* Source: *Nicholls, Hoozemans & Marchand (1999).*

| Region | Flood Incidence | | |
	1990 Exposed population (millions)	1990 Average number of people experiencing flooding (thousands of years)	2080s Increase due to sea-level rise, assuming no adaptation (%)
Atlantic coast	19.0	19	50–9000
Baltic coast	1.4	1	0–3000
Mediterranean coast	4.1	3	260–120 000

expected. Soil properties will deteriorate under warmer and drier climate conditions in southern Europe (Rounsevell, Evans and Bullock 1999). In mountain regions, higher temperatures will lead to an upward shift in biotic zones (Beniston *et al.* 1995). Timber harvests will increase in commercial forests in northern Europe (Sykes & Prentice 1996), but reductions are likely in the Mediterranean, with increased drought and fire risk. Agricultural yields will increase for most crops, due to an increase in atmospheric carbon dioxide concentration (Harrison & Butterfield 1999). However, this effect will be counteracted by risk of water shortages in southern and eastern Europe. Climate change would cause faunal shift, thus affecting marine and freshwater fish and shellfish biodiversity. The insurance industry will face potentially costly climate change impacts due to property damage (Dlugolecki & Berz 2000), but there is correspondingly greater scope for adaptive measures if initiatives are taken soon. The concentration of industries on the coast exposed to high sea-level rise and extreme events necessitates measures to protect or relocate them (Nicholls & de la Vega-Leinert 2000). Recreational preferences are likely to change with higher temperatures, with outdoor activities becoming more popular. However, high temperatures will reduce the traditional peak demand for Mediterranean holidays, and less reliable snow conditions could impact adversely on winter tourism. The risk of flooding (see Table 7.4), erosion, and wetland loss in coastal areas will increase, with implications for human settlements, industry, tourism, agriculture, and coastal natural habitats. There is a range of risks posed to human health by increased exposure to heat episodes (exacerbated by air pollution in urban areas), extension of vector borne diseases, and coastal and riverside flooding. However, current evidence shows that warming would lead to a decrease in wintertime deaths.

7.4.1   Adaptation

The adaptive capacity of socioeconomic systems in western, northern, and southern Europe is relatively high because of economic conditions (high gross domestic product (GNP) and stable growth), a stable population, and well-developed political, institutional, and technological support systems. However, the adaptation potential for natural systems is generally low. The adaptive capabilities in parts of central and eastern Europe are less. Below, we provide some suggestions for adaptation options in the areas of water resources, agriculture, coastal zone management, and ecosystems.

Water. Many places in southern Europe would have to adapt to decreased water availability due to climatic changes. Measures can be taken both on the supply and demand side. Supply-side adaptation options include: (a) changing operating rules; (b) increasing interconnections between sources; (c) finding new sources; (d) improving seasonal forecasting; (e) developing new sources for irrigation (e.g. farm ponds, collecting winter runoff); (f) increasing flood protection; (g) enhancing water level management and increasing dredging for navigation, and (h) installing, or increasing, water storage for power generation.

Demand-side options include: (a) reducing demand (through pricing, publicity, statutory requirements, etc.); (b) increasing irrigation efficiency and changing cropping patterns; (c) anticipating high risk of loss by flooding and reducing exposure by relocation; (d) using smaller ships for navigation, and (e) increasing cooling water use efficiency for power generation.

In *agriculture*, promising short-term options include: (a) changing planting dates and cultivars; (b) changing external inputs (e.g. fertilizers, pesticides); (c) using land-management techniques to maintain soil functions; (d) improving technology for application, better water quality and scheduling; (e) managing soil pH to prevent acidification; (f) conserving soil to prevent erosion; (g) tilling equipment and better timing of field operations to prevent soil compaction; (h) improving irrigation technology and scheduling; (i) practising to conserve soil moisture; (j) using manures, reducing tillage, and managing crop rotation in order to conserve organic matter in the soil; (k) mulching in order to maintain soil temperature; (l) timing manure and sludge applications in order to improve water movement and soil structure, and (m) changing land use for carbon fluxes.

Some other long-term adaptation options include: (a) changes in land use, where crops with high interannual variability in production (e.g. wheat) may be replaced by crops with lower productivity but more stable yields (e.g. pasture); (b) biotechnology and development of 'designer cultivars'; (c) substitution of crops for those that are more drought- and heat-resistant; (d) microclimate modification in order to improve water use efficiency (e.g. windbreaks to reduce evaporation); (e) intercropping in order to increase productivity per unit area, and (f) changes in farming systems to mixed farms with livestock and arable production which have larger resilience to changes in the environment (IPCC 2001).

Several *coastal regions* in Europe are vulnerable to the impacts of sea-level rise. In many places, there are no ready sources of sand available for beach nourishment. Managed realignment of coastal defences will help prevent inundation due to rising sea-levels and flooding. There is a need to develop strategic management approaches that allow both continued human utilization and preservation of coastal ecosystems. Developing methods that balance protection of people and the economy against costs of degradation of the coastal environment will require multidisciplinary research.

Tundra *ecosystems* need to be protected from stresses such as land exploitation and tourism. Establishing buffer areas around wetlands, and restoration of already destroyed wetlands, would minimize stress (Hartig, Grosev & Rosenzweig 1997). Planting of trees that are suitable for a future climate should be done today. However, given the uncertainty of climate change impacts, these measures are unlikely to be taken.

7.5   Latin America and the Caribbean

This region faces many of the same environmental problems as Africa and Asia, but also specifically those such as land tenure, overexploitation of natural ecosystems (e.g. forests and fisheries), and natural disasters such as earthquakes and hurricanes (UNEP 2002). The significantly inequitable income distribution in Latin America contributes to the vulnerability of the poor populations. With the events of the ENSO, Latin America is exposed to significant climate variations (see Table 7.5).

Tropical cyclones, associated heavy rains, flooding, and landslides are very common in Central America and southern Mexico. Glaciers in Latin America have receded since about 1970 (INAGGA-CONAM 1999). Warming in high mountain regions could lead to the disappearance of significant snow and ice surface. Latin America accounts for one of the largest concentrations of biodiversity on Earth

Table 7.5. *Estimated changes projected under IS92 scenario for some countries within Latin America.* Source: *de Siqueira* et al. *(1994); de Siqueira, Salles and Fernandes (1999).*

Region	Temperature	Precipitation
Mexico	Increase	Decrease
Costa Rica		
Pacific sector	+3 °C	−25%
Southeast Caribbean sector		Small increase
Nicaragua		
Pacific sector	+3.7 °C	−36.6%
Caribbean sector	+3.3 °C	−35.7%
Brazil		
Central and south central sector	+4 °C	+10 to +15% for autumn reductions for summer
Central Argentina	Summer: +1.57 °C (+1.08 to 2.21 °C) Winter: +1.33 °C (+1.12 to 1.57 °C)	Summer −12% Winter −5%

(Heywood & Watson 1995), causing the impacts of climate change to increase the risk of biodiversity loss. Global warming will change substantially the availability of fresh water. Potential changes in precipitation and temperature might have dramatic impacts on the pattern and magnitude of runoff, soil moisture, and evaporation, as well as the aridity of some hydrological zones (Mendoza, Villanueva & Adem 1997). Agriculture is the main producing sector for Latin America, and is undoubtedly affected severely, and influenced by, climate variations. Shortened growth seasons, and changes in precipitation, with consequent reductions in crop yields, would have devastating effects. Climate change is expected to lead to changes in soil stocks of carbon and nitrogen. Livestock is raised on rangelands, with no storage of hay or alternative feeds. Thus, prolonged droughts would impact on livestock production. Plantation forestry – a major land use in Brazil – is likely to be affected from decreased precipitation and water availability. A drier climate in plantation areas is likely also to lead to a greater fire hazard. Changes in precipitation and runoff may have significant impacts on mangrove forest communities. Sea-level rise will affect mangrove ecosystems by eliminating their present habitats and creating areas for them elsewhere (Twiley *et al.* 1997). Coastal inundation may affect water availability and agricultural land, exacerbating socio-economic and health problems. Increases in temperature will reduce crop yields by shortening the crop cycle (Rosenzweig & Hillel 1998). Health impacts of climate change would depend on the size, density, location, and wealth of the population (McMichael *et al.* 1996). The ENSO – increases in temperature and precipitation – cause changes in disease vector populations and water-borne diseases, and after the geographical distribution of some diseases. Extreme events tend to increase death and morbidity rates. All these climatic impacts exacerbate current problems and can frustrate efforts to develop sustainably the regional economies.

7.5.1   Adaptation

Latin America suffers from low adaptive capacity, high vulnerability and poor socioeconomic systems, particularly with reference to extreme climate events. Adaptation options relating to effects of sea-level rise on coastal biodiversity include establishing a national legal system and a national strategy for conserving biodiversity that includes terrestrial, marine, and coastal reserves. In Mexico, it was found that increased nitrogen fertilization was the best option to increase maize yields (Conde *et al.* 1997). In Argentina, planting dates were adjusted to take advantage of favourable thermal conditions resulting from fewer late frosts to improve wheat, maize, and sunflower yields (Travasso *et al.* 1999). Genetic improvements could be made in order to obtain cultivars that are better adapted to new growing conditions. In Uruguay and Argentina, increasing photoperiodical sensitivity could lead to a longer growth season for wheat and barley crops (Hofstadter and Bidegain 1997; Travasso *et al.* 1999). Alfalfa, which can withstand higher temperatures, could be used as an alternative forage crop for livestock. Prevention and response mechanisms need to be implemented against disease and other threats to human health.

7.6   North America

Although North America is the engine of globalization and has high levels of development, the region does have several serious environmental problems (UNEP 2002), some of which are likely to be worsened by climatic changes. North America will experience both positive and negative climate change impacts.

Warmer temperatures are likely to result in a seasonal shift in runoff, with more runoff in winter and possible reductions in summer flows, which could affect adversely the availability and quality of water for use during the summer. Evidence suggests that global warming may lead to substantial changes in mean annual stream flows, seasonal distribution of flows, and probabilities of extreme high and low flow conditions (Mearns *et al.* 1995). Water quality change may be driven by changes in hydrological flowpaths associated with changes in precipitation, evapotranspiration, and total flow in streams and rivers. There is evidence that intensity of rainfall events may increase under global warming due to increases in the perceptible water content of the atmosphere. Projected increases in intensity of precipitation could contribute to increased erosion and sedimentation in some areas (Mount 1995) which may increase flooding risks in some watersheds (IPCC 1996). Increased evaporation resulting from warmer water temperatures would be likely to affect future lake levels and outflow (Mortsch *et al.* 2000).

There is a complex link between climatic variations and changes in the processes that influence the productivity and spatial distribution of marine fish populations, as well as uncertainties in the future of commercial fishing patterns (Boesch, Field & Scavia 2000). Climate-related variations in the marine environment – including changes in sea-surface temperatures, nutrient supply and circulation dynamics – play important roles in determining the productivity of several North American fisheries.

Depending on existing conditions, global warming and carbon dioxide enrichment can have positive or negative effects on crop yields. Increased production from direct physiological effects of carbon dioxide and farm and agricultural market level adjustments (e.g. behavioural, institutional, economic) are projected to offset losses (Reilly *et al.* 2000). Warming could affect growing seasons, and summer temperatures and water stress will limit production (Rosenzweig & Tubiello 1997). Distribution and proliferation of weeds, crop diseases and insects are determined largely by climate. Climate change can affect livestock appetite or affect changes in quality and quantity of forage from grasslands and supplies of feed.

Climate change is expected to increase the extent and productivity of forests (Myneni *et al.* 1997). However, extreme and/or long-term climate change scenarios indicate the possibility of widespread decline. Changes in geographic ranges of forest types, increased frequency of fire outbreaks and insect infestations, and changes in carbon storage function, are expected to result from climate change.

There is strong evidence that climate change may also lead to the loss of specific ecosystems types, such as high alpine areas and specific coastal (e.g. salt marsh) and inland wetland types, e.g. prairie 'potholes'. Climate change is expected to have significant impacts on wetland structure and function, primarily through alterations in hydrology, especially the water table (Clair & Ehrman 1998). Coastal and marine biota also are vulnerable to changes in upwelling, current dynamics, freshwater inflow, salinity, and water temperatures (Boesch, Field & Scavia 2000). Climate-related pressures can act directly on wildlife through physiological effects, i.e. changes in growth rates, food demands, abilities to reproduce and survive.

Diseases may expand their ranges in the USA due to the warming of air and water temperatures, and increased runoff from agricultural and urban surfaces. Increased frequency and severity of heat waves may lead to increases in illness and death, particularly among young, elderly, and frail people, especially in large urban centres. Increased frequency of thunderstorms could lead to increased asthma. Frequent floods and other extreme events could lead to increases in deaths and injuries, infectious diseases, and stress-related disorders, as well as other health effects associated with social disruption, environmentally forced migration, and settlement in urban slums. Diseases associated with water may increase with warming of air temperatures, combined with heavy runoff events from agricultural and urban surfaces.

High-value concentrated development in urban centres can lead to high damage costs from extreme events associated climate change. Large cities are considered to be areas of high risk because global warming could lead to problems such as heat stress, water scarcity, and intense rainfall.

Winter recreations as well as businesses and destinations are likely to be impacted negatively by increasing temperatures. Changes in season length, as well as availability and quality of the resource base on which recreational activities depend, will be affected. Changes in magnitude and frequency of extreme events such as hurricanes, avalanches, fires, floods, and loss of beaches, will have considerable implications for tourism and its associated infrastructure.

Insured losses are increasing as the population continues to move to more vulnerable areas (III 1999). Insurers are affected adversely by increased variability or actuarial uncertainty of weather-related events.

7.6.1   Adaptation

Shifting patterns in temperature, precipitation, disease vectors, and water availability will require adaptive response in the region. Communities can reduce their vulnerability to adverse impacts through investments in adaptive infrastructure, e.g. storm protection, water supply infrastructure, and community health services. Rural, poor, and indigenous communities may not be able to make such investments. Furthermore, infrastructure investment decisions are based on a variety of needs other than climate change, including population growth and ageing of existing systems.

Some innovative adaptation technologies are being tested as a response to current climate-related challenges. However, the implementation of these methods needs to be examined. We discuss below some adaptation options that appear particularly relevant in a North American context, in the areas of water resources, fisheries and coastal regions, wetlands and ecosystems, forests and protected areas, health, human infrastructure and settlements, and insurance systems.

Adaptation of *water resources*, to the effects of climate change will occur in conjunction with adjustments and changes in the level and characteristics of water demand. Adaptive responses for changes in runoff patterns include altered management and artificial storage capacity, increased co-ordination in the management of groundwater and surface water supplies, and voluntary water transfers between users. There is a trend towards increased water transfers from irrigated agriculture to urban, and to other higher-value uses. Such reallocations raise questions of priority, and entail adjustment costs that will depend on the institutional mechanisms in place. Conservation is encouraged in order to balance the growing demands and supplies. Changes in reservoir storage policies for holding water in the basins in order to reduce evaporative losses could reduce consumptive use damage (Booker 1995). The roles of water and environmental markets are being explored to help create flexibility of response to uncertain changes in future water availability (Miller, Rhodes & MacDonell 1997). Changes are being made to the design and operation of joint water and energy systems. Current water-management infrastructure has allowed the citizens of North America to make productive use of water and to reduce adverse impacts of extreme high and low flows. Laws and institutions have been created that govern the allocation of water among competing users, and define the rights and obligations of individuals, government entities, and other organizations. These institutional aspects vary by region and have changed over time, reflecting differences in climatic and historical climate and societal values. Under riparian water law and permit systems, governmental authorities may have substantial discretion in regulating water uses under drought conditions (Scott & Coustalin 1995).

In order to adapt to a changing situation in *fisheries and coastal regions*, sustainable fisheries management will require timely and accurate scientific information on environmental conditions affecting fish stocks, as well as institutional and operational capacity to respond quickly to such information. Aquaculture, habitat protection, and fleet reduction, could be encouraged, along with coastal zone planning for high demand areas.

Better control of filling and draining of *wetlands*, preventing additional stress and fragmentation, creating upland buffers, controlling exotic species, protecting low flows and residual water, efforts to restore and create wetlands along with efforts to breed exotic wetland species in captivity, would help reduce the vulnerability of wetlands against climate variability. Creating regional inventories and management plans against risks from climate change could also be undertaken.

In *agriculture*, the ability of farmers to adapt their input and output choices is difficult to forecast, and will depend on market and institutional signals. New practices are to be adopted for agriculture and forestry, taking into account future climatic changes that can affect crops and timber supply. Farm-level adaptation can include: (a) changes in plant harvest dates; (b) crop rotations; (c) selections of crops and crop varieties for cultivation; (d) water consumption for irrigation; (e) use of fertilizers, and (f) tillage practices.

There is moderate potential for adaptation to prevent losses *in forests and protected areas* through conservation programmes to identify and protect threatened ecosystems. Long-term comprehensive systems are needed to monitor forest health and disturbance regimes that can function as early-warning systems for climate change effects on forests. Lands that are managed for timber production are less susceptible to climate change than unmanaged forests, because of the potential for adaptive management. Threatened ecosystems could be protected, while new landscapes might be adapted.

To avoid increasing climate-related *health* problems, socioeconomic factors (e.g. public health measures) play important roles in determining the existence and extents of diseases. Community health needs to be managed for changing risk factors. Behavioral adaptation (e.g. use of air-conditioners and increased intake of fluids) can help people survive increased temperatures. Other adaptation options against heat include: (a) development of community-wide heat emergency plans, and (b) improved heat-warning systems and better heat-related illness management (Patz *et al.* 2000). Adaptive measures to counter health impacts of extreme events could include improved building codes, disaster policies, warning systems, evacuation plans, and disaster relief (Noji 1997). Adaptive measures to reduce water-borne disease include improved water safety criteria, monitoring, treatment of surface water, sewerage, and sanitation systems (Patz *et al.* 2000). For changing levels of pollution, measures such as federal legislation and warning of the general population and susceptible individuals have been taken (Patz *et al.* 2000).

Settlements and communities can reduce their vulnerability to potential adverse impacts from climate change through investments in adaptive infrastructure. Highways, bridges, culverts, residences, commercial buildings, schools, hospitals, airports, ports, drainage systems, communications cables, transmission lines, and other items of infrastructure, have been built based on historical climate experience (Bruce *et al.* 1999). Similarly, land use practices and building codes have been developed to provide effective protection from existing climatic changes. Coastal communities have developed a variety of systems to manage exposure to erosion, flooding, and other hazards from rising sea-levels. Levees and dams often are successful in managing most variations in the weather (Wright 1996). Co-ordinated regional planning and management may allow more efficient adaptation to changing flood risks.

Possible change in frequency/intensity/duration of heavy precipitation events may require changes in land use planning and infrastructure design in order to avoid increased damages arising from flooding, landslides, sewage outflows, and releases of contaminants to natural water bodies.

In the *public and private insurance sectors*, extreme events have caused insurers to pay increased attention to building codes and disaster-preparedness, limiting

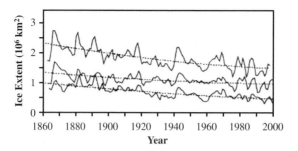

Figure 7.5. Time series of April sea-ice extent in Nordic Sea (1864–1998). Top: Nordic Sea; middle: eastern area; bottom: western area. *Source:* Anisimov *et al.* (2001).

insurance availability or increasing prices, and establishing new risk-spreading mechanisms. Recently, they have begun to use models to predict future climate-related losses. Governments play key roles as insurers or providers of disaster relief, especially in cases in which the private sector deems risks to be unacceptable. There should be increased roles for public emergency assistance and private insurance.

7.7   Polar regions

Climate change in the polar regions is expected to be among the greatest of any region. Increased temperatures have already been observed in major continental areas. Increased precipitation and a spectacular loss of ice-shelves are already emerging signs, although these developments cannot yet be reliably attributed to climate change (Figure 7.5).

The Arctic is extremely vulnerable to climate change, and major physical, ecological, and economic impacts are expected to appear rapidly. Mitigation of the detrimental impacts of thawing in its developed areas (e.g. severe damage to buildings and transport infrastructure) is required (Vyalov, Gerasimov & Fortiev 1998). However, the loss of sea-ice in the Arctic Ocean could open up sea routes for eco-tourism, which could have impacts for trade and local communities.

Most of the Antarctic ice-sheet is likely to thicken due to increased precipitation. Melting of ice-sheets in the west Antarctic and Greenland will contribute to sea-level rise. Increased water and runoff due to melting will lead to an increased flux of fresh water into the Arctic Ocean. Precipitation will increase during the summer, but evaporation and transpiration will also. This, together with earlier snowmelt, will lead to soil moisture deficiency late in the summer (Oechel *et al.* 1997). Changes in hydrology, possible increases in catchment rates, and melting of ice may result in wider dispersion of pollutants from accidents. Changes in ice cover and hydrology of lakes may cause them to become greater sinks for river-borne contaminants.

Climate change is likely to result in alterations to major biomes in the Arctic. Some species may be threatened (e.g. seals, polar bears), (Tynan and DeMaster 1997), whereas others will flourish (caribou, fish), (Gunn & Skogland 1997). Climate change eventually may increase the overall productivity of the natural systems in the polar regions (Henry & Molau 1997). Peat lands may dry out due to increased evaporation and transpiration by plants. Transpiration should increase as non-transpiring lichens and mosses, that currently dominate the tundra regions, are replaced by a denser cover of vascular plants (Rouse *et al.* 1997). Species composition in forests is likely to change, entire forest types may disappear, and new assemblages of species may be established. Melting ice-sheets will expose more bare ground, thus causing changes to terrestrial biology. Warming may lead to a loss of species that cannot tolerate warmer temperatures. However, other species that are able to withstand warmer conditions will reproduce more successfully. Shifts in oceanic circulation associated with global warming are likely to affect the distribution of commercially important fish and their migration routes (Vilhjalmsson 1997). Warmer waters also will increase growth rates of fish. Retreat of ice cover will provide easier access for fisheries. Thinner ice cover would lead to increased solar radiation penetrating the underlying water, thereby increasing photosynthetic production of oxygen and reducing potential winter fishkills. However, longer ice-free seasons would lead to increases in mixing depth and lower oxygen concentrations, and increased stress on cold-water organisms (Rouse *et al.* 1997). Climate warming is likely to alter husbandry practices. Presence of deep snow with an ice surface prevents animals from obtaining forage.

Historically, indigenous groups have shown resilience and ability to survive changes in resource availability, but they may be less well equipped to cope with the combined impacts of climate change and globlization (Peterson & Johnson 1995). Climate change will be disruptive to indigenous peoples, following traditional lifestyles, e.g. hunting patterns and seasonality of snowfall etc. Coastal erosion and retreat due to thawing of ice-rich permafrost are already threatening communities, heritage sites, and oil and gas facilities (Wolfe, Dallimore & Solomon 1998).

Climate change is likely also to lead to increased costs of oil and gas extraction, as new technology would need to be designed to withstand the changes.

7.7.1   Adaptation

Successful future adaptation to change depends on technological advances, institutional arrangements, availability of financing, and information exchange. Stakeholders must be involved in studies from the beginning as well as in discussions of any adaptive and mitigative measures (Weller & Lange 1999). Technologically developed communities are likely to adapt readily to climate change by adopting altered modes of transport and by increased investment in order to take advantage of new commercial trade opportunities. Adaptation to climate change will occur in natural polar ecosystems, mainly through migration and the changing mix of species. Opportunities for adaptation are limited for indigenous human communities following traditional lifestyles. Local adjustments in harvest strategies and in allocation of labour and capital (e.g. boats, snowmobiles, weapons, etc.) is necessary. There is a need to maintain self-esteem, social cohesion and cultural identity of communities.

Extensive trenching may be needed to combat the effects of coastal instability and erosion, especially that caused by melting permafrost (Maxwell 1997). Extreme weather-recording networks and navigational aids are needed with the increase in air and sea travel. There is a need for new building codes, railways, runways, and buildings to cope with the effects of thawing permafrost, snow loads and wind strengths (Maxwell 1997). New designs are needed for onshore oil and gas facilities and winter roads due to permafrost instability and flooding.

7.8 Small island states

Small island states are very vulnerable to climate change due to: (a) their small physical size relative to large expanses of ocean; (b) limited natural resources; (c) relative isolation; (d) extreme openness of small economies; (e) sensitivity to external shocks; (f) exposure to natural disasters and other extreme events; (g) rapidly growing populations with high densities; (h) poorly developed infrastructure, and (i) limited funds, human resources, and skills (Maul 1996). These characteristics limit the capacity of the small island states to mitigate and adapt to climate change and sea-level rise.

Although the contribution of these states to global greenhouse gas emissions is insignificant, projected impacts of climate change and sea-level rise are likely to be serious (Nurse, McLean & Suarez 1998) with high sensitivity to external shocks, and high vulnerability to natural disasters. Economic development and poverty alleviation are the main concerns for many small island states. Thus, with limited resources and low adaptive capacity, these islands face the challenge of meeting social and economic needs of their population in a sustainable manner. At the same time, they are forced to implement appropriate strategies to adapt to increasing threats from climate change, to which they contribute little.

Coastal zone changes experienced already by these states are aggravated by human activities on the coast (Gillie 1997). This will increase vulnerability by reducing natural resilience and increasing the cost of adaptation (Nurse 1992). Serious consideration needs to be given by small island states to the question of whether they have the capacity to adapt to sea-level rise within their own national boundaries, e.g. without relocating populations to other countries. Climate change and sea-level rise will affect shifts in species composition and competition (McIver 1998). Coral reefs, mangroves, and sea grass beds that often rely on stable environmental conditions will be affected adversely by rising air and sea temperature and level (Edwards 1995). Thermal tolerance of reef-building corals are expected to be exceeded with the next few decades. Water resources (Amadore *et al.* 1996), agriculture, and fisheries, are of critical concern due to the limited arable land and water resources. Climate change and sea-level rise will cause unfavourable shifts to the vulnerable biotic composition and adversely affect competition among some species.

Climate change is expected to affect the extensive *biodiversity* directly and indirectly through rising atmospheric carbon dioxide concentrations, which may increase productivity of some species but eliminate others and introduce new species.

Water is in short supply, as many islands rely heavily on rainwater, rivers, surface runoff, and limited groundwater. Soil salinization is common due to sea-level rise and climate change.

Small island states are dependent on *agriculture* for food security and export earnings. Heat stress, changes in soil moisture and temperature, evapotranspiration and rainfall, and extreme events such as floods and droughts, will affect production of some subsistence crops. Climate change is expected to have severe impacts on abundance and distribution of reef-fish population on the islands.

Human systems are likely to be affected by projected changes in climate and sea-levels. Economies of these states could be devastated by extreme events, e.g. cyclone and storm surges, destruction to critical infrastructure, e.g. airports, ports, roads, coastal protection structures, tourism structures, and vital utilities, etc. The tourism industry – a major revenue earner and employment creator for many economies – could be affected adversely by changes in rainfall and temperature regimes, coral bleaching, and loss of beaches. Important traditional assets (e.g. sites of worship near coasts) could be affected by sea-level rise. High incidences of vector- and water-borne diseases have resulted from changes in temperature and rainfall.

7.8.1   Adaptation

Progress in adaptation to climate change will require the integration of appropriate risk-reduction strategies with other sectoral policy initiatives in areas such as sustainable development planning, disaster prevention and management, integrated coastal management, and health care planning.

Because climate change and sea-level rise are inevitable in the future, implementing programmes that lead to wise use of resources is essential. Enhancement and preservation of natural protection (e.g. replanting of mangroves for protection of coral reefs), artificial nourishment of beaches, raising the height of the ground of coastal villages (Nunn & Mimura 1997), removal of material from 'unimportant' islands to build up important ones via sand transfer by pipes and barges, could be done (IPCC 1990). Enhancement of coastal resilience should be undertaken, whereby dynamic systems (e.g. dunes, lagoons, estuaries) are allowed to utilize their natural capacity to grow in response to rising sea-levels (Helmer *et al.* 1996). Adaptation options (e.g. retreat to higher ground) would not be practical, due to the limited availability of space. Hard engineering solutions (e.g. construction of groynes, sea-walls, breakwaters and bulkheads) need to be implemented properly in order to be effective against flooding and coastal erosion where vital infrastructure is at immediate risk (Mimura & Nunn 1998; Solomon & Forbes 1999). Integrated coastal management has been identified as one approach that would allow a sustainable tourism industry. Caribbean islands are placing more emphasis on 'precautionary' approaches. These include: (a) enforcement of land-use regulations; (b) building codes; (c) insurance coverage, and (d) application of traditional responses (e.g. building on stilts, use of expendable, readily available, indigenous building materials), which have proved to be effective responses (Forbes & Solomon 1997; Mimura & Nunn 1998). All small island states that participated in the recent US Country Studies Program concluded that integrated coastal management was the most appropriate adaptation strategy and should form an essential part of their climate change national action plans (Huang 1997).

Establishment of terrestrial, marine, and coastal reserves helps to preserve endangered habitats and ecosystems and contribute to the maintenance of biodiversity, while increasing the resilience of these systems to cope with climate change.

Management interventions for fish stocks include: (a) conservation; (b) restoration; (c) establishing protected areas for critical species; (d) aquaculture; (e) enhancement of vital habitats such as mangroves, coral reefs, and sea grass beds, and (f) implementation of bilateral and multilateral agreements and protocols for the exploitation and management of shared fisheries (IPCC 1998).

Great urgency needs to be placed on improving water-resource management efforts, including inventorying of resources and rational and equitable allocation. Implementation of methods of more efficient rainwater harvesting, desalinization, efficient leak detection and repair, water-saving, and aggressive recycling, is worth considering.

To ensure the sustainability of the tourist industry, Cyprus has recommended a strategy to protect infrastructure combined with planned retreat. The goal would be to maintain the limited beach area by erecting hard structures, enforcing building setbacks, and using artificial nourishment (Nicholls & Hoozemans 1996).

Resettlement within national boundaries could be the only viable option available for human settlement adaptation. In extreme cases, it may be necessary to abandon areas (Nurse, McLean & Suarez 1998), but this would be socially disruptive and require access to substantial resources, which most countries cannot afford.

Health threats could be reduced by: (a) implementing effective health education programmes; (b) preventative maintenance and improvement to health care facilities; (c) cost-effective sewerage and solid waste management practices; (d) disaster preparedness plans, and (e) adoption of early-warning systems for forecasting conditions that are favourable for the outbreak of various diseases. Insecticide-treated nets could be used, and filtering of drinking water done at the individual level in order to prevent the spread of disease.

Raising public awareness and understanding about the threats of climate change and sea-level rise and the need for appropriate adaptation, requires urgent attention. Community-based adaptation approaches would be successful due to the strong social and kinship ties that exist in many small island states. Certain traditional island assets could be useful, e.g. subsistence and traditional technologies and knowledge, in addition to cohesive community structures that in the past have helped to buttress the resilience of the islands against various forms of shock (Kaluwin & Hay 1999). Thus, it would be important for any climate-change adaptation policy to incorporate these traditional coping skills.

References

Amadore, L., Bolhofer, W. C., Cruz, R. V., Feir, R. B., Freysinger, C. A., Guill, S., Jalal, S. K. F., Iglesias, A., Jose, A., Leatherman, S., Lenhart, S., Mukherjee, S., Smith, J. B. and Wisniewski, J. (1996) Climate change vulnerability and adaptation in Asia and the Pacific: workshop summary. *Water, Air and Soil Pollution*, **92**, 1–12.

Ando, M. (1998) Risk assessment of global warming on human health. *Global Environmental Research*, **2**, 69–78.

Arnell, N. W. (1999a) Climate change and global water resources. *Global Environmental Change*, **9**, S51–67.

Arnell, N. W. (1999b) The effect of climate change on hydrological regimes in Europe: a continental perspective. *Global Environmental Change*, **9**, 5–23.

Beniston, M., Ohmura, A., Rotach, M., Tshuck, P., Wild, M. and Marinucci, M. R. (1995) Simulation of climate trends over the Alpine Region. In *Development of a Physically-Based Modeling System for Application to Regional Studies of Current and Future Climate*. Final Scientific Report No. 4031-33250. Bern: Swiss National Science Foundation.

Boko, M. (1992) *Climats et Communautes Rurales du Benin*. Rhythm Climatiques et Rythmes de Developpements. Dijon: Université de Bourgogne.

Boesch, D. F., Field, J. C. and Scavia, D., eds. (2000) *The Potential Consequences of Climate Variability and Change in Coastal Areas and Marine Resources*. Report of the Coastal Areas and Marine Resources Sector Team, US National Assessment of the Potential Consequences of Climate Variability and Change, US Global Change Research Program, National Oceanic and Atmospheric Administration Coastal Ocean Program Decision Analysis Series No. 21. Silver Spring MD: National Oceanic and Atmospheric Administration Coastal Ocean Program.

Booker J. F. (1995) Hydrologic and economic impacts of drought under alternative policy responses. *Water Resources Bulletin*, **31**(5), 889–906.

Bruce, J. P., Burton, I., Egener, I. D. M. and Thelen, J. (1999) *Municipal Risks Assessment: Investigation of the Potential Municipal Impacts and Adaptation Measures Envisioned as a Result of Climate Change*. Ottawa: The National Secretaries on Climate Change.

Clair, T. A. and Ehrman, J. M. (1998) Using neutral networks to assess the influence of changing seasonal climate in modifying discharge, dissolved organic carbon and nitrogen export in eastern Canadian rivers. *Water Resources Research*, **34**(3), 447–55.

Colwell, R. (1996) Global climate and infectious disease: the cholera paradigm. *Science*, **274**, 2025–31.

Conde, C., Liverman, D., Flores, M., Ferrer, R., Araujo, R., Betancourt, E., Villareal, G. and Gay, C. (1997) Vulnerability of rainfed maize crops in Mexico to climate change. *Climate Research*, **9**, 17–23.

Desanker, P. V., Frost, P. G. H., Justice, C. O. and Scholes, R. J., eds. (1997) *The Miombo Network: Framework for a Terrestrial Transect Study of Land Use and Land Cover Change in the Miombo Ecosystems of Central Africa*. Stockholm: International Geosphere-Biosphere Programme.

Dlugolecki, A. and Berz, G. (2000) Insurance. In M. L. Parry, ed., *Assessment of Potential Effects and Adaptations for Climate Change in Europe: The Europe ACACIA Project*. Norwich: University of East Anglia.

Edwards, A. J. (1995) Impact of climate change on coral reefs, mangroves and tropical sea grass ecosystems. In D. Eisma, ed., *Climate Change: Impact on Coastal Habitation*. Boca Raton: Lewis, pp. 209–34.

FAO (1999) *Production Yearbook 1999*. Rome: Food and Agriculture Organization of the United Nations.

Forbes, D. L. and Solomon, S. M. (1997) *Approaches to Vulnerability Assessment on Pacific Island Coasts: Examples from Southeast Viti Levu (Fiji) and Tarawa (Kiribati)*. Miscellaneous Report 277. Suva: South Pacific Applied Geoscience Commission.

Fukushima, Ozaki, T. N., Kaminishi, H., Harasawa, H. and Matushige, K. (2000) Forecasting the changes in lake water quality in response to climate changes, using past relationships between meteorological conditions and water quality. *Hydrological Processes*, **14**, 593–604.

Gillie, R. D. (1997) Causes of coastal erosion in Pacific island nations. *Journal of Coastal Research*, **25**, 174–204.

Greenwood, B. M. (1984) Meningoccocal infections. In D. J. Weatherall, J. G. G. Ledingham, and D. A. Warrell, eds., *Weatherall: Oxford Textbook for Medicine.* Oxford and New York: Oxford University Press, pp. 165–74.

Gunn, A. and Skogland, T. (1997) Responses of caribou and reindeer to global warming. *Ecological Studies,* **124,** 189–200.

Hales, S., Kjellstrom, T., Salmond, C., Town, G. I. and Woodwards, A. (2000) Daily mortality in Christchurch, New Zealand in relation to weather and air pollution. *Australia and New Zealand Journal on Public Health,* **24,** 89–91.

Harris, G., Nilsson, C., Clementson, L. and Thomas, D. (1987) The water masses of the east coast of Tasmania: seasonal and interannual variability and the influence on phytoplankton biomass and productivity. *Australian Journal of Marine and Freshwater Research,* **38,** 569–90.

Harrison, P. A. and Butterfield, R. E. (1999) Modeling climate change impacts on wheat, potato and grapevine in Europe. In R. E. Butterfield, P. A. Harrison and T. E. Downing, eds., *Climate Change, Climate Variability and Agriculture in Europe: An Integrated Assessment.* Environmental Change Unit, Research Report No. 9. Oxford: University of Oxford.

Hartig, E. K., Grosev, O. and Rosenzweig, C. (1997) Climate change, agriculture, and wetlands in Eastern Europe: vulnerability, adaptation, and policy. *Climate Change,* **36,** 107–21.

Hay, S. I., Snow, R. W. and Rogers, D. J. (1998) Predicting malaria seasons in Kenya using multi-temporal metereological satellite sensor data. *Transactions of the Royal Society of Tropical Medicine and Hygiene,* **92,** 12–20.

Helmer, W., Vellinga, P., Litjens, G., Ruijgrok, E. C. M., Goosen, H. and Overmars, W. (1996) *Meegroein met de Zee.* Zeist: Wereld Natuur Fonds.

Henry, G. H. R. and Molau, U. (1997) Tundra plants and climate change: the International Tundra Experiment (ITEX). *Global Change Biology,* **3,** 1–9.

Heywood, V. H. and Watson, R. T., eds. (1995) *Global Biodiversity Assessment.* Cambridge and New York: Cambridge University Press.

Hoegh-Guldberg, O., Berkelmans, R. and Oliver, J. (1997) Coral bleaching implications for the Great Barrier Reef Marine Park. In N. Turia and C. Dalliston, eds., *Proceedings of The Great Barrier Reef Science, Use and Management Conference,* 25–29 November, 1996, Townsville, Australia. Townsville: Great Barrier Reef Marine Park Authority, pp. 210–24.

Hofstadter, R. and Bidegain, M. (1997) Performance of General Circulation Models in southeastern South America. *Climate Research,* **9,** 101–105.

Huang, J. C. K. (1997) Climate change and integrated coastal management: a challenge for small island nations. *Ocean and Coastal Management,* **37,** 95–107.

Huang, Z. G. (1999) *Sea Level Changes in Guangdong and Its Impacts and Strategies,* Guangzhou: Science and Technology Press (in Chinese).

IFRCRCS (1999) *World Disasters Report.* Geneva: International Federation of the Red Cross and Red Crescent Society.

III (1999) *The Insurance Fact Book: 2000.* New York: Insurance Information Institute.

INAGGA-CONAM (1999) Vulnerabilidad de Recursos Hidricosde alta Montana. In *Climate Change 2001: Vulnerability, Impacts, Adaptation to Climate Change*, chap. 14. Cambridge: Cambridge University Press.

IPCC (1998) *The Regional Impacts of Climate Change: An Assessment of Vulnerability. Special Report of IPCC working Group II*. R. T. Watson, M. C. Zinoyowera and R. H. Moss, eds., Cambridge and New York: Cambridge University Press.

IPCC (1996) *Climate Change 1995: Impacts, Adaptation and Mitigation of Climate Change: Scientific-Technical Analyses. Contribution of Working Group II to the Second Assessment Report of the Intergovernmental Panel on Climate Change*. R. T. Watson, M. C. Zinyowera and R. H. Moss, eds., Cambridge and New York: Cambridge University Press.

IPCC (1990) *Strategies for Adaptation to Sea Level Rise*. The Hague: Ministry of Housing.

IPCC (2001) *Climate Change 2001: Vulnerability, Impacts and Adaptation to Climate Change*, Contribution of Working Group II to the Second Assessment Report of the Intergovernmental Panel on Climate Change. Cambridge and New York: Cambridge University Press, pp. 675–77.

Kaluwin, C. and Hay, J. E., eds. (1999) *Climate Change and Sea Level Rise in the South Pacific Region. Proceedings of the Third SPREP Meeting, New Caledonia, August, 1997*. Apia: South Pacific Regional Environment Programme.

Lal, M., Harasawa, H. and Takahashi, K. (2001) *Future Climate Change and its Impacts over Small Island States*. [location of publisher]: Climate Research.

Lengeler, C. (1998) *Insecticide Treated Bednets and Curtains for Malaria Control*. Cochrane Library, Issue 3. Oxford: Oxford University Press.

Loevinsohn, M. E. (1994) Climate warming and increased malaria in Rwanda. *Lancet*, **343**, 714–48.

Maul, G. A. (1996) *Marine Science and Sustainable Development*. Washington: America Geophysical Union.

Maxwell, B. (1997) *Responding to Global Climate Change in Canada's Arctic, Vol. II*. In N. Mayer, ed., *Canada Country Study Climate Impacts and Adaptation*. Downsview ON: Environment Canada.

McIver, D. C., ed. (1998) *Adaptation to Climate Variability and Change*. IPCC Workshop Summary, San José, Costa Rica, 29 March–1 April 1998. Montreal: Atmospheric Environment Service, Environment Canada, p. 55.

McMichael, A. J., Ando, M., Carcavallo, R., Epstein, P., Haines, A., Jendritzky, G., Kalkstein, L., Odongo, R., Patz, J. and Piver, W. (1996) Human population health. In R. T. Watson, M. C. Zinyowera and R. H. Moss, eds., *Climate Change 1995: Impacts, Adaptations, and Mitigation of Climate Change. Contributions of Working Group II to the Second Assessment Report of the Intergovernmental Panel on Climate Change*. Cambridge and New York: Cambridge University Press, pp. 561–84.

Mearns, L. O., Giorgi, F., McDaniel, L. and Shields, C. (1995) Analysis of daily variability of precipitation in a nested regional climate model: comparison with observations and doubled CO_2 results. *Global and Planetary Change*, **10**, 55–78.

Melillo, J. M., Prentice, I. C., Farquhar, G. D., Schuze, E. D. and Sala, O. E. (1996) Terrestrial biotic responses to environmental change and feedbacks to climate. In: J. T., Houghton, L. G. Meira Filho, B. A. Callander, N. Harris, A. Kattenberg and

K. Maskell, eds., *Climate Change 1995: The Science of Climate Change, Contribution of Working Group I to the Second Assessment Report of the Intergovernmental Panel on Climate Change*. New York: Cambridge University Press, pp. 444–516.

Mendoza, M., Villanueva, E. and Adem, J. (1997) Vulnerability of basins and watersheds in Mexico to global climate change. *Climatic Research*, **9**, 139–45.

Miller, K. A., Rhodes, S. L. and MacDonnell, L. J. (1997) Water allocation in a changing climate: institutions and adaptations. *Climate Change*, **35**, 157–77.

Mimura, N. and Nunn, P. D. (1998) Trends of beach erosion and shoreline protection in rural Fiji. *Journal of Coastal Research*, **14**(1), 37–46.

Mortsch, L., Hengeveld, H., Lister, M., Lofgren, B., Quinn, F., Slivitzky, M. and Wenger, L. (2000) Climate change impacts on the hydrology of the Great Lakes – St. Lawrence System. *Canadian Water Resources Journal*, **25**(2), 153–79.

Mount, J. F. (1995) *California Rivers and Streams: Conflict Between Fluvial Process and Land Use*. Berkley: University of California Press.

Myneni, R. B., Keeling, C. D., Tucker, C. J., Asrar, G. and Nemani, R. R. (1997) Increased plant growth in northern high latitudes from 1981 to 1991. *Nature*, **386**, 698–702.

Naylor, R., Falcon, W. and Zavaleta, E. (1997) Variability and growth in grain yields, 1950–94: does the record point to greater instability? *Population and Development Review*, **23**(1), 41–61.

Nelson, F. E. and Anisimov, O. A. (1993) Permafrost zonation in Russia under anthropogenic climate change. *Permafrost Periglacial Processes*, **4**(2), 197–48.

Nicholls, R. J. and Hoozemans, M. J. (1996) The Mediterranean: vulnerability to coastal implications of climate change. *Ocean and Coastal Management*, **31**, 105–32.

Nicholls, R. J., Hoozemans, F. M. J. and Marchand, M. (1999) Increasing flood risk and wetland losses due to global sea level rise: regional and global analyses. *Global Environmental Change*, **9**, S69–87.

Nicholls, R. J. and Mimura, M. (1998) Regional issues raised by sea level rise and their policy implications. *Climate Research*, **11**, 5–18.

Nicholls, R. J. and A. Vega-Leinert, de la (2000) Synthesis of sea level rise impacts and adaptation costs for Europe. In de la A. Vega-Leinert, R. J. Nicholls and R. S. J. Tol, eds., *Proceedings of the European Expert Workshop on European Vulnerability and Adaptation to the Impacts of ASLR*, Hamburg, 19–21 June 2000. Enfield: Flood Hazard Research Centre.

Noble, I. R., Barson, M., Dumsday, R., Friedel, M., Hacker, R., McKenzie, N., Smith, G., Young, M., Maliel, M. and Zammit, C. (1996) Land resources. In *Australia: State of the Environment 1996*. Melbourne: Commonwealth Scientific and Industrial Research Organization, pp. 6.1–6.55.

Noji, E. (1997) The nature of disaster: general characteristics and public health effects. In E. Noji, ed., *The Public Health Consequences of Disasters*. Oxford: Oxford University Press, pp. 3–20.

Nunn, P. D. and Mimura, N. (1997) Vulnerability of South Pacific island nations to sea-level rise. *Journal of Coastal Research*, **24**, 133–51.

Nurse, L. A. (1992) Predicted sea level rise in the wider Caribbean: likely consequences and response options. In P. Fabbri and G. Fierro, eds., *Semi-Enclosed Seas*. Oxford: Elsevier, pp. 52–78.

Nurse, L. A., McLean, R. F. and Suarez, A. G. (1998) Small island states. In: R. T. Watson, M. C. Zinyowera and R. H. Moss, eds., *The Regional Impacts of Climate Change: An Assessment of Vulnerability. A Special Report of the IPCC Working Group II*. Cambridge and New York: Cambridge University Press, pp. 331–54.

Oechel, W. C., Cook, A. C., Hastings, S. J. and Vourlitis, G. L. (1997) Effects of CO_2 on climate change in Arctic ecosystems. In: S. J. Woodin and M. Marquiss, eds., *Ecology of Arctic Environments*. Oxford: Blackwell Science, pp. 255–73.

Patz, J., McGeehin, M., Bernard, S., Ebi, K., Epstein, P., Grambsch, A., Gubler, D., Reiter, P., Romieu, I., Rose, J., Amet, J. S. and Trtanj, J. (2000) The potential health impacts of climate variability and change for the United States: executive summary of the report of the health sector of the US National Assessment. *Environmental Health Perspectives*, **108**(4), 367–76.

Peterson, D. L. and Johnson, D. R. eds. (1995) *Human Ecology and Climate Change*. Washington: Taylor & Francis.

Pimental, D. (1993) Climate changes and food supply. *Forum for Applied Research and Public Policy*, **8**(4), 54–60.

Pittock, A. B., Allan, R. J., Hennessy, K. J., McInnes, K. L., Suppiah, R., Walsh, K. J. and Whetton, P. H. (1999) Climate change, climatic hazards and policy responses in Australia. In T. E. Downing, A. A. Oltshoorn and R. S. L. Tol, eds., *Climate, Change and Risk*. London: Routledge, pp. 19–59.

Reibsame, W. E., Strzepek, K. M., Wescoat, Jr., J. L., Perrit, R., Graile, G. L., Jacobs, J., Leichenko, R., Magadza, C., Phien, H., Urbiztodo, B. J., Restrepo, P., Rose, W. R., Saleh, M., Ti, L. H., Tucci, C. and Yates, D. (1995) Complex river basins. In K. M. Strzepek and J. B. Smith, eds., *As Climate Chnages, International Impacts and Implications*. Cambridge and New York: Cambridge University Press, pp. 57–91.

Reilly, J., Tubiello, F., McCarl, B. and Melillo, J. (2000) Climate change and agriculture in the United States. In *Climate Change Impacts on the United States: The Potential Consequences of Climate Variablity and Change*. Report for the US Global Change Research Program. Cambridge and New York: Cambridge University Press, pp. 379–403.

Rosenzweig, C. and Hillel, D. (1998) *Climate Change and the Global Harvest: Potential Impacts of the Greenhouse Effect on Agriculture*. Oxford: Oxford University Press.

Rosenzweig, C. and Tubiello, F. N. (1997) Impacts of future climate change on Mediterranean agriculture: current methodologies and future directions. *Mitigating Adaptive Strategies in Climate Change*, **1**, 219–32.

Rounsevell, M. D. A., Evans, S. P. and Bullock, P. (1999) Climate change and agricultural soils: impacts and adaptation. *Climatic Change*, **43**, 683–709.

Rouse, W. R., Douglas, M. S. V., Hecky, R. E., Hershey, A. E., Kling, G. W., Lesack, L., Marsh, P., McDonald, M., Nicholson, B. J., Roulet, N. T. and Smol, J. P. (1997) Effect of climate change on freshwaters of Arctic and sub-Arctic North America. *Hydrological Processes*, **11**, 873–902.

Schreider, S. Y., Jakeman, A. J., Pittock, A. B. and Whetton, P. H. (1996) Estimation of possible climate change impacts on water availability, extreme flow events and soil moisture in the Goulburn and Oven Basins, Victoria. *Climatic Change*, **34**, 513–46.

Scott, A. and Coustalin, G. (1995) The evolution of water rights. *Natural Resources Journal*, **35**(4), 831–979.

Sharma N., Damhang, T., Gilgan-Hunt, E., Grey, D., Okaru, V. and Rothberg, D. (1996) *African Water Resources: Challenges and Opportunities for Sustainable Development.* World Bank Technical Paper 33, African Technical Department Series. Washington: The World Bank.

Siegfried, W. R. (1989) Preservation of species in southern African nature reserves. In B. J. Huntley, ed., *Biotic Diversity in Southern Africa: Concepts and Conservation.* Cape Town: Oxford University Press, pp. 186–201.

de Siqueira, O. J. F., Farías, J. R. B. and Sans, L. M. A. (1994) Potential effects of global climate change for Brazilian agriculture: applied simulation studies for wheat, maize and soybeans. In C. Rosenzweig and A. Iglesias, eds., *Implications of Climate Change for International Agriculture: Crop Modeling Study.* Report EPA 230-B-94-003. Washington: US Environmental Protection Agency.

de Siqueira O. J. F., Salles, L. and Fernandes, J. (1999) Efeitos potenciais de mudancas climaticas na agricultura brasileira e estrategias adaptativas para algumas culturas. In *Memorias do Workshop de Mudancas Climatics Globais e a Agropecuaria Brasiloeira.* Campinas, Brazil, pp. 18–19 (in Portuguese).

Singh, A., Dieye, A., Finco, M., Chenoweth, M. S., Fosnight, E. A. and Allotey, A. (1999) *Early Warning of Selected Emerging Environmental Issues in Africa: Change and Correlation from a Geographic Perspective.* Nairobi: United Nations Environment Programme.

Solomon, S. M. and Forbes, D. L. (1999) Coastal hazards and associated management issues on the South Pacific islands. *Oceans and Coastal Management*, **42**, 523–54.

Sutherst, R. W., Yonow, T., Chakraborty, S., O'Donnel, C. and White, N. (1996) A generic approach to defining impacts of climate change on pests, weeds, diseases in Australasia. In W. J. Bouma, G. I. Pearman and M. R. Manning, eds., *Greenhouse: Coping with Climate Change.* Victoria: CSIRO, pp. 190–204.

Sykes, M. T. and Prentice, I. C. (1996) Climate change, tree species distributions and forest dynamics: a case study in the mixed conifer/northern hardwoods zone of Northern Europe. *Climatic Change*, **34**, 161–77.

Travasso, M. I., Magrin, G. O., Rodriguez, G. R. and Boullon, D. R. (1999) Climate change assessment in Argentina, II: adaptation strategies for agriculture. In *Food and Forestry: Global Change and Global Challenges.* Global Change and Terrestrial Ecosystems Focus 3 Conference, Reading.

Tucker C. J., Dregne, H. E. and Newcomb, W. W. (1991) Expansion and contraction of the Sahara Desert from 1980 to 1990. *Science*, **253**, 299–301.

Twiley, R. R., Snedaker, S. C., Yanez-Arancibia, A. and Medina, E. (1997) Biodiversity and ecosystem processes in tropical estuaries: perspectives of mangrove ecosystems. In H. A. Mooney, J. H. Cashman, E. Medina, O. E. Sala, and E. D. Schulze, eds., *Functional Roles of Biodiversity: A Global Perspective.* Chichester: John Wiley and Sons, pp. 327–68.

Tynan, C. T. and DeMaster, D. P. (1997) Observations and predictions of Arctic climatic change: potential effects on marine mammals. *Artic*, **50**, 308–22.

UNEP (1997) *World Atlas of Desertification,* 2nd edn. London: Edward Arnold.

UNEP (2002) *Global Environmental Outlook* 3. London: Earthscan Publications.

Valendik, E. N. (1996) Ecological aspects of forest fires in Siberia. *Siberian Ecological Journal*, **1**, 1–18, (in Russian).

Vilhjalmsson, H. (1997) Climatic variations and some examples of their effects on the marine ecology of Icelandic and Greenland waters, in particular during the present century. *Rit Fiskideildar Journal of the Marine Research Institute, Rykjavik*, **XV**, 9–29.

Vyalov, S. S., Gerasimov, A. S. and Fortiev, S. M. (1998) Influence of Global Warming on the state and geotechnical properties of permafrost. In A. G. Lewkowicz and M. Allard, eds., *Proceedings of the Seventh International Conference on Permafrost*, Yellowknife, 23–27 June 1998. Collection Nordiana, No. 57, Centre d'etudes Nordiques. Quebec: Université Laval, pp. 1097–102.

Walker, H. J. (1998) Arctic deltas. *Journal of Coastal Research*, **14**(3), 718–38.

Walsh, K. and Pittock, A. B. (1998) Potential changes in tropical storms, hurricanes and extreme rainfall events as a result of climate change. *Climatic Change*, **39**, 199–213.

Weller, G. and Lange, M. (1999) *Impacts of Global Climate Change in the Arctic Regions*. International Arctic Science Committee, Center for Global Change and Arctic System Research. Fairbanks AK: University of Alaska, pp. 1–59.

Whitehead, D., Leathwick, J. R. and Hobbs, J. F. F. (1992) How will New Zealand's forests respond to climate change? Potential changes in response to increasing temperature. *New Zealand Journal of Forestry Science*, **22**, 39–53.

WHO (1998) The state of world health. 1997 Report. *World Health Forum*, **18**, 248–60.

Wolfe, S. A., Dallimore, S. R. and Solomon, S. M. (1998) Coastal permafrost investigations along a rapidly eroding shoreline, Tuktoyaktuk, NWT, Canada. In A. G. Lewkowicz and M. Allard, eds., *Proceedings of the Seventh International Conference on Permafrost*, Yellowknife, NWT, 23–27 June 1998. Collection Nordicana 57. Quebec: Centre d'Etudes Nordiques. Université Laval, pp. 1125–31.

WRI (1998) *1998–1999 World Resources Database Diskette: A Guide to the Global Environment*. Washington: World Resources Institute.

Wright, J. M. (1996) Effects of the flood on national policy: some achievements. Major challenges remain. In S. A. Changnon, ed., *The Great Flood of 1993: Causes, Impacts and Responses*. Boulder: Westview Press, pp. 245–75.

Yoshino, M. (1998) Deviation of catch in Japan's fishery in the ENSO years. In *Climate and Environmental Change*. International Geographical Union, Commission on Climatology, Evora, 24–30 August.

8

Mitigating climate change: concepts and linkages with sustainable development

8.1   Review of basic concepts and methodology

Mitigation is defined as an anthropogenic intervention to reduce the sources or enhance the sinks of greenhouse gases. Mitigation is required to meet the ultimate objective of the United Nations Framework Convention on Climate Change (UNFCCC): to stabilize the concentrations of greenhouse gases in the atmosphere. For the most important greenhouse gas – carbon dioxide – this requires that global emissions decrease to a very small fraction of current emissions. The dates upon which global carbon dioxide emissions would have to peak vary along with the targeted stabilization level. For example, for stabilization at 550 ppmv – the goal agreed by the EU – they would have to peak between 2020 and 2030, and for stabilization at 450 ppmv, between 2005 and 2015. The greater the emissions reductions and the earlier they are introduced, the smaller and slower global warming will be. Shorter-lived greenhouse gases such as methane respond more quickly to changing emissions. For gases with atmospheric sinks (e.g. methane and nitrous oxide), stable emissions lead eventually to stabilized concentrations.

A common approach to assessing the potential for mitigation has the following steps.

1. Identify technologies and practices that can reduce emissions or enhance their sinks in a particular sector, within a specific region.
2. Assess the potential to implement these technologies and practices by first identifying the barriers to their implementation (e.g. costs) and then identify the opportunities to overcome these barriers through policies and measures in a cost-effective way.

This approach usually focuses on the short-term feasibility of reaching particular emissions goals (e.g. those of the Kyoto Protocol) through currently available technologies and practices. Nevertheless, sometimes important long-term assessments of mitigation potential are attempted, such as reflected by the Intergovernmental Panel on Climate Change (IPCC) statement that 'most model results indicate that known technological options could achieve a broad range of atmospheric CO_2 stabilization levels, such as 550 ppmv, 450 ppmv or below over the next 100 years or more, but implementation would require associated socio-economic, and institutional changes' (IPCC 2001c).

The above two steps are discussed on some detail in Chapter 9. Here, only a conceptual framework is presented, as proposed by Sathaye et al. (2001). The vertical axis of Figure 8.1 indicates different categories of mitigation potential, while the horizontal axis shows how these potentials may change over time. The upper line is the physical potential, the theoretical upper limit. The technological potential is the maximum potential which could currently be achieved by applying all environmentally sound technologies that are 'known', i.e. have been demonstrated. The technical potential can increase over time due to research and development. At the other, lower, end of the spectrum is the market potential of technologies and practices as they are realized in the market. In the conceptual framework of Figure 8.1, the gap between the market and the technical potential is considered the result of different types of barriers. Starting at the bottom, the gap between the market potential and the economic potential is caused by market failures preventing cost-effective options from being realized, e.g. trade barriers, inadequate information, lack of competition, lack of access to investment funds, etc. These barriers can be removed by reducing or taking away the market failures. But there are also other types of barriers that would prevent implementation of mitigation options, e.g. behavioural issues, values, and cultural norms. If these would be removed (e.g. through changes in life-styles or social systems), the so-called socioeconomic potential would be achieved. The socioeconomic potential can be larger than the economic potential (e.g. people are willing to pay more because something is fashionable, or it meets their desire for a positive contribution to environmental protection) or it can be smaller, e.g. when a technology would be financially profitable for them, but would have an unattractive design. The conceptual approach behind Figure 8.1 is that there are opportunities, through policies and measures, to remove barriers and move the market potential up in the direction of the technological potential. Some actions (indicated in the right-hand column) may be specifically appropriate to remove a particular type of barrier (e.g. information programmes to improve access to up-to-date information), while others are generic and could contribute to various types of barriers simultaneously, e.g. taxes.

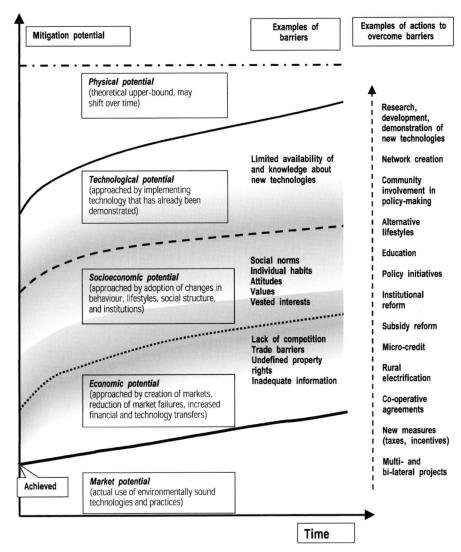

Figure 8.1. Penetration of environmentally sound technologies: a conceptual framework. Various barriers prevent the different potentials from being realized. Opportunities exist to overcome barriers through innovative projects, programmes, and financing arrangements. One action can address more than one barrier. Actions may be pursued to address barriers at all levels simultaneously. Their implementation may require public policies, measures, and instruments. The socioeconomic potential may lie anywhere in the space between the economic and technological potential. *Source*: Sathaye *et al.* (2001).

It should be noted that the above conceptual framework is not undisputed. It represents a model assuming that people respond rationally and individually to economic and technological opportunities, usually in the form of technological devices that are taken as given (see, for example, Wilhite *et al.* 2000). Social science has been involved to explore social and cultural barriers towards the adoption of particular technologies, e.g. in the area of energy efficiency. However, it is then overlooked how the demand for energy services (rather than particular technologies) is conditioned by considerations of convenience, comfort, aesthetics, social norms, etc. Analysing the broader social and cultural contexts that actually form the demand may thus provide a broader set of policy options than the device- and individual-oriented technoeconomic rationality represented in Figure 8.1 (Wilhite *et al.* 2000). This would require a new agenda for social science research on energy demand and the related climate change mitigation options.

The approach does not address the desirability of implementing these technologies and practices from a long-term climate change perspective. Analysing both desirability and feasibility of options can be done using different decision-analytical frameworks. Usually, applying different frameworks provides different answers to different questions. We discussed in Chapter 3 which analytical frameworks have been applied to the analysis of sustainable development, notably cost–benefit analysis and multicriteria analysis. These frameworks also have been applied to the evaluation of climate change mitigation options. In addition, cost-effectiveness analysis – the so-called tolerable windows/safe landing approaches and multiscenario approaches – have been used to analyse mitigation options, often in relation to climate change constraints. Below, we describe briefly key aspects of their application in the area of climate change, including application issues. For a more detailed discussion, refer to Munasinghe *et al.* (1996) and Jepma and Munasinghe (1998).

8.1.1   *Cost–benefit analysis and multicriteria analysis*

In applying cost–benefit to climate policy, the costs of mitigating climate change are assessed as well as the benefits of this mitigation in terms of avoided climate damage, and incremental changes to the costs of adaptation. Usually, monetary valuation of both climate response actions and damage is attempted. Chapter 3 discusses how the net present value can be calculated as the difference between the total benefits and costs, discounted over time, which should be positive for a strategy to be acceptable. Ideally, the optimal level of response would be where the marginal costs of mitigation and adaptation equal those of avoided climate damage. While the approach is applied most often at the project level, it is also applied to large-scale, aggregate problems such as climate change.

While the cost–benefit approach usually attempts to use a single (monetary) value for comparison, multicriteria analysis (or multi-attribute analysis) defines a framework for evaluating multiple objectives in parallel (see Chapter 3). This approach acknowledges that in the real world, decisions are always based on multiple criteria rather than only one criterion (like minimizing monetary costs in cost–benefit analysis). For example, a typical set of criteria for evaluating public policy might include flexibility, urgency, low cost, irreversibility, consistency, economic efficiency, profitability, political feasibility, health and safety, legal and administrative feasibility, equity, environmental quality, private versus public sector, and unique or critical resources. Different policy strategies can be evaluated using such criteria in a qualitative matrix. Such a matrix can reveal both synergies and trade-offs between the various criteria. In addition, various quantification methods have been developed and applied over the years for different problems. Although in the broader environment-related policy process this approach is implicitly followed, it has not been applied in any systematic, quantified fashion to the scientific assessment of climate change.

8.1.2   Cost-effectiveness analysis

Chapter 3 discusses the cost-effectiveness approach in a generic way, focusing on broad sustainable development issues. In this chapter, the cost-effectiveness approach implies determining a particular goal for either short-term or long-term greenhouse gas emissions limitation and then analysing the cheapest and most efficient way to achieve this goal. An example of a short-term application is the analysis of the achievement of the Kyoto Protocol targets for industrialized countries for the first commitment period. An example for the longer term is the analysis of a cost-minimizing pathway leading to stabilized greenhouse gas concentrations in the atmosphere. In order to perform such an analysis, a considerable amount of information and assumptions are needed. This includes: (a) the development or selection of a reference pathway for socioeconomic developments; (b) the availability of mitigation technologies over time; (c) their costs as affected by assumptions such as discounting, and (d) the efficiency of policies, measures, and instruments. In some sense, the cost–effectiveness approach is a special form of the cost–benefits approach, taking the achievement of the selected goal as the benefit. Usually, cost-effectiveness analysis takes the benefits of achieving this goal, as well as the costs, into account: if the costs are considered to be unacceptably high, the goal is relaxed, taking a higher risk in terms of impacts.

8.1.3   Tolerable windows/safe landing approach

The tolerable windows and safe landing approaches are methodologies developed specifically to analyse one particular climate-change-related question:

how do short-term emissions requirements depend on long-term climate objectives? First, a long-term goal is chosen in terms of climate change (e.g. rate of global temperature change and/or sea-level rise) and subsequently a so-called ('safe') emissions corridor are derived using global integrated models. The approach has been inspired by the 'critical loads' and 'critical levels' approaches used in the assessment of acid deposition strategies, in which short-term emissions constraints for acidifying compounds (sulphur dioxide and nitrogen oxides) are derived from acid deposition levels and concentrations, and which are considered to be tolerable from an ecological perspective. The safe-landing approach has been applied to determine emissions corridors for the near term (in a particular year, e.g. 2010) as constrained by assumptions for a tolerable (rate of) temperature change, sea-level rise and feasible rates of global emissions reductions. The methodologically more sophisticated tolerable windows approach provides a continuous 'tolerable window' for emissions over time. These methods have been used to support the negotiations of the Kyoto Protocol, and could be used again to consider issues related to the 'adequacy of commitments' and possible future emissions in succeeding commitment periods. They have been used to demonstrate the importance of inertia in the climate and socio-economic systems, e.g. delaying emissions limitations seriously reduces the magnitude of the tolerable corridors for particular selected long-term goals.

8.1.4   Multiscenario approaches

For all the above methodologies, there is a need for the development or selection of a baseline scenario against which the costs and benefits of policies can be measured in order to apply them to climate change response. Although it has been established already that there is nothing like a policy-free, 'business-as-usual', scenario, and the dependency of the results of any analysis in this area depends on the selected baseline, most analyses have since continued to adopt single baselines, e.g. the IS92A case from the IPCC IS92 scenario range. This not only hides the uncertainty in the results caused by this assumption, but also gives the wrong impression that probability can be attached to scenarios. Therefore, researchers increasingly explore a multiscenario analysis (Toth *et al.* 2001). In this approach, a – sometimes very large – number of fundamentally different plausible scenarios are generated, using computer models. Rather than weighing the results probabilistically or aggregating them, the analyst uses them as tools to transmit information about the deep uncertainties about future socio-economic developments and the way these developments might be influenced by policies.

8.1.5   *Robust decision-making*

All of these approaches have to deal with large uncertainties. This is why decision-makers are interested in 'robust' options. However, 'robustness' can be interpreted in various ways. It can apply to optimal levels of mitigation in view of uncertainties about climate change and decisions about its acceptability as uncertainties may be reduced over time. Various issues play a role here. For example, inertia of the socio-economic system is important, but largely unknown in detail. At the beginning of this section, we noted that in order to stabilize atmospheric concentrations at 450 ppmv or below, global emissions would have to peak within the next 10 years. If this does not happen, it would imply that the option to achieve this target, may it be considered politically desirable in the future, would be foregone if emission reductions do not start very soon. Hua-Duong, Grubb and Hourcade (1997) address the same problem with an integrated assessment model. They find that the optimal emissions profile in a situation in which the decision about a stabilization target is taken in 2020 (with the possibility of a 450 ppmv goal) would be considerably lower than the optimal emissions profile in case a 550 ppmv target is decided upon now, and hence, a considerable amount of 'hedging' would be appropriate. However, another way of framing this issue can lead to different insights. Manne and Richels (1995) assign low but non-zero probability to a very high-consequence scenario and a high probability to an average baseline scenario. Again, the uncertainties in 2020 are assumed to be resolved. In this case, the economically efficient solution would require only mimimal hedging as compared to the reference case. This analysis, corroborated by other modelling teams participating in the Energy Modelling Forum, takes into account both costs and benefits of emissions abatement. While these results appear to be contradictory, they contain similar messages (Toth *et al.* 2001): costs and benefits of near-term actions have to be weighed against those of delayed emissions controls, and issues like the uncertainty with respect to future climate goals and the inertia of the socioeconomic systems have to be incorporated in the decision process. The apparent contradiction originates primarily from the gap in knowledge with respect to the inertia of the socioeconomic system (i.e. the feasibility of achieving rapid changes in greenhouse gas emissions), as it is influenced not only by economic, but also cultural and institutional, forces.

'Robustness' also can refer more specifically to the choice of technologies that would mitigate greenhouse gas emissions under a spectrum of possible futures, reflected by alternative development pathways either considered to be likely or preferable by different stakeholders. For the energy sector, for example, increased usage of natural gas and biofuels for the near to mid term are options characteristic of many stabilization scenarios (Morita *et al.* 2001), while options such as enhanced nuclear capacity or carbon capture and storage are characteristic only for a small number of mitigation scenarios reflecting particular world views. While the latter options may be less *robust* according to this definition of robustness, they may be the preferred options in particular situations or regions. Yet another definition of *robust* refers to options that satisfy a variety of criteria, such as those of sustainability, feasibility, and cost-effectiveness in the energy sector, climate change being a subcriterion of the broader sustainability criterion.

8.1.6   *Applications issues in evaluating climate policies*

There are many pitfalls when applying various evaluation techniques. First, there are large uncertainties in valuing various aspects of both costs and benefits, the assignment of monetary values to some issues is questioned, and distributional issues are important. Second, there are many types of costs that are important during implementation. Third, the selection of an appropriate discount rate is controversial. Fourth, the choice of the emissions baseline, and the flexibility to carry out mitigation, will affect costs. Furthermore, many options have a wide variety of costs and benefits beyond those directly related to climate change, as discussed in Chapter 4 in the context of the many linkages between climate change responses and broader (sustainable) development strategies. Quantifying such 'ancillary' costs and benefits, or assigning a portion of total costs and benefits to climate change, involves many uncertainties and sometimes arbitrary decisions (see also Sections 5.3 and 10.4).

Measurement, aggregation and distribution

There is some controversy about the measurement of environmental and social impacts, equity, and sustainability in monetary terms. This is because of the difficulties as to what monetary values should be attached to biophysical and social changes (see Chapter 3). Furthermore, even if all costs and benefits could be monetized, the aggregate costs are not the only useful consideration in evaluating alternative policy instruments. The distribution of these costs and benefits across businesses, regions, individuals, and over time, is also important.

Implementation costs

All climate change policies necessitate some costs of implementation, i.e. costs of changes to existing rules and regulations, making sure that the necessary infrastructure is available, training and educating those who are to implement the policy as well those affected by the measures, etc. Unfortunately, such costs are not covered fully in conventional cost analyses. Implementation costs in this context are meant to reflect the more permanent institutional aspects of putting a programme into place, and differ from those costs conventionally considered as transaction costs. The latter, by definition, are temporary costs. Considerable work needs to be done to quantify the institutional and other costs of programmes, so that the reported figures are a better representation of the true costs that will be incurred if programmes are implemented. The full analysis of implementation costs (including environmental and social linkages), necessitates an economywide analysis that involves, for example, the use of computable general equilibrium models and multisectoral macroeconomic models (Munasinghe 1996).

Discounting

There are broadly two approaches to discounting: (a) an ethical or prescriptive approach based on what rates of discount should be applied, and (b) a descriptive approach based on what rates of discount people (savers as well as investors) actually apply in their day-to-day decisions (Munasinghe *et al.* 1996). For mitigation analysis, the country must base its decisions at least partly on discount rates that reflect the opportunity cost of capital. Rates ranging from 4 to 6 per cent probably would be justified in developed countries. The rate could be 10–12 per cent or even higher in developing countries (Watts 1999). The long-term nature of the climate change problem is the key issue in discounting. The benefits of reduced greenhouse gas emissions vary with the time of emissions reduction and the actual atmospheric greenhouse gas concentration profile from the time such reduction begins – up to 100 years or more afterwards. The bigger challenge is the dispute that climate change mitigation projects should face different discount rates, unless the mitigation project is of very long duration (Arrow *et al.* 1996). Rates that decline over time are being used increasingly, thereby giving more weight to benefits that occur in the long term (Weitzman 1998). We note that these rates do not reflect private rates of return, which typically must be greater, e.g. at around 10–25 per cent to justify a project (see Section 5.3).

Baseline

The baseline case gives the emissions of greenhouse gases in the absence of climate change interventions being considered, and is critical to the assessment of incremental costs of climate change mitigation. This is because the definition of the baseline scenario determines the potential for future greenhouse gas emissions reduction, as well as the costs of implementing these reduction policies. Changes from the baseline scenario frequently are put into categories of incremental 'costs' and 'benefits'. The baseline scenario also has a number of important implicit assumptions about future economic policies at the macroeconomic and sectoral levels, including sectoral structure, resource intensity, prices, and thereby technology choices.

Flexibility

For a wide variety of options, the costs of mitigation depend on the regulatory framework adopted by national governments to reduce greenhouse gases. In general, the more flexibility the framework allows over time and space, the lower the costs of achieving a given reduction. More flexibility at a given moment can reduce costs, e.g. in the form of more trading partners. Also flexibility over time can reduce costs, e.g. banking of emissions allowances. The opposite is expected with inflexible rules and few trading partners. Flexibility enhances the ability to reduce carbon dioxide emissions at the lowest cost, either domestically or internationally.

8.2   Long-term mitigation and stabilization scenarios

8.2.1   Introduction

In Chapter 2 we discussed socioeconomic scenarios and associated greenhouse gas emissions profiles. In these scenarios, based on assumptions with respect to demographics, economic development, and technological change, explicit climate policies did not play a role. We have seen that the world can develop according to a wide variety of possible futures. This is one of the most significant uncertainties with respect to future climate change. However, the resolution of this type of uncertainty is dependent upon human choices and actions, unlike a scientific uncertainty (like the exact magnitude of the climate sensitivity). Some of the scenarios discussed in Chapter 2, such as the B1 scenario of the IPCC *Special Report on Climate Change* or the 'Great Transitions' scenario of the Global Scenario Group, lead to global reductions of greenhouse gas emissions in the second half of this century and eventually could lead to stabilization of carbon dioxide concentrations. Several questions could be raised with respect to the desirability and feasibility of mitigation:

1. What could be the desirable level of greenhouse gas concentration stabilization?
2. Through which technologies and practices could these levels be reached for different baseline scenarios?
3. What would be the implications for the near term?

In the following section, these questions are addressed.

8.2.2   Desirable levels of greenhouse gas concentrations

The desirable level of greenhouse concentrations is unknown, or rather, has not been agreed upon. In a qualitative sense, it is determined by the ultimate objective of the UNFCCC of avoiding dangerous interference with the climate system (UNFCCC Article 2). The formulation of this objective is complex, not unambiguous, and hence subject to different interpretations. For example, does dangerous interference with the climate system also refer to dangerous interference with the biosphere and society as impacted upon by climate change? As we have noted in Chapter 1, the conditions specified in Article 2 suggest it does: the level and timescale for stabilization of greenhouse gas concentrations in the atmosphere should allow ecosystems: (a) to adapt naturally to climate change; (b) ensure that food production would not be threatened, and (c) that economic development proceeds in a sustainable manner. These three conditions are not necessarily compatible: if we would interpret the first condition in a way that would aim at allowing *all* ecosystems to adapt naturally to climate change, the most vulnerable ecosystems (e.g. in the Arctic or alpine regions) would set a very tight standard. Indeed, according to the IPCC, adverse impacts on ecosystems have been identified already (McCarthy *et al.* 2001), suggesting that these ecosystems currently are unable to adapt naturally to climate change. Nevertheless, very low emissions concentrations targets (e.g. below 450 ppmv, or a return to pre-industrial levels) would require very deep and global emissions reductions, but also reductions to be achieved very soon. This could lead to high costs and seriously affect development opportunities, especially for developing countries, and would thus be inconsistent with the third criterion of Article 2. The IPCC (2001d) states clearly, therefore, that the determination of what constitutes dangerous interference with the climate system is a value judgement arrived at through sociopolitical processes rather than scientific decision. Scientific, technical, and economic research can provide information that can support decision-making here.

It is also unlikely that such a key decision as the selection of a climate protection target can be taken at any time in any definite way. Decision-makers will have to make a series of decisions sequentially under uncertain conditions, as information gradually improves. Additional complexity in evaluating Article 2 is caused by the fact that the level of dangerousness would be dependent not only on the level of climate change impacts, but also on the adaptive and mitigative capacity of societies and ecosystems. It may be evident that the level of dangerousness depends on the degree to which ecosystems and societies can adapt to the changes. It may be less evident that the level also depends on the degree to which society mitigates climate change. The reason for this is that the stress posed on ecosystems and societies by climatic changes depends not only on their eventual magnitude, but also on the rate of change. Theoretically, the more rapid and rigorous greenhouse gas emissions reductions are implemented, the lower the rate of climate change, and the lower the vulnerability. In addition, the higher the mitigative capacity of societies, the higher their adaptive capacity. Interpreting Article 2 from a strong sustainability perspective (see Chapter 2), any damage by climate change should be avoided, regardless of the costs. From a weak sustainability perspective, the costs of avoiding these damages through mitigation would be taken into account, in line with the third condition included in Article 2. Protecting sensitive ecosystems may require a target level of 450 ppmv of carbon dioxide or below but, as will be illustrated in Chapter 10, below 550 ppmv the costs of mitigation may rapidly and possibly exponentially increase with a decreasing target level, because the required rapid emissions reductions would require premature retirement of capital goods. Without any agreement on the target level, a range of stabilization levels should be considered. In this chapter, we will consider scenarios leading to stabilization of carbon dioxide concentrations at 450, 550, 650 and 750 ppm.

The limited scientific inquiry has attempted primarily to provide information on the adaptive capacity of ecosystems. In the early 1990s, suggestions for climate targets in terms of the (rate of) global mean temperature change were formulated (Rijsberman & Swart 1990; Vellinga & Swart 1991). Based on anecdotal information on migration rates of plant species, it was suggested that to limit risks of climate change, the rate of change of the global mean temperature should not go beyond 0.1 °C per decade. The Wissenschaftlicher Beirat der Bundesregierung globale Umweltveränderungen (German Advisory Council on Global Change) bases climate constraints on geohistorical arguments leading to a recommended tolerable global temperature change of 2 °C compared to pre-industrial times, and a maximum rate of change of 0.2 °C per decade. The European Commission agreed that the global average temperature should not exceed the pre-industrial level by more than 2 °C and that therefore carbon dioxide concentration levels lower than 550 ppmv should guide global emissions reduction efforts (European Commission 1996). Taking into account that global temperature has already increased by about 0.5 °C and that further increases due to historical emissions have been committed, this goal is consistent with rates of temperature change below 0.15 °C per decade. The reason these early recommendations were not supported further or adjusted by ecologists is due mainly to their generalized nature: in reality, vulnerability of ecosystems is dependent on local conditions, of which temperature change is only one. As discussed in Chapter 1, the relationship between global mean temperature and local temperature changes is very uncertain. In a workshop on the issue, ecologists concluded (intuitively) that 0.1 °C decade would be an upper value:

historically observed migration rates of vegetation species (20–100 km per century) would translate into a number closer to 0.01 °C per decade, much lower than the 0.15 °C per decade observed over the last decades (Leemans & Hootsmans 2000). However, ecosystem resilience could allow periods with high rates, possibly for some decades. Because of the complexities involved, indicators other than absolute values of temperature change have been suggested (e.g. the amount of time a particular temperature range is exceeded). Different threshold values of temperature range exceedance could be determined for different ecosystems (Leemans & Hootsmans 2000). An additional complexity is that rapid climate change could disrupt long-established relationships between species, which may affect their capability to adapt naturally to climate change. The IPCC (2001d) introduced a new insight into this debate: the implications of very long-term, possibly dramatic and irreversible impacts of climate change, such as the melting of the Greenland Ice-cap (see Box 8.1). While these issues address 'the avoidable impacts' of climate change and relate to the three conditions of UNFCCC Article 2, the complex relationship between the climate parameters, which are often regionally different, has led researchers and policy-makers to focus on the stabilization of the atmospheric concentration of greenhouse gas emissions in the atmosphere, reflected in the first part of Article 2.

8.2.3   Pathways towards stabilization

Although the UNFCCC refers to stabilization of aggregate greenhouse gas concentrations, scientific, technical, and economic analyses have so far focused only on carbon dioxide (see Box 8.2). In the next chapter, the economic consequences of stabilizing carbon dioxide concentrations at various levels is discussed. Here, we discuss the implications of stabilization for different socioeconomic and technical development pathways. The IPCC assessed the implications of stabilizing carbon dioxide concentrations in different possible future worlds, by considering how the carbon dioxide concentration could be stabilized taking the different *Special Report on Emission Scenarios* (*SRES*) discussed in Chapter 2 as baseline (Morita *et al.* 2001). Nine modelling teams participated in the so-called 'post-*SRES*' effort. In Figure 2.9, the emissions ranges of the post-*SRES* stabilization analyses are superposed on the emissions ranges of the *SRES* base cases.

The carbon dioxide concentration level is determined more by the cumulative emissions over time than by the emissions in any particular year, and therefore different emissions profiles leading the same stabilization level are possible: the rate of change of emissions is important as well as the year in which the concentration is stabilizing. This explains the range of emissions in Figure 2.9

Box 8.1 The 'Ultimate Objective' of the UNFCCC and long-term
irreversible impacts

Projected climate changes during the twenty-first century have the potential
to lead to future large-scale and possibly irreversible changes in the systems
resulting from impacts at continental and global scales (IPCC 2001b).
Examples include the slowing of the North Atlantic ocean circulation
(thermohaline circulation), melting of the Greenland ice-cap and
disintegration of the west Antarctic ice-sheet, and a 'run-away' greenhouse
effect through accelerated releases of carbon from permafrost regions and
methane from hydrates as a result of global warming. The probability of
such changes in this century is very low, and if they will ever occur, is
dependent mostly on the magnitude, rate, and duration of climatic changes.
An interesting exception appears to be the melting of the Greenland Ice-cap.
According to the results of ice-sheet models assessed by IPCC (2001a), a local
warming exceeding 3 °C, if sustained for millennia, would lead to virtually a
complete melting of this ice-cap and an associated sea-level rise of about 7 m.
A local warming of 5.5 °C sustained for one millennium would probably
result in a sea-level rise of 3 m. Local warming over Greenland may be 1–3
times the global average. Figure 8.2 shows the dependency of both 2100 and
equilibrium temperature changes on a number of stabilization levels. This
graph suggests that the risk of the eventual melting of the Greenland ice-cap
can be reduced by stabilising carbon dioxide concentrations at lower levels,
but does not exclude this eventual melting even for stabilization at 450 ppm.
What does this imply for decision-making? According to IPCC (2001d):
'decision making has to deal with uncertainties including the risk of
non-linear and/or irreversible changes and entails balancing the risks of
either insufficient or excessive action, and involves careful consideration of
the consequences (both environmental and economic), their likelihood, and
society's attitude towards risk'. The ultimate objective of the UNFCCC does
not provide clear guidance on how such long-term, large-scale and possibly
irreversible changes can be evaluated. To what extent policy-makers take
these risks into account depends on their attitude towards risks (risk-taking
or risk-adverse) and their willingness to consider the very long-term and
near-term decision-making.

Temperature change relative to 1990 (°C)

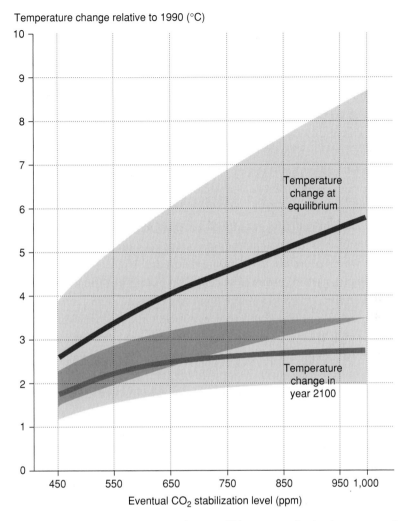

Figure 8.2. Stabilizing CO_2 concentrations would lessen warming but by an uncertain amount. Temperature changes compared to 1990 in (a) 2100 and (b) at equilibrium are estimated using a simple climate model. Lowest and the highest estimates for each stabilization level assume a climate sensitivity of 1.7 and 4.2°C, respectively. The center line is an average of the lowest and the highest estimates. For a color version of the Figure see the original publication or website http://www.grida.no/climate/ipcc_tar/vol4/english/figspm-7.htm. *Source:* IPCC, 2001d.

for any given stabilization target, dependent on the models used and assumptions made. From the perspective of internationally co-ordinated policy-making, two important pieces of information are interesting: when do the global emissions have to start declining in order to make stabilization at a particular level possible?, and: when should global emissions fall below the 1990 levels? According to

---

Box 8.2   Stabilizing greenhouse gas concentrations or stabilizing carbon dioxide concentrations?

For convenience and simplicity, in practice policy-makers and researchers alike focus on stabilization of the carbon dioxide concentration rather than on the aggregate of all greenhouse gases. Theoretically, the inclusion of other greenhouse gases (e.g. those in the Kyoto Protocol) would allow for more flexibility in meeting overall stabilization goals. Such goals could be expressed in terms of stabilization of radiative forcing in the atmosphere rather than concentrations. Sometimes, the concept of stabilization of greenhouse gas equivalent concentrations in the atmosphere is used, in which case the radiative forcing of non-carbon dioxide concentration changes is used to calculate the equivalent concentration of carbon dioxide which would cause the same radiative forcing. Wigley (2001) shows that the difference between carbon dioxide and its equivalent concentrations could be significant. Using an average non-carbon dioxide emissions scenario, Wigley calculates that stabilizing its equivalent concentrations at 550 ppm would imply a stabilization at about 400 ppm.

Carbon dioxide emissions have to decrease in order to enable stabilization of its atmospheric concentration. This generally implies that non-carbon dioxide emissions from the energy sector should also decrease. In addition, such developments in the energy sector in scenario analyses are usually matched by similar efforts to reduce greenhouse gas emissions in other sectors. This again leads to decreasing total emissions of non-carbon dioxide greenhouse gases, including those with a short atmospheric lifetime, leading to actually decreasing concentrations of those gases, with associated decreased radiative forcing. Taking such developments into account, carbon dioxide emissions reductions could be less than under a carbon dioxide-only regime, as long as non-carbon dioxide emissions continue to decrease, unless one would like to go beyond stabilization and, for example, return to pre-industrial temperature levels. Clearly, research addressing both near-term and long-term optimal combinations of control across the various greenhouse gases in the context of stabilizing equivalent carbon dioxide concentrations or radiative forcing would be very useful.

Table 8.1, for stabilizing at 450 ppm, global emissions would have to peak within the next 5–15 years. Parallel numbers for 550, 650, and 750 ppmv carbon dioxide stabilization are: 20–30 years, 30–45, and 80–180 years, respectively. Global emissions would have needed to fall below 1990 levels by 2000–40 for 450 ppm, 2030–100 for 550 ppm, 2055–145 for 650 ppm, and 2080–180 for 750 ppm (IPCC

2001d). The large ranges are caused by uncertainties in the carbon cycle as well as different possible emissions profiles.

Notwithstanding the wide range of emissions in both baseline and stabilization scenarios, two conclusions can be drawn immediately from the Figure 2.9. First, not surprisingly, the effort needed to achieve stabilization becomes greater as the stabilization target is tightened (the effort represented by the gap between the baseline and various stabilization profiles in any scenario group). Second, and maybe more interestingly, the effort to stabilize atmospheric carbon dioxide concentrations is very dependent upon the characteristics of the baseline scenario, i.e. on the socioeconomic development pathway. For example, the gap to be closed between the baseline scenario in the A2 scenario group is much larger than that to be closed in the B1 case, and probably more difficult to close. There are several reasons why it is difficult to close the gap in A2: (a) technological change is slow; (b) the energy system continues to be based on fossil energy; (c) the support for mitigation is low because of the dominance of the material economic growth objectives, and (d) the opportunities for international co-ordination of carbon dioxide mitigation measures are modest in the regionalized world. In B1, the situation is very different: (a) rapid technological development in an environmentally sound way; (b) changes in lifestyle and economic structure; (c) popular support for environmental measures in an environmentally conscious world, and (d) rapid and internationally co-ordinated transfer of environmentally sound – predominantly renewable – technologies in a globalizing world society. In this scenario, stabilization of the carbon dioxide concentration at 550 ppm is in reach by the end of the century without additional climate policies, and the gap to be closed to stabilize at 450 ppm is relatively small. This points very strongly at the importance of linking climate change mitigation measures with broader socioeconomic objectives: if society would develop in a way that would be compatible with social, economic, and environmental sustainability objectives (as in B1), it would be much easier to stabilize carbon dioxide concentration and hence avoid dangerous interference with the climate system.

It should be noted, though, that the B1 scenario represents but one particular world development in a more sustainable direction. One can imagine many different world developments that would also be compatible with sustainable development objectives, or maybe even more so. For example, the 'Great Transitions' scenario discussed in Chapter 2 is different from B1 in that it assumes fundamental changes (transitions) in societies' values and associated behaviour, leading to improvements in social and environmental conditions beyond those in B1. Another example is the A1T scenario, in which greenhouse gas emissions are low because of very rapid technological developments that are concentrated in non-fossil energy sources. In this scenario, further reduction of carbon dioxide emissions in order to stabilize the greenhouse gas concentration at low levels can

Table 8.1. *Some considerations for possible carbon dioxide stabilization goals.*

450 ppm	• Lowest environmental risks, often positive side-effects
	• Consistent with Kyoto targets, provided deep reductions afterwards
	• Accumulated carbon dioxide emissions[a] this century 365–735 Gt carbon
	• Global emissions peak in 2005–2015
	• OECD countries 30–40 per cent reduction by 2030[b], 70–80 per cent by 2050 as compared to 1990
	• Developing countries' emissions to diverge from baseline soon
	• Costs can be very high, dependent on baseline (see Chapter 10)
550 ppm	• Environmental risks decreased significantly
	• Consistent with 550 ppmv carbon dioxide stabilization target of European Union
	• Compatible with Kyoto target, provided gradual further reductions afterwards
	• Accumulated carbon dioxide emissions this century 590–1135 Gt carbon
	• Global emissions peak in 2020–30
	• Developing countries' emissions diverge from baseline after some decades
	• Costs can be very high, dependent on baseline (see Chapter 10)
650 ppm	• Environmental risks in this century slightly decreased
	• Kyoto targets strict, only slow further declines needed
	• Accumulated carbon dioxide emissions this century 735–1370 Gt carbon
	• Global emissions peak in 2030–45
	• Developing countries only to diverge from baseline in second half of the twenty-first century
	• Costs in various scenarios modest (see Chapter 10)
750 ppm	• Environmental risks not significantly decreased in this century
	• Kyoto targets stricter than needed
	• Accumulated carbon dioxide emissions this century 820–1500 Gt carbon
	• Global emissions peak in 2080–180
	• Developing countries only to diverge from baseline in second half of the twenty-first century
	• Costs in various scenarios low (see Chapter 10)

[a] Accumulated carbon dioxide emissions for baseline scenarios A1B, 1415 Gt carbon; A1T, 985 Gt carbon; A1FI, 2105 Gt carbon; A2, 1780; B1, 900 Gt carbon; B2, 1080 Gt carbon.

[b] Typical value from the scenario literature; extensive emissions trading with other regions can lead to lower OECD reduction requirements.

Source: Authors' interpretation from findings of IPCC (2001c, d).

Table 8.2. *Groups of narrative 'global futures' scenarios.*

Scenario group	Scenario subgroups	No. of scenarios
1. Pessimistic scenarios	*Breakdown*: collapse of human society	5
	Fractured world: deterioration into antagonistic regional blocs	9
	Chaos: instability and disorder	4
	Conservative: world economic crash is succeeded by conservative and risk-averse regime	2
2. Current trends scenarios	*Conventional*: no significant change from current and/or continuation of present-day trends	12
	High growth: government facilitates business, leading to prosperity	14
	Asia shift: economic power shifts from the West to Asia	5
	Economy paramount: emphasis on economic values leads to deterioration in social and environmental conditions	9
3. High-tech optimist scenarios	*Cybertopia*: information and communication technologies facilitate individualistic, diverse, and innovative world	16
	Technotopia: technology solves all or most of humanity's problems	5
4. Sustainable development scenarios	*Our common future*: increased economic activity is made to be consistent with improved equity and environmental quality	21
	Low consumption: conscious shift from consumerism	16

Source: Morita *et al.* (2001).

be achieved by technologies that are not preferred in B1, such as carbon dioxide scrubbing and storage, and nuclear energy. Preliminary analysis (see Chapter 10) suggests that in this case the costs of stabilizing the carbon dioxide concentrations (e.g. at 550 ppm or below) can be relatively high, but in a very high-income scenario this may not be prohibitive.

Although, for the purpose of analysis, the implications of stabilization of greenhouse gas concentration quantified scenarios are indispensable, an assessment of the qualitative 'global futures' scenarios can provide additional insights about variables that escape quantification, e.g. issues of governance, security, and social equity. Morita *et al.* (2001) made an inventory of more than hundred of such scenarios, which they grouped into four typical categories: (a) pessimistic; (b) current trend; (c) high-tech optimist, and (d) sustainable development (see Table 8.2). Many of these scenarios are normative, describing futures desirable (e.g. 'a better world') or undesirable (e.g. 'collapse') in the eyes of the authors.

Table 8.3 shows how futurists (often implicitly) assume various socioeconomic factors to be associated with either rising or falling greenhouse gas emissions, the latter a requirement for concentration stabilization. Not surprisingly, scenarios with decreasing greenhouse emissions – which are smaller in number – are often characterized by: (a) low population growth; (b) rapid technological innovation and diffusion; (c) increased usage of renewable resources; (d) generally improved environmental quality, and (e) increasing social and economic equity. This suggests that socioeconomic development consistent with such trends is likely to be climate-friendly, too. The assessment suggests that a comprehensive evaluation of greenhouse gas stabilization scenarios could take a wider range of factors into account than just those factors that are usually quantified in model analysis. Designing greenhouse gas policies in a way that would be resilient against a wide variety of possible future conditions makes them more robust. The global futures scenarios support the findings from the *SRES* report that increasing levels of income and population do not necessarily lead to increasing greenhouse gas emissions. Technology and lifestyle choices and economic structural changes are considered by the futurists to be able to reduce greenhouse gas emissions without necessarily affecting welfare. Low emissions futures can be associated with policies and actions beyond climate policy (Morita *et al.* 2001).

Lifestyle choices could help reduce greenhouse gas emissions, but regardless of the effectiveness of such choices, technological innovation is vital in any attempt to stabilize greenhouse gas concentrations. This is illustrated by the IPCC statement quoted above, in which it is stated that greenhouse gas concentrations can be stabilized at levels of 550, 450 ppmv, or even below with currently known technologies – 'known' referring to technologies existing in operation or at pilot plant stage today. The choice of such known options is already very wide indeed, and the future market potential of the various options, uncertain. Therefore, an interesting question is: which technologies would be robust in the sense that they would be attractive in different world views and for reaching different climate goals? Morita *et al.* (2001) explored this question by analysing the set of post-*SRES* scenarios, which combine different baseline scenarios for socioeconomic development (A1, A2, B1, B2) with different stabilization levels (450, 550, 650, 750 ppm, respectively). Table 8.4 shows which options are applied by the scenario modellers to arrive at a required stabilization level of 550 ppm of carbon dioxide.

According to the scenario analysis represented in Table 8.1, demand reduction (through efficiency increases, structural economic changes) is a key robust option in any scenario. This option is particularly dominant for the first half of the twenty-first century. In the second half, different (energy supply) options are increasingly needed to achieve the emissions reductions necessary to stabilize

Table 8.3. *Factors associated with changing greenhouse gas emissions in narrative global futures scenarios.*

Factor	Rising greenhouse gases	Falling greenhouse gases
Economy	Growing, post-industrial economy with globalization, (mostly) low government intervention, and generally high level of competition	Some scenarios show rising GDP, others show economic activity limited to ecologically sustainable levels; generally high level of government intervention
Population	Growing population with high level of migration	Growing population that stabilizes at relatively low level; low level of migration
Governance	No clear pattern in governance	Improvements in citizen participation in governance, community vitality and responsiveness of institutions
Equity	Generally declining income equality within nations and no clear pattern in social equity or international income equality	Increasing social equity and income equality within and among nations
Conflict/security	High level of conflict and security activity (mostly), deteriorating conflict resolution capability	Low level of conflict and security activity, improved conflict resolution capability
Technology	Low level of technology, innovation and technological diffusion	High level of technology, innovation and technological diffusion
Resource availability	Declining renewable resource and water availability; no clear pattern for nonrenewable resource and food availability	Increasing availability of renewable resources, food and water; no clear pattern for nonrenewable resource
Environment	Declining environmental quality	Improving environmental quality

Source: Morita *et al.* (2001).

atmospheric concentrations. In all scenarios, the gradual penetration of renewable energy is important; in the first half, biomass plays a key role, followed by a further large-scale introduction of other renewable sources in the second. Options such as nuclear energy and capture and storage of carbon dioxide are only important in a few scenarios, reflecting that these options are controversial and only considered

Table 8.4. *Sources of emissions reductions for 550 ppmv carbon dioxide stabilization across the nine 'post-SRES' models.*

2050	A1B	A1FI	A1T	A2	B1	B2
Substitution among	0.0–1.2	0.6–3.3	0.0–0.11	0.13–0.96	0.0–0.11	−0.02–0.54
fossil fuels	(0.55)	(2.2)	(0.1)	(0.6)	(0.02)	(0.25)
Switch to nuclear	−0.42–0.56	0.33–1.1	−0.03–0.12	0.18–0.98	0.0–0.25	0.12–0.54
	(0.45)	(0.53)	(0.04)	(0.97)	(0.17)	(0.48)
Switch to biomass	−0.15–1.7	0.26–1.3	−0.1–0.14	0.64–2.3	0.0–0.73	−0.14–0.87
	(1.13)	(0.76)	(0.02)	(1.14)	(0.34)	(0.31)
Switch to other	−0.1–3.1	0.26–4.91	−0.15–0.05	0.0–1.1	0.0–0.51	0.1–0.85
renewables	(1.12)	(0.88)	(−0.05)	(0.55)	(0.29)	(0.38)
Carbon dioxide	0.0–2.5	0.0–5.75	0.49–0.58	0.0–0.5	0.0–0.0	0.0–2.4
scrubbing and	(0.0)	(3.09)	(0.53)	(0.0)	(0.0)	(0.14)
removal						
Demand reduction	1.0–3.5	1.66–8.5	0.03–0.99	0.83–5.7	0.0–1.1	−0.2–2.4
	(2.45)	(4.6)	(0.51)	(2.3)	(0.23)	(1.15)
2100						
TOTAL reduction	(5.46)		(1.12)	(6.7)	(1.25)	(3.5)
Substitution among	−0.1–2.2	0.2–11.8	0.1–0.1	2.4–5.4	0.0–0.2	0.6–2.7
fossil fuels	(0.97)	(1.82)	(0.09)	(2.95)	(0.09)	(1.35)
Switch to nuclear	0.3–6.4	−2.4–1.9	0.0–2.0	0.3–1.7	0.0–3.1	−0.2–5.1
	(0.55)	(1.20)	(1.03)	(1.18)	(0.02)	(2.28)
Switch to biomass	−0.8–1.5	−0.2–5.5	−0.2–0.3	1.1–3.8	0.0–4.3	−1.9–1.5
	(1.03)	(2.50)	(0.07)	(1.84)	(0.04)	(0.63)
Switch to other	0.1–2.5	0.6–15.1	−0.1–0.0	2.2–6.7	0.1–0.3	0.1–3.2
renewables	(1.51)	(2.70)	(−0.05)	(3.33)	(0.28)	(2.07)
Carbon dioxide	0.0–4.7	0.0–23.8	0.5–1.6	0.0–5.8	0.0–1.1	0.0–3.0
scrubbing and	(0.00)	(0.39)	(1.06)	(0.0)	(0.0)	(0.63)
removal						
Demand reduction	0.5–6.6	1.9–17.7	0.0–0.2	5.2–15.6	0.1–0.3	0.7–3.5
	(0.94)	(10.4)	(0.11)	(10.21)	(0.08)	(1.64)
	7.1–11.9	21.7–30.5	0.3–4.4	21.7–26.9	0.2–9.6	6.0–10.6
	(9.16)	(21.1)	(2.31)	(22.81)	(0.39)	(8.14)

Minimum–maximum and (median) in 2050 and 2100 (gigatonnes of carbon)

Note: Emission reductions are estimated by subtracting the mitigation value (in gigatonnes of carbon) from the baseline value (in gigatonnes of carbon) of each scenario.

Relative valuation signs added by the authors: ++: very important option in this scenario.

+ important option in this scenario; 0: not very important option in this scenario.

Source: For 2100, Morita *et al.* (2001); 2050 added on the basis of data provided by T. Morita.

to provide important opportunities in some worldviews (e.g. both options in the technology-oriented A1T, nuclear in B2), and in some regions. The various options are discussed in more detail in Chapter 9.

8.2.4   *Near-term implications of greenhouse gas stabilization*

For today's decision-makers in governments and the private sector, near-term objectives are more important than the century timescale discussed above. However, because of the ultimate objective of climate change response – stabilization of greenhouse gas concentrations – the long-term goals provide a framework for near-term decision-making.[1] As long as the long-term goals have not been agreed upon, climate change response is a matter of sequential decision-making: a stepwise approach allowing for adjusting response measures as more information becomes available. The relevant question is: what is the best course of action for the near term given the expected long-term climate change and the accompanying uncertainties? The issue of timing of emissions reductions is a hotly debated issue, in which a wide variety of, often conflicting, arguments is used (see Table 8.5). Some arguments tend to favour very early stringent abatement measures while others support a more gradual approach to emissions mitigation. However, since the earlier discussions in the IPCC *Second Assessment Report* based on the so-called 'WRE' scenarios after Wigley, Richels and Edmonds (1996) that tended to focus on 'early action' versus 'delayed response', there are more nuances in the debate now. It is now acknowledged that the question is not so much about delaying action, but about the type of action needed now. 'Action' does not necessarily imply immediate emissions reductions, but can involve different types of action, including stimulation of technological innovation through enhanced research, development and demonstration of environmentally sound technologies; hence, the distinction in Table 8.5 between modest and stringent abatement. In order to stabilize the atmospheric concentration of carbon dioxide, large future emissions reductions will be needed: eventual carbon dioxide emissions would need to decline to a very small fraction of current emissions (IPCC 2001a). This would require a fundamental restructuring of the energy system, either by switching fuels or by moving in a direction of a 'clean fossil fuel' system in which system carbon dioxide would be captured and stored underground or in the deep ocean. Either way, investing in technological innovation is a robust option. Greenhouse gas mitigation is likely to be easier and less costly if a long-term policy perspective is provided by governments to the

---

[1] In some countries, this is acknowledged. For example, The Netherlands Environmental Policy Plan contains agreed indicative carbon dioxide emission reduction targets of 30–50 per cent by 2030, in addition to the near-term target related to the Kyoto Protocol (VROM 2001).

Table 8.5. *Balancing the near-term mitigation portfolio.*

Issue	Favouring modest early abatement	Favouring stringent early abatement
Technology development	• Energy technologies are changing and improved versions of existing technologies becoming available, even without policy intervention • Modest early deployment of rapidly improving technologies allows learning-curve cost reductions, without premature lock-in to existing, low-productivity technology • The development of radically advanced technologies will require investment in basic research	• Availability of low-cost measures may have substantial impact on emissions trajectories • Endogenous (market-induced) change could accelerate development of low-cost solutions (learning by doing) • Clustering effects highlight the importance of moving to lower emission trajectories • Induces early switch of corporate energy research and development from fossil frontier developments to low carbon technologies
Capital stock and inertia	• Beginning with initially modest emissions limits avoids premature retirement of existing capital stocks and takes advantage of the natural rate of capital stock turnover • It also reduces the switching cost of existing capital and prevents rising prices of investments caused by crowding-out effects	• Exploit more fully natural stock turnover by influencing new investments from the present onwards • Limiting emissions to levels consistent with low carbon dioxide concentrations, preserves an option to limit carbon dioxide concentrations to low levels using current technology • Reduces the risks from uncertainties in stabilization constraints and, hence, the risk of being forced into very rapid reductions that would require premature capital retirement later.

Social-effects and inertia	• Gradual emission reduction reduces the extent of induced sectoral unemployment by giving more time to retrain the workforce and for structural shifts in the labour market and education • Reduces welfare losses associated with the need for fast changes in people's lifestyles and living arrangements	• If lower stabilization targets would be ultimately required, stronger early action reduces the maximum rate of emissions abatement required subsequently and reduces associated transitional problems, disruption, and the welfare losses associated with the need for faster later changes in lifestyles and living arrangements
Discounting and intergenerational equity	• Reduces the present value of future abatement costs (ceteris paribus), but possibly reduces future relative costs by furnishing cheap technologies and increasing future income levels	• Reduces impacts and (ceteris paribus) reduces their present value
Carbon cycle and radiative change	• Small increase in near-term, transient carbon dioxide concentration • More early emissions absorbed, thus enabling higher total carbon emissions this century under a given stabilization constraint (to be compensated by lower emissions thereafter)	• Small decrease in near-term, transient carbon dioxide concentration. • Reduces peak rates in temperature change
Climate change impacts	• Little evidence about damages from multidecade episodes of relatively rapid change in the past	• Avoids possibly higher damages caused by faster rates of climate change

Source: IPCC 2001c.

private sector. As recognized by the IPCC (2001d), unplanned and unexpected policies ('quick fixes') with sudden short-term effects may cost much more than planned policies that could have more gradual economic benefits experienced well in advance.

An important question relating to the adequacy of the Kyoto Protocol commitments is: are the Kyoto targets adequate if we would like to meet the ultimate objective of the UNFCCC? This question cannot be answered easily at this stage, since we do not know how this ultimate objective will be quantified in terms of stabilization levels, while at the same time a wide range of pathways are possible which may lead to any chosen stabilization level. In addition, the Kyoto Protocol only requires Annex I countries to reduce their emissions, leaving open the emissions paths of the other countries. The adequacy of the Kyoto Protocol to stabilize carbon dioxide concentrations depends on post-Kyoto commitments of both Annex I and non-Annex I countries. If the emissions in developing countries are never constrained, while those in developed countries are stablized at 2010 levels, the effect on global mean temperature would be marginal. However, if the Kyoto Protocol would lead to further emissions controls, this would change. We may deduce some general insights from modelling analyses with respect to the question: does Kyoto at least put us on the right track to stabilization of concentration? Figure 2.9 shows the emissions reductions from 1990 levels in Annex I countries[2] for different stabilization levels, as modelled in the so-called 'post-*SRES*' analysis (Morita *et al.* 2001). There is a wide range of results, depending on the particular modelling approach, the assumptions with respect to the timepath of emissions reductions, and with reference to the different developments in developed and developing countries. Nevertheless, some general tendencies are suggested via Figure 2.9. For a stabilization target of 450 ppm, the Kyoto targets are on the lower end of what may be required, while after 2010 further reductions are likely to be needed. For 550 ppm, the Kyoto targets are well within the range of emissions reductions likely to be needed, while afterwards only gradual further reductions may be required. For 650 ppm or higher stabilization targets, the Kyoto targets are on the higher end of what may be needed, and there would be no need for further reductions for some time, according to this analysis.

Another type of analysis may provide different answers. For example, the tolerable windows or safe-landing approach discussed in Section 8.1 take a different, 'backcasting' approach that can lead to more stringent near-term emissions restrictions, especially if constraints are being posed on 'tolerable' rates of temperature change (see Box 8.3).

---

[2] It should be noted that these analyses do assume various emissions pathways for non-Annex I regions, affecting the results for Annex I regions.

Box 8.3   Near-term implications of EU long-term climate targets: the safe-landing approach applied

The safe-landing analysis (see Section 8.1 for a discussion of the methodology) has been used to explore the implications of the long-term climate goals – stabilization of the carbon dioxide concentration at 550 ppm and global mean temperature not exceeding 2 °C as compared to pre-industrial levels – of the European Commission for the near term (Swart *et al.* 1998). Taking only the temperature goal in 2100 as limiting factor, a 'best guess' climate sensitivity, stable sulphur dioxide emissions, and a maximum feasible annual emissions reduction rate of 2 per cent per year, the whole range of IPCC IS92 scenarios fall within the calculated *global* carbon dioxide emissions corridor, suggesting that no emissions reductions before 2010 are needed. However, if we introduce additional constraints, this changes. For example, by adding a constraint on sea-level rise of 20 cm, as recommended by the Aliance of Small Island States group of countries, the corridor would narrow. The same applies if a constraint on the rate of temperature change is added, e.g. 0.15 °C per decade (consistent with the Commission's goal) or 0.1 °C per decade (suggested on the basis of ecosystem vulnerability). The resulting corridor appears to be very dependent on the assumed constraint on the rate of temperature change and the number of decades for which this rate is tolerated.

In the context of the UNFCCC and Kyoto Protocol, for the near term the *Annex I* emissions are vital. If it is assumed that non-Annex I emissions develop as in the IS92A scenario, the Annex I emissions in 2010 should be less than 5 per cent above the 1990 levels (see Figure 8.3 (a) for assumed constraints). Figure 8.3 (b) shows the dependency of the emissions corridor on the constraint on the rate of change: for 0.1 °C per decade, allowed to be exceeded for two decades, the top of the corridor in 2010 moves down from 6.1 to 3.3 gigatonnes (Gt) of carbon per year, an impracticable 40 per cent emissions reduction. Evidently, lower or higher emissions in non-Annex I countries would lead to a wider or narrower Annex I emissions corridor, respectively. This analysis suggests that it is very important to understand better the relationship between climate change impacts and the rate of temperature change, clearly a gap in knowledge. Another important factor is the assumed maximum feasible rate of emissions reduction. In the above calculations, 2 per cent per year was assumed, but reducing this rate to 1 per cent per year narrows the emissions corridor to 2010 by more than 10 per cent. In this context, it is also important to consider intergenerational issues.

If future emissions were to develop according to the top of the simulated corridors, they would have to be reduced afterwards to meet the global climate goals, with the maximum assumed global rate of emissions reduction. This leaves no room for future tightening of the climate goals. In this case, the calculated emissions corridor suggests that Annex I emissions would have to fall below zero before 2030, unless non-Annex I emissions would stay well below IS92a emissions. These analyses illustrate the importance of taking into account the inertia in climate and socioeconomic systems when analysing near-term mitigation requirements in view of long-term climate goals.

The so-called 'tolerable windows approach' (e.g. Toth *et al.* 2001) is similar to the safe landing approach, and its application leads to similar findings: near-term emissions will be constrained significantly if long-term climate change goals are taken into account, the more so as the feasible rate of future emissions reductions would be lower. The tolerable windows approach discussed above can also provide some insights related to equitable distribution of the burden of climate stabilization. If it is assumed that the Annex-I emissions follow the IS92a baseline until 2010 and the emissions of non-Annex I countries follow this IS92a pathway until their emissions reach those of Annex I levels on the basis of their 1992 population, a very tight emissions corridor results, with annual emissions reduction rates in the order of 3–4 per cent per year (Toth *et al.* 2001). It should be noted that the climate protection goals assumed in this analysis are more relaxed than in the safe-landing analysis discussed above, while higher feasible rates of emissions reductions are also assumed.[1]

> [1] An optimistic 'continuity rate' (i.e. the speed at which fossil fuels could be phased out in approximate synchrony with the turnover of capital stocks avoiding premature retirement of capital) could reduce fossil-fuel related carbon dioxide emissions by 2.5–3.0 per cent per year. Efficiency improvements (historically in the order of 1 per cent per year) theoretically could be boosted to simular improvement rates. With a world economic growth rate of 3 per cent per year, for example, this would lead to a maximum feasible emissions reduction rate of 2–3 per cent per year. Economic, political, and social realities will reduce the feasibility of such potentially achievable reduction rates (Swart *et al.* 1998).

8.3   Links with development, equity, and sustainability issues

In Section 8.2, we saw that sustainable development and climate change response are closely linked. The UNFCCC and the Kyoto Protocol address this link in various ways. The UNFCCC recognizes that countries have the right to – and

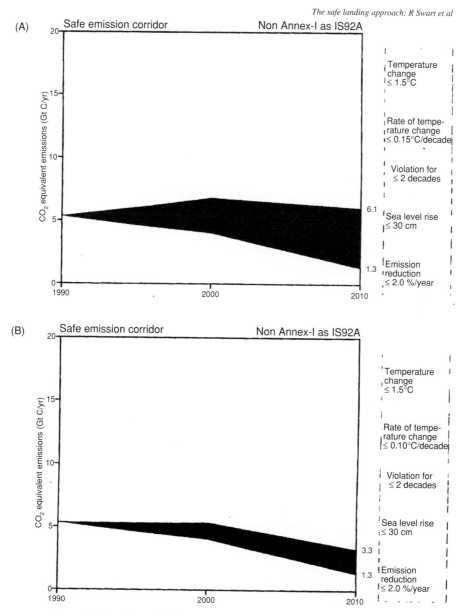

Figure 8.3. Two Annex-I emissions corridors consistent with the EU long-term climate objective with different constraints on the decadal rate of temperature change. *Source:* Swart *et al.* (1998).

should – promote sustainable development, and that countries in their efforts to protect the climate system shall do this on the basis of equity – and in accordance with their common but differentiated responsibilities (Article 3). As discussed above, the ultimate objective includes a condition that economic development should be allowed in a sustainable manner (Article 2). The Clean Development Mechanism of the Kyoto Protocol has been established not only to give industrialized countries more flexibility for meeting their quantified emissions limitations and reductions requirements, but also to help developing countries achieve sustainable development and contribute to the ultimate objective of the Convention. Scenario analysis described above suggests that in a world in which concerted efforts are being made in the direction of a more equitable and sustainable development, stabilization of greenhouse gas concentrations would be easier. Conversely, stabilization of greenhouse gas concentrations and associated reduction of greenhouse gas emissions would require changes in production and consumption patterns that would have multiple benefits.

The development, implementation, and transfer of environmentally sound technologies usually is pursued for reasons unrelated to climate change, e.g. the need to reduce local air pollution problems or the benefits of increased energy efficiency. The implications of general social and economic strategies are discussed in Chapter 3. Climate change mitigation strategies affect a set of issues broader than climate change. If stringent greenhouse gas mitigation actions are pursued too quickly, the economic consequences may be significant, affecting opportunities for sustainable economic growth as considered in the ultimate objective of the UNFCCC. If introduced gradually, however, consequences may be positive. For example: (a) increasing energy efficiency reduces energy costs and fuel import dependency; (b) a shift to non-fossil energy sources helps in abating local air pollution and regional acidification of soils, agriculture, and ecosystems, and (c) stopping deforestation and regrowth of forests can help protect local forest assets and biodiversity. Such 'ancillary' benefits are discussed in Chapter 10.

In general, decisions will be taken for many reasons together, climate change being just one of them. In that sense, it may be better to talk about 'co-benefits' rather than 'ancillary' benefits. Increasingly, integrated policies are pursued, often combining air pollution and greenhouse gas mitigation strategies (see, for example, USEPA 2000). Usually, like explicit climate strategies, such combined strategies are pursuing incremental abatement, often through technological means.

Two fundamental questions now are: (a) what is the extent to which the climate problem can be mitigated through incremental technological changes, and (b) to what extent are fundamental economic structural and behavioural changes required? As noted above, according to the IPCC, no fundamentally new tech-nologies are required to stabilize the carbon dioxide concentration at 450 ppm or below, but a disclaimer adds that this is only true provided that associated socioeconomic and institutional changes would occur. There is no clear separa-tion between technological changes and socioeconomic and behavioural changes, since one depends on the other. The scenario literature suggests that – looking at the climate problem in isolation – technology may be able to do the job, but structural and behavioural changes would make this job easier. Conversely, technological innovation is a characteristic of all plausible climate change mit-igation scenarios analysed. The Global Scenario Group explore this question in some depth, in the broader context of sustainable development. They assert that in a 'conventional development world' – in which socioeconomic develop-ments would continue without major surprises or fundamental transformations in the basis for human civilization – it would be possible to meet a wide range of sustainable development goals with strict, but incremental, policies. These policies address not only environmental objectives (e.g. greenhouse gas stabi-lization, amelioration of water scarcity, abatement of desertification, protection of forests and biodiversity, and abatement of toxic substances) but also social goals, e.g. narrowing the income gap and halving world hunger (Gallopin *et al.* 1997; Raskin *et al.* 1998). Such developments would continue to be based on glob-alization, industrialization, and market forces. As the anti-globalization move-ments have demonstrated in recent years, such developments are not without major tensions. Movements towards social inequity and environmental degra-dation as a side-effect of the conventional development pathway have to be corrected continuously, making the world's socioeconomic and environmental systems increasingly brittle. The situation could be compared with running up a downward-moving escalator leading to an ever-increasing need to be better equipped (Raskin *et al.* 2002). Other futures may be possible, which would involve innovative social and economic arrangements and fundamental changes in values, protecting natural systems, equitable distribution of resources, efficiency next to sufficiency, and a strong sense of solidarity ('Great Transitions': see Gallopin *et al.* 1997; Raskin *et al.* 2001). In such futures geared towards sustainable development, decoupling between welfare and resource throughput would not only be achieved primarily through a 'technology wedge' (as in the conventional development sce-narios), but also by a 'lifestyle wedge' that is governed by changes in taste and behaviour.

8.3.1   *Equity aspects of climate change mitigation strategies*

Stabilization of greenhouse gas concentrations eventually requires decreasing global emissions, and hence at some point in the future, all regions would have to participate in an emissions control scheme. First steps have been taken for Annex I countries in the Kyoto Protocol, but although the USA has required 'meaningful participation' by developing countries, no commitments have been agreed to date with non-Annex I countries; nor is it expected that this would happen in the near future, because of the arguments that the Annex I countries not only caused most of the problem until now, but also have the most resources to deal with it. For example, in 1988 the richest 20 per cent of the population of the world accounted for 82.7 per cent of the global income, closely tied to greenhouse gas emissions, and the poorest 20 per cent for only 1.4 per cent (60 times less).

Nevertheless, from a scientific perspective, it is interesting to consider the issue of 'common but different responsibilities' and 'burden sharing' in the context of meeting the ultimate objective of the UNFCCC.

A wide variety of perspectives on equity can be distinguished (Banuri *et al.* 2001). A *rights-based* approach would be based on equal rights of people to the global commons, and can be translated into equal per capita emissions rights. This is the backbone of the 'contraction and convergence' approach that has been embraced in the climate change policy debate by quite a number of nongovernmental and political groups, 'contraction' implying a global reduction of carbon dioxide emissions consistent with the carrying capacity of the atmosphere, and 'convergence' referring to converging per capita incomes in different regions at some future date. A *liability approach* departs from the assumption that people should not be harmed by the actions of others without adequate compensation. The issue of an adaptation fund in the UNFCCC negotiations is related to this perspective. A *poverty-based approach* focuses on the need to protect the poor from adverse consequences, e.g. through decreasing their vulnerability by capacity-building activities and livelihood enhancement. An *opportunity-based* approach would be based on the rights of people to enjoy a standard of living enjoyed by the well-to-do of the world's population, an approach related to the current limitation of emissions control commitments to industrialized countries. One related proposal is to have non-Annex I countries 'graduate' to a status requiring emissions control when they reach a particular level of income or per capita emissions. This would lead to a model of increasing levels of participation in a global emissions limitations regime, or a 'multistage' approach, and could be considered as a *responsibility approach* to climate change response (Berk *et al.* 2001): countries are all responsible for managing the problem, at least when they reach a certain level of development. Toth *et al.* (2001) discuss an even larger set of principles, and find that applying different principles to the problem of climate change mitigation would lead to very different outcomes. The application of a particular principle and the possibility of achieving a particular climate goal are linked. For a graduation-based approach, it could well be possible that lower levels of carbon dioxide concentration stabilization would not be possible if the point of graduation would be too high (Berk *et al.* 2001).

References

Arrow, K. J., Cline, W. R., Maler, K.-G., Munasinghe, M., Squitieri, R. and Stiglitz, J. E. (1996) Intertemporal equity, discounting, and economic efficiency. In J. P. Bruce, Lee Hoesung and E. F. Haites, eds. (1996) *Climate Change 1995: Economic and Social Dimensions of Climate Change.* Cambridge: Cambridge University Press.

Banuri, T., Weyant, J., Akumu, G., Najam, A., Pinguelli Rosa, L., Rayner, S., Sachs, W., Sharma, R. and Yohe, G. (2001) Setting the stage: climate change and sustainable development. In B. Metz, O. Davidson, R. Swart and J. Pan, eds., *Climate Change 2001: Mitigation.* Cambridge: Cambridge University Press.

Berk, M., van Minnen, J. G., Metz, B. and Moomaw, W. (2001) *Climate Options for the Long term (COOL) – Final Report.* Bilthoven: National Institute for Public Health and Environment.

European Commission (1996) *Communication on Climate Change, Council Conclusions.* Brussels: European Commission.

Gallopin, G., Hammond, A., Raskin, P. and Swart, R. (1997) *Branch Points: Global Scenarios and Human Choice.* Boston MA: Global Scenario Group, Stockholm Environment Institute.

Hua-Duong, M., Grubb, M. J. and Hourcade, J.-C. (1997) Influence of socio-economic inertia and uncertainty on optimal CO_2 abatement. *Nature*, **390**(4), 426–46.

IPCC (2001a) *Climate Change 2001: The Scientific Basis.* Cambridge: Cambridge University Press.

IPCC (2001b) *Climate Change 2001: Impacts, Adaptation, and Vulnerability.* Cambridge: Cambridge University Press.

IPCC (2001c) B. Metz, O. Davidson, R. Swart and J. Pan, eds. *Climate Change 2001: Mitigation.* Cambridge: Cambridge University Press.

IPCC (2001d) *Climate Change 2001: Synthesis Report.* Cambridge: Cambridge University Press.

Jepma, C. and Munasinghe, M. (1998) *Climate Change Policy: Facts, Issues and Analyses.* Cambridge: Cambridge University Press.

Leemans, R. and Hootsmans, R. (2000) *Assessing Ecosystem Vulnerability and Identifying Climate Protection Indicators.* Report 410200039. Bilthoven: Nationaal Onderzoek Programma Mondiale Luchtverontreiniging en Klimaatverandering Dutch National Research Programme on Global Air Pollution and Climate Change.

Manne, A. and Richels, R. (1995) The greenhouse debate: economic efficiency, burden sharing and hedging strategies. *Energy Journal*, **16**(4), 1–37.

IPCC (2001), O. F. Canziani, N. A. Leary, D. J. Dokken and K. S. White, eds., *Climate Change 2001: Impacts, Adaptation, and Vulnerability*. Contribution of Working

Group II of the Intergovernmental Panel on Climate Change to the *Third Assessment Report*. Cambridge: Cambridge University Press.

Morita, T., Robinson, J., Adegbulugbe, A., Alcamo, J., Herbert, D., Lebre-Larovere, E., Nakicenovic, N., Pitcher, H., Raskin, P., Riahi, K., Sankovski, A., Sokolov, V., de Vries, B. and Yamaji, K. (2001) Greenhouse gas emission mitigation scenarios and implications. In B. Metz, O. Davidson, R. Swart and J. Pan, eds., *Climate Change 2001: Mitigation*. Cambridge: Cambridge University Press.

Munasinghe, M. (1996) *Environmental Impacts of Macroeconomic and Sectoral Policies*. Solomons, Washington, and Nairobi: International Society for Ecological Economics, The World Bank, and United Nations Environment Programme.

Munasinghe, M., Meier, P., Hoel, M., Wong, S. and Aaheim, A. (1996) Applicability of techniques of cost–benefit analysis to climate change. In J. P. Bruce, H. Lee and E. H. Haites, eds., *Climate Change 1995: Economic and Social Dimensions*. Geneva: Intergovernmental Panel on Climate Change, Chap. 5.

Raskin, P., Gallopin, G., Gutman, P., Hammond, A. and Swart, R. (1998) *Bending the Curve: Towards Global Sustainability*. Global Scenario Group. Boston MA: Stockholm Environment Institute.

Raskin, P., Banuri, T., Gallopin, G., Gutman, P., Hammond, A., Kates, R., and Swart, R. (2002) *Great Transitions, The Promise and Lure of the Times Ahead*. Boston: Global Scenario Group, Stockholm Environment Institute.

Rijsberman, F. and Swart, R. J. (1990) *Targets and Indicators of Climate Change*. Stockholm: Stockholm Environment Institute.

Sathaye, J., Bouille, D., Biswas, D., Crabbe, P., Geng, Luis, Hall, D., Imura, H., Jaffe, A., Michaelis, L., Peszko, G., Verbruggen, A., Worrell, E. and Yamba, F. (2001) Barriers, opportunities, and market potential of technologies and practices. In B. Metz, O. Davidson, R. Swart and J. Pan, eds., *Climate Change 2001: Mitigation*. Cambridge: Cambridge University Press.

Swart, R., Berk, M. M., Janssen, M., Kreileman, E. and Leemans, R. (1998) The safe landing analysis: risks and trade-offs in climate change. In J. Alcamo, R. Leemans and E. Kreileman, eds., *Global Change Sceanrios of the 21st century. Results from the IMAGE 2.1 Model*. London: Pergamon and Elsevier Science, pp. 193–218.

Toth, F. L., Mwandosya, M., Carraro, C., Christensen, J., Edmonds, J., Flannery, B., Gay-Garcia, C., Lee, Hoesung, Meyer-Abich, K. M., Nikitina, E., Rahman, A., Richels, R., Ye, R., Villavicencio, A., Wake, Y. and Weyant, J. (2001) Decision-making frameworks. In B. Metz, O. Davidson, R. Swart and J. Pan, eds., *Climate Change 2001: Mitigation*. Cambridge: Cambridge University Press.

USEPA (2000) *Developing Country Case Studies: Integrated Strategies for Air Pollution and Greenhouse Mitigation*. Colorado: National Renewable Energy Laboratory.

Vellinga, P. and Swart, R. (1991) The greenhouse marathon: a proposal for a global strategy. *Climatic Change*, **18**, 7–12.

Watts, W. (1999) *Discounting and Sustainability*. Brussels: European Commission.

Weitzman, M. (1998) *Gamma Discounting for Global Warming*. Discussion Paper. Harvard: Harvard University Press.

Wigley, T. M. L. (2001) Stabilization of CO_2 and other greenhouse gas concentrations. *Climatic Change.*

Wigley, T. M. L., Richels, R. and Edmonds, J. A. (1996) Economic and environmental choices in the stabilization of atmospheric CO_2 concentrations. *Nature*, **379**, 240–43.

Wilhite, H., Shove, E., Lutzenhiser, L. and Kempton, W. (2000) The legacy of twenty years of energy demand management: we know more about individual behaviour but next to nothing about demand. In E. Jochem, J. Sathay and D. Bouille, eds., *Society, Behaviour and Climate Change Mitigation.* Dordrecht: Kluwer Academic.

9

Mitigation measures: technologies, practices, barriers, and policy instruments

9.1   Scope for greenhouse gas emissions reduction technologies

9.1.1   Introduction

In Chapter 8, we outlined concepts and methods relating to climate change mitigation. One approach to climate mitigation, as adopted by the Intergovernmental Panel on Climate Change (IPCC) (Anderson *et al.* 2001; Bashmakov *et al.* 2001; Kauppi *et al.* 2001; Moomaw *et al.* 2001; Sathaye *et al.* 2001), is followed in this chapter. Currently, many technologies exist that can help in mitigating greenhouse gas emissions (see Section 9.1.2), but there are barriers that prevent them from being implemented fully (see Section 9.1.3). There are also opportunities to implement such technologies, and where barriers exist, they may be removed by generic or targeted policies, measures and instruments (Jepma & Munasinghe 1998).

9.1.2   Buildings and settlements

The building sector contributes almost one-third to global energy-related carbon dioxide emissions. This section describes the many options for the reduction of greenhouse gas emissions from buildings and settlements. They include options to increase the energy efficiency of appliances, equipment, and building structures. Because of population and income growth, as well as other social changes such as decrease of household size in industrialized countries, energy use related to buildings and the associated emissions is growing rapidly. Energy demand in this sector after 1990 has grown by about 2.5 per cent per year, notwithstanding efficiency increases for equipment and appliances. At the same time, the buildings sector offers some of the most cost-effective opportunities for reductions of greenhouse gases, using currently available technologies and practices.

The most significant greenhouse gas for this sector is carbon dioxide, while in developing countries, inefficient cooking stoves emit methane, and in cooling and heating equipment halocarbons such as hydrofluorocarbons (HFCs) play a role. The latter are discussed in Section 9.1.8. Mitigation options mentioned by the IPCC include (Moomaw *et al.* 2001):

1. *Use of the most energy-efficient equipment and appliances available.* The energy intensity of appliances such as refrigerators, washing machines, dishwashers, air-conditioning, heating and ventilation systems, lamps, etc. has increased significantly over the last decades. The benefit in terms of greenhouse gas emissions has been limited by a tendency towards bigger and/or more powerful appliances, and by the addition of new types of appliances such as clothes dryers and computers, among many others. For individual appliances, energy efficiency increases of 30–40 per cent are currently possible (in a range of 10–70 per cent). In developing countries, improved biomass and woodburning stoves can decrease greenhouse gas emissions.

2. *Integrated building design.* This option capitalizes on the opportunities provided by building siting and interactions between building components. For example, insulation of windows (double- or even triple-glazed), walls, and roofs allows for reduced size and capacity and plumbing for space heating and cooling equipment. Efficiency increases from more than 10 to more than 70 per cent have been reported.

3. *Reducing standby losses.* Many devices used in households continue to use a significant amount of energy while in standby mode (estimated from 5 per cent in the USA to 12 per cent in Japan). From limiting appliances to be in the standby mode, through technological advances these losses can be limited to 1 W, significantly reducing energy consumption. If, in the USA, this 1 W target would be met, US $2 billion would be saved annually on the US energy bill.

4. *Solar energy for buildings.* In many circumstances, solar water heating can be applied, replacing fossil-fuel heating. In rural areas without gridded electricity, photovoltaic systems can be attractive alternatives. In areas with a central electricity network, excess energy from such systems could be fed into the grid. Although prices are currently high, they are decreasing and sales are increasing rapidly around the world.

5. *Distributed power generation for buildings.* Relatively small electricity generating systems installed in relative proximity of building sites reduce transport losses and provide opportunities for combined heat and power, and better co-ordination of demand and supply. These advantages more

than compensate for the lower conversion efficiency as compared to large centralized plants. They can be powered by renewables (e.g. photovoltaic systems) or fossil fuels, e.g. microturbines.

In general, the options above apply to both the residential and commercial buildings sectors. In addition to these primarily technological solutions, changes in behaviour by households and firms can lead to energy conservation and reduced greenhouse gas emissions from buildings. These options are discussed in Section 9.3.

9.1.3   Transport

Road transport

The economic sector with the most rapidly rising greenhouse gas emissions worldwide is transportation, and already contributes more than 20 per cent to global energy-related carbon dioxide emissions. The most important is road transport (73 per cent of the sector's contribution by 1996). The numbers of motor vehicles in the USA has been growing by 2.5 per cent per year since 1970, and almost twice as fast in the rest of the world (4.8 per cent). While, in the 1980s, the average fuel efficiency of passenger cars has declined significantly, it has stabilized now (see Figure 9.1). This is not the result of a slowing down of technological energy-efficiency improvements; on the contrary, those have been very rapid. However, this development was counteracted by the fact that people purchased heavier and more powerful vehicles. Because of the almost full reliance of road transport on liquid fossil fuels and an infrastructure fully designed for this purpose, limiting emissions of carbon dioxide and other greenhouse gases[1] in the transport sector is a major challenge. The availability of relatively low-cost liquid fossil fuels and the rising need for greater mobility will accelerate demand for transportation services and associated infrastructure.

From the perspective of global warming, these trends may be bad news, but fortunately there is also good news from the technology front. Future emissions could be reduced by both efficiency improvements and fuel shifts. Table 9.1 shows greenhouse gas emissions per kilometre for a series of technologies. Moomaw *et al.* (2001) discuss five categories of improvements.

1. *Hybrid electric vehicles.* Hybrid cars have been introduced into the market in various industrialized countries, e.g. the Toyota Prius in Japan and the USA, the Honda Insight. These vehicles combine an internal combustion engine with an electric drive unit and battery. Features can include the recapture of braking energy, downsized engine because the battery

---

[1] Cars equipped with catalytic convertors produce nitrous oxide; methane can result from incomplete combustion.

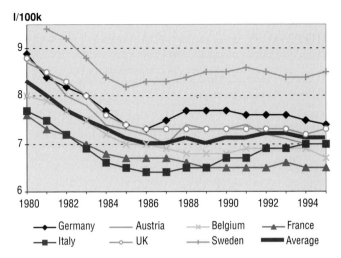

Figure 9.1. Weighted average fuel consumption of new passenger cars. *Source:* Moomaw *et al.* (2001).

can be used as a power booster, avoidance of idling losses, and increased efficiency by avoiding low-efficiency modes using the electric drive unit. Efficiency increases can be as high as 50 per cent at relatively high speeds and more than 100 per cent for slow city traffic.

2.   *Lower-weight structural materials.* In the past, expensive production techniques, and features such as low repairability and low-weight vehicles of impaired safety, were commonplace, but recent developments using plastics and aluminium rather than steel have demonstrated the feasibility of weight reductions of 20–30 per cent or more, with associated efficiency improvements.

3.   *Direct injection petrol and diesel engines.* Direct injection lean-burn petrol engines are already available in Japan and Europe, where they can lead to efficiency increases of between 12 and 20 per cent. For truck diesel engines, the efficiency gains of direct injection engines are even greater, but the cost increases are considerable.

4.   *Automotive fuel cells.* Manufacturers in various regions have announced the introduction of fuel cell cars around 2005, 10 years earlier than estimated only 5 years ago. Evidently, the advantages in terms of greenhouse gas emissions depend on the fuel used to drive the fuel cell. While hydrogen would be the cleanest and most efficient fuel, this would require the build-up of a completely new, or parallel, distribution infrastructure. As Table 9.1 shows, even petrol-powered fuel cell cars may achieve a 50 per cent greenhouse gas emissions reduction, with even higher numbers for methanol.

Table 9.1. *Greenhouse gas emissions from advanced vehicle technologies and alternative fuels.* Source: *Moomaw et al. (2001).*

| | Carbon dioxide equivalent g km^{-1} | | | | | | |
| | Fuel cycle stage | | | Greenhouse gas | | | |
	Feedstock	Fuel	Operation	Carbon dioxide	Nitrous oxide	Methane	Total
Petrol (reformulated)	15.6	52.7	228.9	282.2	5.7	9.4	295.6
Petrol direct injection (DI)	12.6	42.1	184.3	225.6	5.7	7.7	237.6
Propane (from natural gas)	19.0	13.6	197.6	217.5	5.5	7.3	228.9
Compressed natural gas (CNG)	30.7	21.3	174.6	206.2	3.1	17.3	225.3
Diesel DI	10.6	27.2	161.6	191.7	3.3	4.5	198.4
20% biodiesel DI	11.7	32.7	132.7	169.1	3.7	4.3	176.1
Grid-hybrid (RFG)	9.8	63.5	88.8	152.7	4.1	5.3	161.2
Hybrid (RFG)	8.6	27.5	123.3	148.3	5.7	5.4	158.5
Electric vehicle (EV, US mix)	12.3	145.2	0.0	152.1	0.6	4.8	156.6
Fuel cell (petrol)	7.8	26.1	112.6	140.8	1.4	4.4	145.7
Hybrid (CNG)	19.1	13.2	110.5	127.6	2.7	12.5	142.0
Fuel cell (methanol NG)	8.1	17.9	83.1	105.0	1.2	3.0	108.5
Fuel cell (hydrogen from methane)	11.0	97.3	0.0	103.1	0.2	5.0	107.6
Electric vehicle (CA mix)	10.4	51.1	0.0	58.5	0.2	2.8	61.1
Fuel cell (solar)	0.0	20.3	0.0	18.9	0.2	1.2	20.2

100 yr global warming potentials

Carbon dioxide	Nitrous oxide	Methane
1	310	21

5. *Use of biofuels.* If biofuels are produced sustainably, using them to power cars – possibly blended with petrol – would have a great advantage over fully petrol-powered cars in terms of greenhouse gas emissions. Low oil prices have prevented penetration of these fuels in most parts of the world, and government-stimulated programmes for biofuel usages (e.g. the Brazilian ethanol programme) have been decreased in recent years. This does not exclude a shift to biofuels in the future for transportation purposes. In many medium- to long-term greenhouse gas mitigation scenarios, the use of biofuels plays a key role in the transition to a low-carbon energy system. Regions like Latin America, eastern Europe, and Africa may become major biofuel exporters. In addition, biofuels production could be considered for processing on abandoned agricultural lands in Europe.

Many of the above options also are relevant for truck freight, most often powered by diesel engines. The total efficiency gains for trucks, combining all options, could be 60 per cent over current levels (Moomaw *et al.* 2001).

The above options are all addressing the link between transport and its environmental consequences from the perspective of vehicle technology. However, if one would look at the issue from a perspective beyond vehicle technology, a wider set of options can be identified (see, for example, Nijkamp 2001). Transportation, or mobility, has important spatial implications as well as social and behavioural aspects. Spatial organization, transport activities, infrastructure, and ecosystems are closely interlinked. Furthermore, transport is not a goal in itself, it is a 'derived demand' (Nijkamp 2001), because people go to work, pursue recreation, buy food and amenities, and visit family and friends. A wider set of options therefore includes:

1. *Delinking transport volumes and economic development.* This could be achieved by reducing home–work distances, land-use management strategies (e.g. parking area regulations, restrictions on low-density urban design), investment in transit, pedestrian and bicycle infrastructure, or pursuing more regionalized production systems.
2. *Development of new logistic systems.* This could involve not only stimulating higher occupancy rates of vehicles or shared car ownership, and increasing load factors of trucks, but also shifts from road to rail and boat freight transport.
3. *Promotion of substitutes for physical transport.* This category of options includes the, currently, largely untapped possibilities for using information and communication technologies to avoid transportation, e.g. teleworking, teleconferencing, and teleshopping.

For developing regions, some special options may be relevant. For example, paving roads and more regular vehicle maintenance can reduce energy consumption and greenhouse gas emissions. Most of the above options have important ancillary benefits for non-climate problems and, hence, contribute to sustainable development. In fact, in many if not most cases, these ancillary benefits will be the prime movers of policies. Examples are: (a) reduction of local and urban air pollution (sulphur, hydrocarbons, soot, carbon monoxide, etc.); (b) reduction of regional air pollution (e.g. acidifying substances) and (c) noise reduction, enhancement of transport safety, addressing traffic congestion, and reducing road damage (see, for example, Barker *et al.* 2001). According to the Organization for Economic Co-operation and Development (OECD 1999, quoted in Nijkamp 2001), the non-internalized environmental and social costs of transport are estimated to amount to at least 5 per cent of gross domestic product (GDP) for industrialized countries. Many of the options above would reduce these costs, and many may be no-regrets options.

Other transport modes

Other important transport modes are air, rail, and shipping transport. Of these, air traffic – with 12 per cent of world transport energy use by 1996 – is the most rapidly growing sector. Notwithstanding many recent improvements, there are several important opportunities to improve the energy efficiency of aircraft by 40 per cent or more, such as engine characteristics, advanced aerodynamics, and material substitution (Moomaw *et al.* 2001). These technologies could have a major impact on aircraft energy consumption and greenhouse gas emissions over the next decades. There is less optimism about fuel shifts. Technological break-throughs would be needed to make alternative fuels competitive with traditional jet fuel. Options for the future are liquid methane and hydrogen produced from nuclear or renewable energy sources (Penner *et al.* 1999). As for road traffic, there are also non-technological options to reduce the contribution of air traffic to green-house gas emissions. These include increased load factors, reduced airport congestion, slowing growth via market instruments (e.g. taxes), and enhanced procedures for take off and landing. Penner *et al.* (2001) specifically mention improvements in air traffic management and other operational procedures (reduction in fuel burn 8–18 per cent), as well as a series of regulatory, economic, and other options. The latter include: (a) more stringent aircraft engine emissions regulations; (b) removal of subsidies and incentives that have negative environmental consequences; (c) environmental levies; (d) emissions trading; (e) voluntary agreements; (f) research programmes, and (g) substitution of air transport by rail and coach.

For marine and river transport, diesel-powered ships are the most important. Moomaw *et al.* (2001) quote significant improvements in performance: thermal efficiency or marine propulsion (5–10 per cent), propeller design and maintenance (2–8 per cent), hydraulic drag reduction (10 per cent), ship size, speed, increased load factors, and new propulsion systems (12–64 per cent). For rail transport, it is interesting to note that conventional trains are the most efficient mode of mass transportation, but the new high-speed trains are much less energy-efficient, while technological progress is slower than in road transport. In the future, the gap in energy efficiency between road and rail transport may narrow significantly.

9.1.4   Industry

Again, in the manufacturing sector, carbon dioxide from energy combustion is the most important source of greenhouse gas emissions. Industry contributes more than 40 per cent to global carbon emissions. In some industries (e.g. cement production, refineries, and ammonia, steel and aluminum production) carbon emissions can be better qualified as process emissions. In addition, some specific industries emit nitrous oxide, and ozone-depleting substances or their replacements, and sulphur hexafluoride (see Section 9.1.8). Although in the last twenty years, many advances have been made in energy efficiency improvements in industry, industry is still a major contributor to global greenhouse gas emissions (about one-third of the greenhouse gases covered by the Kyoto Protocol). Fortunately, there are many remaining options further to increase efficiency and reduce greenhouse gas emissions. Nevertheless, it is difficult to estimate the future potential, because many inter-related factors play a role: (a) the volume of production; (b) the structure of production in the sector, and (c) the efficiency of the manufacturing processes (Moomaw *et al.* 2001). It is not clear at all how the sector will evolve in the future. For example, within the sector, there appears to be a structural shift to more energy intensive subsectors. However, energy intensity has declined per unit of product or per unit of value added. From a broader perspective, after the oil crisis of the 1970s, there has been some delinking between material throughput and associated greenhouse gas emissions on the one hand, and economic growth on the other (this structural change of the economy towards a more service-oriented society is sometimes also called 'dematerialization'). More recently, however, there are signs in many industrialized countries of a relinking, possibly linked to the above-mentioned structural shift within the sector itself.

There are many options to reduce greenhouse gas emissions from manufacturing. They differ very much per type of industry and per region. Options are largest in developing countries and economies in transition with often outdated equipment. In industrialized countries, the economies of the USA and Australia are less efficient than those in Japan and western Europe, but gains can be made in all regions. Assessments usually focus on the heavy, energy-intensive industry, e.g. oil-refining, production of metals (e.g. iron and steel, aluminium), bulk chemicals (e.g. petrochemicals, fertilizers, chlorine), paper, and cement. There are also many options in the light industries, but these are more difficult to capture because of the very wide range of processes. The IPCC categorizes the options into the following categories (Moomaw *et al.* 2001).

1. *Energy efficiency improvements.* The main way to reduce greenhouse gas emissions in the manufacturing sector is through energy efficiency increases. Table 9.1 gives examples of available options and a qualitative assessment of their potential and costs. Implementation of process control and energy management systems, process integration and cogeneration of heat and power feature as key options in all industry branches.

2. *Fuel switching.* The difference in carbon intensity between similar industries in different countries suggests that a 10–20 per cent carbon dioxide emissions reduction could be achieved by switching between fossil fuels in some circumstances, dependent upon local practical possibilities.

3. *Renewable energy.* Local production of heat and power with renewables can decrease greenhouse gas emissions from industry (see Section 9.1.5).

4. *Carbon dioxide removal.* For large energy-intensive industries, recovery of carbon dioxide from stack gases and disposal in depleted oil or gas reservoirs, aquifers, or the deep ocean, can be a serious option. The understanding of this option has improved significantly over the last few years (see also Section 9.1.5). In some industries (e.g. ammonia production and refineries) carbon dioxide is produced as a by-product of hydrogen production, and here, carbon dioxide recovery is easier than from stack gases.

5. *Material efficiency improvements.* Increasingly it is acknowledged that the production and usage of materials offers interesting opportunities for greenhouse gas emissions reduction: (a) economic use of materials in product design; (b) material substitution; (c) product or material recycling; (d) quality cascading, and (e) good housekeeping. Making producers responsible for each product over a greater part of its lifetime (e.g. the obligation to take it back after usage) can stimulate such options. Integrated materials-energy strategies (including waste management, materials efficiency, material substitution, substitution of petrochemical basis chemicals, end-of-pipe options) could reduce greenhouse gas emissions in western Europe by about one-third (with taxes of €50/t to or below) to two-thirds (with taxes of (€100/t or above), (Kram *et al.* 2001).

6. *Nitrous oxide emissions from industrial processes.* An important source is adipic acid production. Large emissions reductions through thermal or catalytic destruction have been agreed voluntarily by several major producers. Because nitrous oxide concentrations in the waste gases of nitric acid

production – the second important nitrous oxide source – are lower, emissions reduction is somewhat more difficult than in the adipic acid industry. Nevertheless, significant reductions are possible, e.g. through catalytic destruction.

7. *Perfluorocarbon (PFC) emissions from aluminium and semiconductor production.* A major source of PFCs is aluminium smelting, where PFCs are released from the carbon in the anode used for the electrolytic reduction of alumina. Different processes have different emissions and, through new technologies, retrofitting, and improved plant operation, large emission reductions have been achieved already and are expected to continue, often through industry–government collaborative programmes. Since the 1990s, PFCs are used in the semiconductor industry. Options for emission reductions include process optimization, alternative chemicals, recovery and recycling, and effluent abatement.

8. *Hydrofluorocarbon-23 (HFC-23) emissions from hydrochlorofluorocarbon-22 (HCFC-22) production.* Hydrochlorofluorocarbon 22 will only be phased out for consumer uses in the context of the Montreal Protocol in some decades (to 2040), and production as feedstock chemical is expected to continue. Therefore, emissions reduction options are relevant. They include optimization of the HCFC-22 production process in developing countries, and thermal destruction technologies. Other ozone-depleting substances and their replacements are discussed in Section 9.1.8.

9. *Sulphurhexafluoride from magnesium production and insulated switchgear.* Sulphurhexafluoride is a very powerful greenhouse gas and is included in the Kyoto Protocol. It is a very long-lived substance and therefore needs careful attention. It is used primarily in insulating equipment for high-voltage transmission units, where it is a very efficient insulator, allowing for much more compact equipment. In addition to low-emissions technology, a particularly important option is improved handling during installation, maintenance, and decommissioning, including recycling and avoiding release. A less important source is the use as anti-oxidation protective cover applied in the magnesium industry. Currently, the main emissions reduction options for this application are more careful handling and its replacement by the toxic and corrosive sulphur dioxide. For small emission sources (e.g. air in sport shoes and luxury car tyres) such substitution seems unnecessary.

Of the above options, those addressing carbon dioxide and energy consumption generally also have many ancillary benefits, while those addressing non-carbon dioxide greenhouse gases primarily are addressing climate change.

Table 9.2. *Overview of important examples of industrial energy efficiency improvement technologies and indications of associated emissions reductions potentials and costs. The scale is not linear. Costs vary from region to region. Source: Moomaw et al. (2001).*

Sector	Technology	Potential in 2010	Emission reduction costs	Remarks
All industry	Implementation of process control and energy management systems	▮	–	Estimate: 5 per cent saving on primary energy demand worldwide
	Electronic adjustable speed drives	▮	++	In industrial countries ~30 per cent of industrial electricity demand is for electric drive systems
	High-efficiency electric motors	▮	+	Not known for developing countries
	Optimized design of electric drive systems, including low-resistance piping and ducting	▮	+++	
	Process integration, e.g. by applying pinch technology	▮	+	Savings vary per plant from 0 to 40 per cent of fuel demand; costs depend on required retrofit activity
	Cogeneration of heat and power	▮	–	
Food, beverages and tobacco	Application of efficient evaporation processes (dairy, sugar)	▪	+	
	Membrane separation	▪	++	
Textiles	Improved drying systems, e.g. heat recovery	▪	++	
Pulp and paper	Application of continuous digesters (pulping)	▪	+	Applicable to chemical pulping only; energy generally supplied as biofuels
	Heat recovery in thermal mechanical pulping	▪	+++	Energy generally supplied as biofuels
	Incineration of residues (bark, black liquor) for power generation	▪	+	

Sector	Measure		Saving	Note
	Pressing to higher consistency, e.g. by extended nip press (paper making)	■	–	Not applicable to all paper grades
	Improved drying, e.g. impulse drying or condensing belt drying	■	–	Pre-industrial stage; results in a smaller paper machine (all paper grades)
	Reduced air requirements, e.g. by humidity control in paper machine drying hoods	■	+	
	Gas turbine cogeneration (paper making)	■	–	
Refineries	Reflux overhead vapour recompression (distillation)	■	+	
	Staged crude preheat (distillation)	■	+	
	Application of mechanical vacuum pumps (distillation and cracking)	■	+	
	Gas turbine crude preheating (distillation)	■	–	Applicable to 30% of the heat demand of refineries
	Replacement of fluid coking by gasification (cracking)	■	+	
	Power recovery, e.g. at hydrocracker	■	–	
	Improved catalysts (catalytic reforming)	■	+	
	Autothermal reforming	■	–	*
Fertilizers	Efficient carbon dioxide separation, e.g. by using membranes	■	+	* Saving depends strongly on opportunities for process integration of old and new techniques
	Low-pressure ammonia synthesis	■	+	* Site-specific: an optimum has to be found between synthesis pressure, gas volumes to be handled, and reaction speed

(cont.)

Table 9.2. (*cont.*)

Sector	Technology	Potential in 2010	Emission reduction costs	Remarks
Petrochemicals	Mechanical vapour recompression, e.g. for propane/propene splitting	■	+	
	Gas turbine co-generation	■	−	Not yet demonstrated for furnace heating
	De-bottlenecking	■■	−	Estimate: 5 per cent saving on fuel demand
	Improved reactors design, e.g. by applying ceramics or membranes	■	+	Not yet commercial
	Low pressure synthesis for methanol	■	+	* Site-specific: an optimum has to be found between synthesis pressure, gas volumes to be handled, and reaction speed
Other chemicals	Replacement of mercury and diaphragm processes by membrane electrolysis (chlorine)	■	+	* In some countries, e.g. Japan, membrane electrolysis is already the prevailing technology
	Gas turbine co-generation	■■	−	
Iron and steel	Pulverized coal injection up to 40 per cent in the blast furnace (primary steel)	■■	−	Maximum injection rate is still topic of research
	Heat recovery from sinter plants and coke ovens (primary steel)	■	+	
	Recovery of process gas from coke ovens, blast furnaces and basic oxygen furnaces (primary steel)	■■	−	
	Power recovery from blast furnace off-gases (primary steel)	■	+	
	Replacement of open-hearth furnaces by basic oxygen furnaces (primary steel)	■■	−	* Mainly former Soviet Union and China

Table 9.2. (cont.)

	Technology			
	Application of continuous casting and thin slab casting	■	−	*	Replacement of ingot casting
	Efficient production of low-temperature heat (heat recovery from high-temperature processes and co-generation)	■	++		Heat recovery from high temperature processes is technically difficult
	Scrap preheating in electric arc furnaces (secondary steel)	■	+		
	Oxygen and fuel injection in electric arc furnaces (secondary steel)	■	−		
	Efficient ladle preheating	■			
	Second-generation smelt reduction processes (primary steel)	■	−		First commercial units expected after 2005
	Near net-shape casting techniques	■	−		Not yet commercial
Aluminium	Retrofit existing Hall–Héroult process, e.g. alumina point-feeding, computer control	■	−/+		
	Conversion to state-of-the-art Point-feeder, pre-baked technology	■	+		
	Wetable cathode	■	+++		Not yet commercial
	Fluidized bed kilns in Bayer process	■	++		
	Co-generation integrated in Bayer process				
Cement and other non-metallic minerals	Replacement of wet process kilns	■	−/+	*	
	Application of multistage preheaters and precalciners	■	+		No savings expected in retrofit situations

(cont.)

Table 9.2. (cont.)

Sector	Technology	Potential in 2010	Emission reduction costs	Remarks
	Utilization of clinker production waste heat or cogeneration for drying raw materials	■	−	
	Application of high-efficiency classifiers and grinding techniques	■	+	
	Application of regenerative furnaces and improving efficiency of existing furnaces (glass)	■	+	Costs of replacing recuperative furnaces by regenerative furnaces are high (++)
	Tunnel and roller kilns for bricks and ceramic products	■	−	*
Metal-processing and other light industry	Efficient design of buildings, air-conditioning and air-treatment systems, and heat supply systems	■	−	*
	Replacement of electric melters by gas-fired melters (foundries)	■	−	*
	Recuperative burners (foundries)	■	−	*
Cross-sectoral	Heat cascading with other industrial sectors	■	+	
	Waste heat utilization for non-industrial sectors	■	+	

Potential: ■ = 0–10 Mt carbon; ■ = 10–30 Mt carbon; ■ = 30–100 Mt carbon; ■ = > 100 Mt carbon. Annualized costs at discount rate of 10 per cent: − = benefits are larger than the costs; + = US $0–100 per tonne of carbon; ++ = US $100–300 per tonne of carbon; +++ > US $300 per tonne of carbon; An asterisk (∗) indicates that cost data are valid only in regular replacement or expansion.

9.1.5   Energy

A key source of greenhouse gas emissions is the energy sector, notably the production, transport, conversion, and use of fossil fuels. This sector contributes more than one-third to global carbon-derived emissions. The energy efficiency increases of economies, usually in the order of 1 per cent per year, are insufficient to delink greenhouse gas emissions from economic growth. In addition, there are two broad categories of options to reduce the carbon intensity of the economy, and the energy system. First, fuel shifts to sources with lower specific carbon emissions (e.g. from coal to oil to gas) or from fossil fuels to renewable and nuclear energy. Second, fossil fuel use could continue, but the carbon dioxide could be captured after combustion and disposed of underground or in the ocean. Conversion of fossil fuels would be done in the cleanest and most efficient fashion, e.g. 'clean coal technologies'. Sometimes, the second option is called *clean fossil*. A combination of these two options is possible, and even considered likely judging from recent mitigation scenarios (see Chapter 2). Sometimes, it is argued that the timing of a necessary transition in the world energy system is determined by fossil fuel scarcity rather than environmental concerns such as climate change, taking into account the vast reserves of coal, and unconventional oil and gas (see Box 9.1).

Box 9.1   What limits fossil fuel use first: fuel scarcity or carbon
dioxide stabilization goals?

In the political and public debates, it is still often suggested that scarcity of fossil fuel resources, rather than environmental problems such as climate change, are limiting factors for the energy sector. Figure 2.3 and Table 9.3 show that this is incorrect, at least in physical terms: scarcity of fossil fuel resources will not limit carbon emissions in this century. This is caused primarily by the large coal and unconventional gas and oil deposits. Figure 2.3 also shows the carbon associated with the various IPCC scenarios discussed in Chapter 2. The carbon in proven gas and oil reserves (the part of the resources of which exploitation is economically feasible) and conventional oil resources, however, is less than the cumulative carbon that can be emitted under scenarios that lead to stabilization at 450 ppm or above. This does suggest that there is a choice to be made: as these reserves and resources are being depleted, additional resources have to be made available – either coal and unconventional oil and gas, or non-fossil fuels. Both tracks would require significant additional investments. As the IPCC concludes, the choice of energy mix and associated investment will determine whether – and if so, at what level and cost – greenhouse gas emissions can be stabilized (Metz *et al.* 2001). Currently, most research and development is directed as traditionally, towards the first, fossil, track: discovering and developing more conventional and unconventional fossil fuel resources.

Table 9.3. *Aggregate fossil fuel energy occurrences and uranium in exaJoules.* Source: *Moomaw* et al. *(2001).*

	Consumption				Resources	Additional
	1860–1998	1998	Reserves	Resources[a]	base[b]	occurrences
Oil						
Conventional	4 854	132.7	5 899	7 663	13 562	
Unconventional	285	9.2	6 604	15 410	22 014	61 000
Natural gas[c]						
Conventional	2 346	80.2	5 358	11 681	17 179	
Unconventional	33	4.2	8 039	10 802	18 841	16 000
Clathrates						780 000
Coal	5 990	92.2	41 994	100 358	142 351	121 000
Total fossil occurrences	13 508	319.3	69 214	142 980	212 193	992 000
Uranium – once through fuel cycle[d]	1 100	17.5	1 977	5 723	7 700	2 000 000[e]
Uranium – reprocessing & breeding[f]			120 000	342 000	462 000	>120 000 000

[a] Reserves to be discovered or resources to be developed as reserves.

[b] Resources base is the sum of reserves and resources.

[c] Includes natural gas liquids.

[d] Adapted from OECD/NEA and IAEA 2000. Thermal energy values are reactor technology dependent and based on an average thermal energy equivalent of 500 TJ per tonne of uranium. In addition, there are secondary uranium sources such as fissile material from national or utility stockpiles, reprocessing former military materials, and from re-enriched depleted uranium.

[e] Includes uranium from sea-water.

[f] Natural uranium reserves and resources are about 60 times larger if fast-breeder reactors are used (Nakicenovic *et al.* 1996).

We summarize some characteristics of the options in the energy sector from IPCC (Moomaw *et al.* 2001):

1. *Efficiency improvements in fossil fuel power generation.* In general, and particularly in developing countries, there is a large differential between the actual efficiency of power plants and the most advanced technologies available. Average efficiency may shift from 30 to more than 60 per cent efficiency over the next decades, while combined heat and power technologies will reduce energy losses further. For example, advanced pulverized-coal plants can achieve efficiencies beyond 45 per cent. Combined cycle gas turbines, generating electricity through gas as well as waste heat usage, can reach efficiencies of up to 60 per cent and possibly higher ones in the future. Integrated gasification combined cycle technology can use coal or liquid fuels as feedstock and can reach similar efficiencies as combined cycle gas turbines. From the gasification process, carbon dioxide could be recovered. Combined heat and power can reach 90 per cent efficiency levels, and be applied where there is a demand for heat, e.g. in industrial areas or in combination with district heating. Finally, in the longer term, fuel cells are expected to enter niche markets, based on small unit size, potentially low emissions, low maintenance, and less noise.

2. *Shift to biomass.* In many mitigation and stabilization scenarios (see Chapter 2), the use of biofuels is a popular option in the short to medium term to replace fossil fuels in an intermediary stage towards an energy system based on a wider set of renewable fuels. According to many energy analysts, Latin America and Africa especially have a huge potential to become the 'green OPEC' of the world. However, others add words of caution, pointing out the many possible conflicts with food production and nature conservation, as well as other environmental and social problems that could occur if the biomass production does not take place in a sustainable fashion. Bioenergy includes crops specifically produced for energy purposes, agriculture and forestry residues, and fuels derived from waste (water) streams. Bioenergy from wastes in agriculture and forestry especially has good economic potential. Energy crops have higher costs, and recent programmes (e.g. the sugarcane/ethanol programme in Brazil) have not been sustained, due to low oil prices. Commercial biofuels have only been successful in some countries it government incentives (e.g. tax breaks, subsidies, etc.) were implemented. In many developing countries, 'traditional' biomass is the main energy source. Upgrading this to more 'modern' biomass technologies offers potential for significant efficiency

increases, but also requires higher up-front investments. Biomass can be transformed into gaseous or liquid fuel more easily than coal.

3.   *Shift to hydropower.* Presently the most important renewable energy source in the world is hydropower. In industrialized regions it has been developed up to (or possibly sometimes beyond) the limit of social and environmental constraints, e.g. 65 per cent of the technical potential in western Europe, and 76 per cent in the USA. In developing countries, there is currently a large untapped potential, assuming a 40–60 per cent usage of the technical potential. In many regions, there is increasing resistance to large dams, because of: (a) the associated loss of agricultural land, or valuable ecosystems; (b) the ensuing evacuation of local people (e.g. 1.2 million in the case of China's Three Gorges Dam), and sometimes (c) high transmission costs. From the perspective of the greenhouse effect, it is relevant to mention the methane emissions caused by decaying organic matter after filling of dams. Taking these problems into account, it seems more attractive to focus on medium- to small-scale hydropower plants, which often may be more consistent with sustainable development objectives.

4.   *Shift to wind power.* Wind power is at present a small power source from a global perspective, but its application is growing fast in various regions, and costs are declining. In areas with high winds, the option can be competitive with other power sources. The main problem is the intermittent supply, which leads to increased costs if back-up power has to be ensured.

5.   *Shift to solar power.* Solar Power is, in principle, the energy source with the greatest potential. Even the lowest potential estimate exceeds current global energy use by a factor of 4. However: (a) the costs are still very high; (b) there is large variation over space and time, and (c) often large surface areas are needed. Therefore, in most long-term mitigation scenarios, solar energy is only contributing seriously to energy requirements by the second half of the twenty-first century. Although technology is improving costs declining, and application increasing, it is still a minor source of energy at the world level, even as compared to other renewables such as wind, biofuels, and hydropower. Important markets for solar photovoltaics will be in rural areas not linked up to a grid network. Efficiencies are improving gradually, with experimentation in new materials: silicon cells, about 20, thin film technologies, below 6–8, and indium gallium selenide cells, 16–18 per cent. Another important and increasingly applied use for solar energy is in rooftop solar collectors for water heating purposes.

6.   *Shift to nuclear.* One of the more controversial options for mitigating greenhouse gas emissions is the expansion of nuclear energy for the

replacement of fossil fuels. This option is feasible only in countries in which it is not limited by public concerns. Because of largely unresolved problems with safety, waste disposal, and nuclear proliferation in many regions, the public has lost confidence in the technology. However, the majority of current plants are suggested to be competitive on a marginal cost basis in a deregulated market (Moomaw *et al.* 2001). This is not true in places with gas supply infrastructures already in place. Lifetime extension can be an economically attractive option. Reducing waste volumes by reprocessing fuel is possible, but incurs proliferation risks. The future for nuclear power depends on the possibility that an economically competitive technology is developed with inherent safety features, safeguards against proliferation, and a socially acceptable solution for the waste problem. Many of these problems can be addressed by technological solutions, but these have economic consequences. Several new reactor types have such advanced safety features. The exclusion of nuclear energy from the Clean Development Mechanism as agreed at the Conference of the Parties (COP7) in Marrakech suggests that the solution is not yet considered to be consistent with sustainable development objectives.

7. *Carbon capture and disposal.* In the last few years, the understanding of the possibilities of capture of carbon dioxide from fossil fuel combustion and its disposal have increased. It appears that the costs of this option are competitive with many others, so in many stabilization scenarios it plays a significant role. Unlike others, this 'end-of-pipe' option offers the attractive feature that one can continue with current consumption and production patterns based on fossil fuel. If one considers these patterns unsustainable without climate change and disagrees with an end-of-pipe solution that basically generates another waste product, it is not an attractive option. It is therefore no surprise that the option generally is considered positively by the fossil fuel industry, and negatively by some environmental nongovernmental organizations. The carbon dioxide can be separated from waste gases after the combustion of fossil fuels in power plants using solvents, or separated during the gasification of fossil fuels in a catalytic shift process. In the latter case, the concentration of carbon dioxide is higher, so separation is easier. Nevertheless, in both cases there is an energy penalty: the efficiency of power plants goes down. After separation, the carbon dioxide is pressurized and transported to the disposal site, again leading to energy losses. Possibilities for disposal are depleted oil and gas reservoirs, or coalfields, underground saline reservoirs, or the deep ocean. In the first case, enhanced oil or gas recovery will lead to lower

costs (or in some cases even benefits). There is confidence that the carbon dioxide can be stored safely for a very long period, both underground and in the ocean. Experience already exists with the former, but for the latter more research is needed. Producing hydrogen from fossil fuel capture and storing the carbon dioxide can be seen as a way to facilitate a transition to a hydrogen economy, possibly in the future fuelled by renewables rather than fossil fuels.

9.1.6   Agriculture

The contribution of agriculture to global warming

There are three ways in which agriculture contributes to greenhouse gas emissions (Moomaw *et al.* 2001). First, through deforestation and land-use changes, and through fossil fuel consumption on farms,[2] agriculture contributes between 21 and 25 per cent of global carbon dioxide emissions. Second, various agricultural activities such as rice paddies, land-use changes, biomass burning, enteric fermentation of livestock, and animal waste cause 55–60 per cent of the global emissions of methane. Thirdly, 65–80 per cent of global emissions of nitrous oxide are caused by agriculture, from nitrogenous fertilizers, and animal wastes. Compared to carbon dioxide emissions from fossil fuels, the exact amounts of these emissions are more uncertain. The energy emissions are gradually increasing worldwide on a per hectare basis, but declining on a per tonne of product basis. At the 1998 World Food Summit it was agreed to strive for a halving of the undernourished population of the world by 2015. Although improved access to food is one of the key strategies to achieve this purpose, increasing total production is important, too. From the perspective of sustainable development, one can imagine many trade-offs between food security (with increasing agricultural production) and environmental protection. In general, increased production can be achieved by increases in arable land area and in agricultural productivity. Most future projections assume that the major part of the demand increases will be met by intensity increases. For example, the Earth Summit's goal may require a 4–7-fold increase in energy consumption by the agricultural sector, with associated greenhouse gas emissions. At the same time, fertilizer consumption is very likely to increase in many regions, with not only the associated risk of nitrate pollution of soil and groundwater resources but also emissions of nitrous oxide. Still, significant amounts of natural lands are likely to be converted into arable land,

---

[2] This is only about 3 per cent of total consumer energy demand, yet there are many indirect emissions of greenhouse gases. Emissions from fertilizer production are usually assigned to the industrial sector, and emissions related to the transport of products to markets assigned to the transport sector.

sometimes at the expense of valuable natural ecosystems, e.g. tropical forests. Land-use changes and deforestation are discussed in Section 9.2. Here, we discuss options related to the on-farm energy input as well as emissions of nitrous oxide and methane.

On-farm energy inputs

With reference to the energy input at the farm level, there is a significant difference between the highly intensive agriculture in industrialized countries and the low-input agriculture in the developing countries. In the industrialized countries, the public has become increasingly concerned about animal welfare due to problems in Europe with mad cow disease, pig fever, and foot-and-mouth disease, and the risks of chemical inputs such as fertilizers and pesticides which threaten drinking water quality in many regions. This has led to a trend towards less intensive agriculture, although very slowly. In developing countries, the need to feed a growing number of people, who in many regions enjoy increasing incomes, leads to the need to rapidly boost productivity, in usually the same way as it was pursued in the industrialized countries, i.e. through increased inputs. Nevertheless, there are various options to reduce energy input and carbon emissions from agriculture. Selected examples from Moomaw *et al.* (2001) are:

1. *Enhanced soil carbon uptake.* Strategies to increase soil organic matter and sequester carbon can go hand-in-hand with strategies to control erosion, salinization, land degradation, and desertification, and thus contribute to broader sustainable development objectives.
2. *Conservation tillage.* One important example of enhanced carbon uptake is conservation tillage, which also has energy benefits. Conventional tillage consumes 60 per cent of tractor fuels in industrialized countries. Minimum or zero cultivation techniques can save fuel, conserve soil moisture, and reduce soil erosion, but require greater weed control.
3. *Tractor operation and selection.* Through education of drivers, energy can be saved in tractor operation, e.g. matching machinery to the task, reducing soil compaction, etc.
4. *Irrigation scheduling.* If water would be applied at the best times and places according to local weather and soil conditions, energy for pumping can be saved.
5. *Reduction of post-harvest crop -losses.* Particularly in tropical countries, post-harvest losses of perishable crops are significant (from 10 per cent of cereals to 25 per cent for fruit and animal products). Storage and transport in sealed and cooled spaces can reduce losses, but increase energy consumption.

Some possibilities are uncertain or controversial. For example, greenhouse agriculture can be energy-efficient in terms of energy per unit product if areas are large enough and well controlled, but can also lead to increased energy requirements if these conditions do not apply. To meet world food demand while avoiding further conversion of natural lands, or the environmental problems associated with high-input agriculture or livestock management, sometimes a shift to more fish is suggested. Unfortunately, the potential for growth in world fisheries is very limited. Marine food resources are already being overexploited, causing damage to marine ecosystems and the long-term sustainability of the seafood industry. Aquaculture production reduces that pressure, but at the same time poses new environmental challenges, especially related to natural coastal wetlands that are often converted to aquaculture. A final major uncertainty is the potential role of biotechnology in agriculture, in particular genetic engineering. Biotechnology has the potential to increase yields, reduce chemical input requirements and improve nutritional values. At the same time, risks include decreases in crop diversity, degrading ecosystems by diffusion of pest-resistant genetically modified organisms, and dependency of farmers on agribusiness (Raskin *et al.* 2001).

Methane emissions

The main sources of methane in agriculture are wet rice production and ruminants. There has been much research to increase the understanding of methane production in rice paddies in recent years, but it remains difficult to reduce emissions – which are determined by local weather, soil, and management conditions – from this source. Emissions per hectare can be reduced through intermittent flooding and usage of inorganic, rather than organic, fertilizers. More options are available to reduce emissions from the second source: ruminant livestock, e.g. improved diets for cattle or chemical, antibiotic, or biological additives. Here, it is important to note that improved diets will enhance animal productivity, would lead to lower methane emissions per unit of product, but not necessarily emissions per animal. In different regions, there can be different trends in animal product consumption. In industrialized countries, the growth in demand for animal products has decreased, and can even turn into a decline because of consumer public health concerns. In many developing countries, demand for animal products is expected to increase with increasing incomes (Rosegrant, Agcaoili-Sombilla & Perez 1995). More vegetarian diets, or a shift from red to white meat can affect significantly methane emissions, and also land use. It is an open question to what extent such trends can or should be changed (see also Section 9.3 for a discussion on social and behavioural options to mitigate greenhouse gas emissions).

Nitrous oxide emissions from fertilized soils

Notwithstanding the potential decline in nitrous oxide emissions in industrialized countries due to enhanced effectiveness of nitrogenous fertilizer application and stabilization or decline of the amount of fertilizer applied, the need for productivity increases in the developing countries is generally assumed to dominate and lead to increasing nitrous oxide emissions over this century (Alcamo & Swart 1998). Uncertainties about the actual emissions of nitrous oxide from fertilization are large, varying between 1 to more than 2.5 per cent losses of nitrogen. The uncertainties are caused, among other factors, by the dependence of the emissions of local soil, management, climate conditions, and fertilizer type. Only recently, experts realized that emissions are not limited to the time and place at which the nitrogen is added, but can occur from nitrification and denitrification processes over a long time period and over a larger area, as the nitrogen compounds are transported via groundwater. Nevertheless, researchers are confident that emissions can de reduced, maybe by up to 30 per cent on a global scale, by adjustments of nitrogen fertilizer application techniques, slow-release fertilizers, organic manures, and nitrification inhibitors.

9.1.7 Waste

There are five ways in which waste management affects greenhouse gas emissions (Moomaw *et al.* 2001): (a) methane emissions from landfill; (b) substitution of fossil fuels by waste combustion; (c) material and energy savings in extractive and manufacturing industries by recycling; (d) carbon storage in forests because of decreased demand for virgin paper through recycling, and (e) energy use related to transport of waste.

Developing countries following consumption patterns of the industrialized countries lead to increasing waste volumes. Most of the waste is disposed through landfill sites and open dumps. Due to anaerobic decomposition of organic materials, methane is produced as a component of landfill gas, along with other compounds which, a.o., can cause odour and groundwater pollution problems. A good way to minimize emissions from landfill is to capture the gas and use it for electricity generation. In many circumstances, this can be a cost-effective solution. In the USA, more than half of the landfill methane can be captured profitably – regulations require at least 40 per cent to be captured (Moomaw *et al.* 2001). However, this strategy sometimes may be at odds with another, i.e. reducing landfill waste through recycling. Recycling of paper and other organic waste reduces the amount of landfill gas produced, decreasing the profitability of this option. Nevertheless, in general, recycling is an attractive option to reduce waste streams. According to one study, if the level of recycling in the whole of the USA would equal that of

Seattle, the national greenhouse gas emissions would be reduced by 4 per cent. In the poor countries recycling is very much part of the informal economy.

Another option is the aerobic composting or anaerobic digestion of food and other organic waste, which would have to be separated from other waste. The composted waste materials then can be reused as fertilizer. A particularly attractive option in many circumstances is the anaerobic digestion of manure and wastewater sludge, producing biogas. The biogas can be used to generate electricity or fuel vehicles, replacing fossil fuels. Storage of manure as well as wastewater treatment facilities can cause significant greenhouse gas emissions. A final option to reduce greenhouse gas emissions from waste management discussed here is to improve the efficiency of waste incineration, e.g. by fluidized bed combustion, gasification, or pyrolysis of organic waste (producing combustible fuels), and co-incineration of waste with fossil fuels (Moomaw *et al.* 2001). Waste can thus substitute for fossil fuels as a heat and power source.

The options discussed above take the production of waste as a given. Considering the waste issue from a wider perspective, and taking into account the pressure on the environment that affects sustainable development opportunities, a different possibility is the reduction of primary waste streams. This is the most environmentally sound and cost-effective option. Policies aimed at minimizing waste production could address packaging practices and lifetime product responsibility laws, reduction of material losses in production processes, and even at the household level, e.g. minimizing food waste. In developing countries, much less waste is produced, and focused policies may help to avoid reaching the same level of waste production as in industrialized countries.

9.1.8   *Ozone-depleting substances and their replacements*

A special category of greenhouse gases is that of the ozone-depleting substances. The problems of stratospheric ozone depletion and global warming are interlinked in various ways. There are some links in the physical and chemical dynamics of the atmosphere system, e.g. the increased penetration of ultraviolet radiation in the troposphere due to stratospheric ozone depletion enhances the abundance of hyroxyl radicals, affecting the lifetime of radiatively active greenhouse gases, e.g. methane (Ramaswamy *et al.* 2001). From a mitigation point of view, it is important to note that many of the ozone-depleting substances are also greenhouse gases with a high global warming potential. The phasing out of CFCs and their initial replacements, HCFCs, in order to abate the depletion of the stratospheric ozone layer (see Figure 9.2 for estimates of halocarbon consumption), is thus also very positive from a global warming point of view. However, halocarbon substitutes for HCFCs such as HFCs and PFCs – while not having an ozone depleting component – do have a significant global warming potential, although this is often

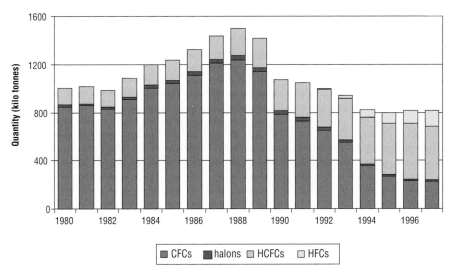

Figure 9.2. The estimated global consumption of CFCs, halons, HCFCs and HFCs. *Source:* Anderson *et al.* (2001).

lower than that of the ozone-depleting substances they replace (Anderson *et al.* 2001). When these HFCs were starting to enter the market in the early 1990s, very high future emissions were projected because they were suggested to replace a large part of the HCFCs to be phased out, and in addition to satisfy the rapidly increasing demand for services requiring such substances, e.g. cooling, insulation, and foam-blowing, etc. More recently, experts consider these early projections to be overestimates because, in many of the applications, other alternative substances and practices would dominate the HCFC replacement. Such assessments do not take into account possible new uses of the substances. Because of the uncertainties with reference to the future emissions of these compounds, the potential to mitigate them should be evaluated carefully. Hydrofluorocarbons, PFCs and sulphurhexafluoride are compounds included in the Kyoto Protocol. It should be noted that there is a time-lag between the consumption of halocarbons and their emissions, because in many applications the major part of the substances will only be released at the end of the lifetime of the equipment in which they are used.

The most important source of HFCs are in cooling and heating applications: refrigeration, air-conditioning and heat pumps. According to Anderson *et al.* (2001), lifetime emissions of HFCs can be reduced through improved design of devices, tighter components to reduce leakage, and recovery and recycling during servicing and disposal, at moderate to low costs. In addition, the emissions can be reduced by using substitutes for halocarbons (e.g. hydrocarbons, ammonia, and carbon dioxide) or completely different technologies. In this area, it is important to note that in order to estimate carefully the benefits of a particular solution in terms of global warming, a life-cycle analysis is needed that also takes into account the energy efficiency of the alternative solutions. It has been claimed that these benefits would be reduced because of an efficiency penalty if using solutions other than halocarbons. It should be noted that some of the applications in this area currently show a very rapid growth, notably automobile air-conditioning. Other applications are in refrigerators and freezers, residential and commercial air-conditioning and heating, food-processing, cold storage, and transport.

Similar points could be made for the second important source: the blowing of insulating foam. There are various alternative compounds to avoid the use of HFCs (e.g. pentanes and carbon dioxide) or one may opt for non-foam insulating options such as mineral fibres and vacuum panels. Furthermore, in this case the alternative solutions may have a insulating value different from the HFC-blown foam solution (Anderson *et al.* 2001).

Finally, sources of HFCs and PFCs are solvents and cleaning agents, medical and other aerosol products, fire protection, and non-insulating foams. Again, similar solutions are available for these applications: increased containment, recovery, destruction, and substitution by non-fluorocarbon fluids and 'not-in-kind' technologies (Anderson *et al.* 2001). An exception is formed by some medical and fire protection applications, e.g. metered dose inhalers for the 300 million asthma patients around the world, and fire protection equipment in aircraft and military vehicles, respectively. Dry powder inhalers are a HFC-free alternative which, however, are not satisfactory to all patients.

Sulphurhexafluoride is a compound with a very high global warming potential (see Table 1.1). It is emitted primarily from the production and usage of insulation for high-voltage electricity (75 per cent) and from magnesium production (7 per cent). Options to reduce emissions from electrical insulation applications include improved production, maintenance, handling, testing and decommissioning procedures, as well as the adoption of low-emissions technologies (Moomaw *et al.* 2001). Emissions from magnesium production can be reduced by switching (back) to sulphur dioxide, accompanied by upgrading of handling of the gas, and operational changes. There are many smaller applications, e.g. in sports shoes.

9.1.9   Geo-engineering

A rather controversial issue in the climate change debate is geo-engineering, defined as the effort to stabilize the climate system by managing the energy balance of the Earth, thereby overcoming the enhanced greenhouse effect (Kauppi *et al.* 2001). As such, geo-engineering addresses the symptoms rather than the causes of the problem. Sometimes, other activities are grouped under the term 'geo-engineering', e.g. efforts to enhance biological uptake of carbon dioxide in oceans and freshwater reservoirs, or even, in terrestrial systems, its capture from fossil fuel combustion, and storage. The latter two options are discussed in sections 9.1.5 and 9.2, respectively. As for the ocean uptake, it is important to note the differences with terrestrial systems: while the net primary productivity of marine and terrestrial ecosystems are similar (50 and 60 gigatonnes (Gt) of carbon per year, respectively), the marine biota contain only 3 Gt carbon versus 2500 Gt carbon in the terrestrial systems, the latter characterized by storage in dead and living vegetation, and soils. It has been proposed to upgrade the 'biological pump' – which moves carbon through the small biotic marine reservoir into the large abiotic reservoir of inorganic carbon dissolved in the oceans – by adding limiting micronutrients (e.g. iron) where the ocean productivity is limited by shortage of such micronutrients, e.g. in the Southern Ocean. Indeed, both model and small-scale experiments have confirmed this possibility. However, the feasibility of this approach depends on the likelihood that the process will require iron to be added over long periods of time, and also on the possible longer-term consequences, e.g. for local and regional ecosystems that have not yet been investigated fully.

The term 'geo-engineering' is most often associated with futuristic proposals to influence the energy balance of the Earth by reflecting incoming solar radiation back into space. Space mirrors, artificial aerosols (e.g. alumina particles), and reflecting balloons have been proposed. The background to the latter proposals is that the effect of a carbon dioxide doubling could be offset by increasing the albedo of the Earth by only 1.5–2 per cent (Kauppi *et al.* 2001). Aluminium scatterers and balloons have been suggested as solutions that would have low costs as compared to other mitigation options, and lower risks than the geo-engineering solutions proposed earlier. However, it should be noted that the ways such techniques would influence the radiative balance of the Earth would differ fundamentally from those of the greenhouse gases, and the solutions could have many as yet uncertain environmental, ethical, and even legal, implications.

9.1.10   Synthesis

A main finding of the IPCC *Third Assessment Report* of technological options is that technological advancement has been rapid since earlier assessments. Examples were given in Sections 9.1.2–9.1.9. Although no comprehensive global assessment has been done, Moomaw *et al.* (2001) give their expert opinion on the basis of a large number of partial (sectoral and regional) studies (see Table 9.4). This table shows a large potential, about half of which is related to energy efficiency improvements in buildings, transport and industry. Half of the potential could also be achieved by 2020 with net economic benefits, profits by energy saving exceeding capital, operating, and maintenance costs. However, here some

Table 9.4. *Estimates of potential global greenhouse gas emissions reductions in 2010 and 2020. Source: Metz et al. (2001).*

Sector		Historic emissions in 1990 (Mt carbon equivalent per year)	Historic carbon equivalent annual growth rate in 1990–95 (per cent)	Potential emission reductions in 2010 (Mt carbon equivalent per year)	Potential emission reductions in 2020 (Mt carbon equivalent per year)	Net direct costs per tonne of carbon avoided
Buildings[a]	Carbon dioxide only	1650	1.0	700–750	1000–1100	Most reductions are available at negative net direct costs
Transport	Carbon dioxide only	1080	2.4	100–300	300–700	Most studies indicate net direct costs less than US $25/t carbon but two suggest net direct costs will exceed US $50/t carbon
Industry	Carbon dioxide only	2300	0.4			
energy efficiency				300–500	700–900	More than half available at net negative direct costs
material efficiency				±200	±600	Costs are uncertain
Industry	Non-carbon dioxide gases	170		±100	±100	N$_2$O emissions reduction costs are US $0–10/t carbon equivalent
Agriculture[b]	Carbon dioxide only	210				Most reductions will cost between US $0–100/t carbon equivalent with limited opportunities for negative net direct cost options
	Non-carbon dioxide gases	1250–2800	n.a.	150–300	350–750	
Waste[b]	Methane only	240	1.0	±200	±200	About 75 per cent of the savings as methane recovery from landfills at net negative direct cost; 25 per cent at a cost of US $20/t carbon equivalent

					Comments
Montreal Protocol replacement applications		0	n.a.	±100	n.a.	About half of reductions due to difference in study baseline and SRES baseline values. Remaining half of the reductions available at net direct costs below US $200/t carbon equivalent
Non-carbon dioxide gases						
Energy supply and conversion[c]	Carbon dioxide only	(1620)	1.5	50–150	350–700	Limited net negative direct cost options exist; many options are available for less than US $100/t carbon equivalent
Total		6900–8400[d]		1900–2600[e]	3600–5050[e]	

[a] Buildings include appliances, buildings, and the building shell.

[b] The range for agriculture is mainly caused by large uncertainties about methane, nitrous oxide and soil-related emissions of carbon dioxide. Waste is dominated by landfill methane and the other sectors could be estimated with more precision as they are dominated by fossil carbon dioxide.

[c] Included in sector values above. Reductions include electricity generation options only (fuel switching to gas/nuclear, carbon dioxide capture and storage, improved power station efficiencies, and renewables).

[d] Total includes all sectors reviewed in Chapter 3 for all six gases. It excludes non-energy-related sources of carbon dioxide (cement production, 160 Mt carbon; gas flaring, 60 Mt carbon; and land use change, 600–1400 Mt carbon) and energy used for conversion of fuels in the end-use sector totals (630 Mt carbon). Note that forestry emissions and their carbon sink mitigation options are not included.

[e] The baseline SRES scenarios (for six gases included in the Kyoto Protocol) project a range of emissions of 11 500–14 000 Mt carbon equivalent for 2010 and of 12 000–16 000 Mt carbon equivalent for 2020. The emissions-reduction estimates are most compatible with baseline emissions trends in the SRES B2 scenario. The potential reductions take into account regular turnover of capital stock. They are not limited to cost-effective options, but exclude options with costs above US $100/t carbon equivalent (except for Montreal Protocol gases) or options that will not be adopted through the use of generally accepted policies.

disclaimers are in order. First, a discount rate of 5–12 per cent has been used in most of the studies used to arrive at this potential, consistent with the range of discount rates used in the public sector, but below rates often used in the private sector. Second, transaction costs have generally not been comprehensively incorporated, nor have the macroeconomic costs of large-scale introduction of these technologies. Third – and important from the sustainable development point of view – ancillary benefits (e.g. in terms of abatement of local air pollution) are generally not taken into account in arriving at these cost estimates. With the same type of assumptions, the other half of the potential could be achieved at costs below US $100 per tonne of carbon. A fuller discussion of the costs (and ancillary benefits) of mitigative options can be found in Chapter 10.

9.2   Scope for biological greenhouse gas mitigation methods

9.2.1   *Potential of biological options*

Biological mitigation options here relate to the potential of the terrestrial biosphere to sequester carbon from the atmosphere and store it, e.g. in forests or agricultural soils, often in this context called 'carbon sinks'. The IPCC estimates that about 100 Gt of cabon can be sequestered cumulatively by biological mitigation options over the next half century (Kauppi *et al.* 2001). This accounts for about 10–20 per cent of the carbon emissions from the combustion of fossil fuels. Therefore, on the one hand these options should not be considered as *the* solution to the problem, since they can only make a partial and often temporary[3] contribution but, on the other hand, the options deserve special attention because the potential is considerable, especially in particular regions. In addition, there is a very evident link between these options and wider sustainable development strategies, warranting a separate section in this chapter. The sinks issue is a politically controversial one (see Box 9.2).

It is important to note that the above potential is the estimated potential of human-, or management-induced carbon sequestration activities. It does not include the result of the various natural or indirect factors that influence the carbon fluxes to and from the terrestrial biosphere. Currently, the biosphere is believed to be a net sink of carbon, as a result of carbon and nitrogen fertilization (plant growth enhancement because of elevated carbon dioxide concentration levels and nitrogen deposition levels due to human action), land-use

---

[3] Temporary because plants or trees only sequester carbon up to a certain saturation point, while they can also release it again, e.g. through fire, if trees are cut, or soils intensively cultivated.

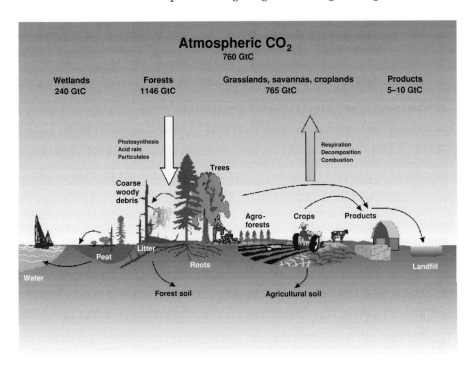

Figure 9.3. Stocks and flows of carbon in different ecosystems, their components, and human activities. Not shown are estimates of carbon stocks in tundra (127 GtC), deserts and semi-deserts (199 GtC) and oceans (approx. 39 000 GtC). *Source:* Kauppi *et al.* (2001).

changes in the past (clear-cutting followed by regrowth, especially in boreal and temperate zones), and climate change. When forests mature and decomposition levels increase with temperature changes in the second half of this century, this sink may decline or even turn into a source (see, for example, Watson *et al.* 2001). This can be counteracted through the management activities referred to above, and which will be discussed in this section. Figure 9.3 gives an overview of the stocks and flows of carbon in the atmosphere and terrestrial biosphere, and provides a framework for the discussion on opportunities for carbon sequestration.

There are three ways of biologically mitigating carbon emissions. First, existing carbon reservoirs (e.g. forests) can be protected to avoid being compromised, e.g. tropical deforestation can be slowed or stopped. Second, existing carbon reservoirs can be expanded or new ones created, e.g. through forest plantations or soil management practices. Third, fossil fuel emissions can be reduced by substituting fossil fuels by biological products e.g. biomass for fuel, or wood as a base material for products. These three options are discussed below.

Box 9.2 Sinks in the Kyoto Protocol

The UNFCCC negotiations related to the inclusion of sinks (or rather land-use and land-use change and forestry (LULUCF)) in the Kyoto Protocol have demonstrated that these options are very controversial from a political perspective. This has not so much to do with the scientific uncertainties with reference to the sequestration potential as such, but with: (a) the ability to measure sequestration adequately; (b) the stability of carbon reservoirs; (c) the possible positive or negative side-effects, and (d) aspects related to sovereignty. First, the uncertainties regarding the exact amount of carbon sequestered by a particular project are generally greater than for a fossil-fuel-related project, and depend very much on definitions, e.g. of 'forest'. Second, for sequestration by trees, for example, there is no guarantee at the start of a project that the trees will not be cut or burnt later, which raises a liability issue. In addition, there is the matter of leakage. For example, a forest conservation programme may avoid local deforestation, but socioeconomic pressure may lead to deforestation elsewhere, which may not have taken place without the conservation programme. Third, pursuing carbon sequestration by forest plantations can have negative effects on biodiversity, e.g. in a situation in which natural forests would be replaced by single-species plantations, or by introducing exotic species. Fourth, if the Kyoto mechanisms – notably the Clean Development Mechanism – would allow sinks, projects established by industrialized countries in developing countries to meet emissions targets could infringe upon the sovereignty of developing countries over their own lands. Including sinks would lower the costs of domestic carbon emissions reductions. Those countries in which there are many carbon sequestration options (i.e. where there is much space or where sequestration is already ongoing or planned for reasons other than climate change mitigation (e.g. in the USA and Australia)), have supported the inclusion of sinks more than those with fewer options and more concerns for the risks, e.g. western European countries.

Rules for LULUCF activities were negotiated in subsequent sessions of the Conference of Parties, not only including definitions for LULUCF activities eligible under the Protocol but also separating sinks into different categories. Firstly, emissions or sequestration from afforestation, reforestation, and deforestation activities are included under Article 3.3. Reporting on sinks under this article is compulsory, and the carbon change resulting from these activities must be accounted for towards emissions reduction targets during the commitment period. Any potential net emissions, which might occur from higher deforestation than afforestation, will be offset through forest management activities, up to a total level of 9 (Mt) of carbon per year for each country for the 5-year commitment period. Secondly, countries agreed in Marrakech that the 'additional activities' under Article 3.4 will include activities such as forest management and agricultural activities. The remaining forest management activities, which have not been used to offset the net emissions from Article 3.3 activities, can be accounted for to help meet emission reduction targets. However, first they have to be subjected to a 85 per cent discount and an individual cap for each party, listed in an appendix in the Marrakech Accords (UNFCCC 2002). The additional agricultural activities, which can be accounted under Article 3.4, are cropland management, revegetation, and grazing land management. Reporting on Article 3.4 is optional: countries have to decide if they apply any or all the activities under Article 3.4. Nevertheless, they have to declare which activities they are going to use at least 2 years prior to the beginning of the commitment period. Finally, sinks have eventually been accepted under the Clean Development Mechanism and Joint Implementation. Under the Clean Development Mechanism, only afforestation and reforestation projects will be considered eligible. Greenhouse gas removals from such projects may be used only to help meet emission targets up to 1 per cent of a party's baseline for each year of the commitment period. On the other hand, both Article 3.3 and Article 3.4 activities are eligible under Joint Implementation, but countries have to include Joint Implementation projects involving forest management in the cap on forest management under Article 3.4.

Conservation and limiting carbon losses

There are three ways of avoiding carbon losses: (a) slowing and halting deforestation; (b) reducing impact logging, and (c) conserving soil carbon in non-forest ecosystems, particularly in agriculture. Deforestation, primarily in the tropics, currently releases between 1.1 and 1.7 Gt of carbon per year to the atmosphere, with a best guess of about 1.6 Gt of carbon per year (Bolin *et al.* 2000). The estimates are very uncertain, as deforestation activities show large variations over time and regions. Detailed studies in countries that take more factors into account tend to give somewhat lower net emissions than large-scale aggregate estimates. There have been several long-term scenario studies with very different outlooks, the difference being caused by different assumptions and methods with reference to the development of the driving forces. There are two groups of long-term scenarios (Alcamo & Swart 1998). According to one perspective, decreasing rates of population growth, continuous increases in agricultural activities, and increased emphasis on forest conservation lead to decreasing rates of deforestation and associated net carbon emissions from now onwards. According to another perspective, continued pressure on the forests (e.g. by logging enterprises and by landless farmers) will lead to increased rates of deforestation in the next decades. The two perspectives converge in the latter half of the twenty-first century, when the exhaustion of forest resources coupled with stabilizing population numbers and increased agricultural productivity would lead to low, or even negative, rates of deforestation and associated low carbon emissions.

The above numbers suggest that there would be a great potential to cut global carbon emissions if deforestation could be reduced, halted, or maybe even reversed. Here, one would have to consider the driving forces, which are very complex, often interlinked, and vary in different regions. They include primary factors such as: (a) population and income growth; (b) economic, political, and institutional conditions (e.g. the external debt problem), and (c) values and attitudes. However, one also should consider derived factors such as increased demand for land for agriculture and livestock management, and for forest products, notably timber and firewood. Understanding these factors at the local level is a prerequisite for effective conservation of tropical forests. In general, conversion of forests for croplands and pasture is a very important factor in many regions (Kauppi et al. 2001). Unsustainable harvest of timber is only a main cause of deforestation in a number of subregional areas, such as South East Asia and some West African countries. However, it should be taken into account that logging companies, which may only have a relatively small direct impact, often pave the way for landless farmers to enter the forest and finish the job. Moreover, fuelwood harvesting and charcoal production have been important causes of deforestation only in a limited number of areas, especially in Africa. Conversion of forests for pasture, related to cattle density, has been a major cause, primarily in Latin America. Furthermore, in many countries, colonization of forest areas is actively stimulated for political reasons, e.g. providing land to poor farmers, pursuing economic development of the full national land area, organized migration of people from overpopulated areas to low-density – and often ecologically fragile – areas, or control and security in sometimes disputed border areas. In many developing countries, deforestation is stimulated by a combination of factors, including: (a) inequalities in access to land and associated landlessness; (b) insecure land tenure; (c) large-scale expansion of commercial, and often export-oriented, agriculture; (d) erosion of traditional systems of resource management and community control, and (e) migration of poor people.

These causes are very similar to those in the industralized regions in the past, when forests were cleared to provide access to lands for farming and livestock management, but also to provide wood for the manufacture of merchandise and construction of warships. Figure 9.4 shows the historic patterns of land-use change over the last two millennia in three main regions. In Europe, decrease of forest lands has changed into an increase over the last few centuries. In North America, this occurred as well, at a somewhat later stage and with a lower minimum. This suggests that, as countries become more affluent, stable and technologically advanced, the reasons for deforestation disappear – similar to the 'technological leapfrogging' discussed in Chapter 4 (Figure 4.5) – thereby allowing developing countries metaphorically to 'tunnel' through the curve of per capita greenhouse gas emissions versus development level. Figure 9.4 suggests that in developing countries there may be important options to avoid the level of forest destruction of the currently industrialized countries. We should add a word of caution here, since one of the factors allowing the industrialized countries to conserve and even expand their forest area is that they import forest produce from the developing countries, an option the developing countries do not have.

Clearly, forest conservation should be viewed from the wider perspective of sustainable development, carbon sequestration being only one of the environmental, social, and economic objectives of land-use management options. This requires a longer-term perspective as well. Evidently, this is difficult to realize for countries eager to earn some money now (e.g. for debt relief) which they can get relatively easily from wood exports, or for landless farmers who do not have alternative options for survival. Nevertheless, the logs of century-old trees can be sold only once, and when the poor forest soils are quickly exhausted, the loggers have to move on.

Conserving forests has many advantages independent of climate change considerations. One of the most important is erosion and flood control through reduced runoff. Not only does decreased erosion have local advantages (e.g. reduced desertification, soil degradation, and water scarcity) but downstream there are also advantages: dams used for hydropower will be less silted, and flooding problems will be reduced. From a biodiversity viewpoint, forest conservation advantages are obvious, since the converted area will usually have lower biodiversity. The main disadvantage is that conserving the forest prevents local people from using it as a resource, forcing them to move to other forest areas unless alternative means to earn a living are provided. In conclusion, we may say that conservation of natural forests generally is the easiest and most cost-effective way of storing carbon.

A second approach to conserving carbon in existing ecosystems is reducing carbon losses during disturbance. A project in Malaysia suggests that as much as 50 per cent of the damage done under normal operating practices can be avoided through so-called 'reduced impact logging'. This includes changes of conventional

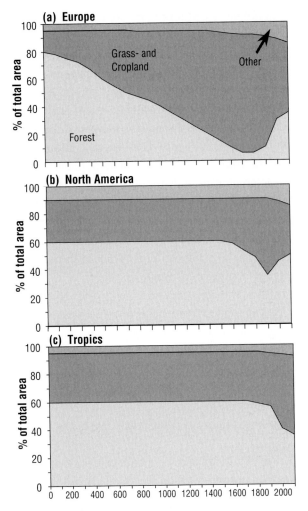

Figure 9.4. Historical changes in land-uses in three world regions. *Source:* Kauppi *et al.* (2001).

bulldozer harvesting practices, directional logging, pre-felling climber cutting, skid trail design, and post-harvest log landing rehabilitation (Kauppi *et al.* 2001).

A third way of conserving carbon in existing ecosystems is the avoidance of carbon losses from agricultural lands, e.g. through reduced tillage or no-tillage techniques. Such techniques not only increase soil carbon and climate benefits, but also improve soil quality and water-holding capacity and productivity while reducing erosion. They could be combined with methods such as crop rotation and effective fertilization practices, to optimize carbon sequestration further.

9.2.2   *Sequestration by reforestation and afforestation*

The second strategy for sequestering carbon is that of adding to the carbon pools rather than protecting existing pools. This can be done through forestry activities as well as via agricultural management techniques. First, forest plantations can be established. Already 61.3 million hectares were under forest plantations worldwide by 1990, and this area is increasing by, on average, 3.2 million hectares each year, balancing part of the carbon losses from deforestation. Detailed studies for major tropical countries in the early 1990s that include forest regeneration and reafforestation, show that net emissions for most countries are lower than those from aggregate estimates (Makundi *et al.* 1998).

Terrestrial systems provide an active mechanism for biological removal of carbon dioxide from the atmosphere. They act as reservoirs of photosynthetically-fixed carbon by storing it in various forms in plant tissues, dead organic material, and soils. Terrestrial systems also provide a flow of harvestable products that not only contain carbon but also compete in the market-place with fossil fuels, construction material (e.g. cement) and other materials (e.g. plastics) that also have implications for the global carbon cycle.

Second, the amount of carbon stored in existing forests can be increased through various management techniques, including: (a) protection against fire, diseases, herbivores, insects, and other pests; (b) changing rotation; (c) controlling stand density; (d) enhancing available nutrients; (e) water table control; (f) species and genotype selection; (g) biotechnology; (h) reducing regeneration delays; (i) reduced-impact logging; (j) managing logging residues; (k) recycling wood products, and (l) improving forest product manufacturing efficiency (Kauppi *et al.* 2001). Such opportunities only exist, in practice, in a small percentage of the total forest area (e.g. 10 per cent), but may be increased in the future. The amounts of carbon sequestered are lower than from forest conservation of forest plantations, but still can be significant in many regions. Figure 9.5a shows that the significance of these options is very different across regions, particularly with respect to the overall national emissions. Figure 9.5b (from the same study, Nabuurs *et al.* 2000) also shows that management practices to achieve these potentials vary in the different countries, emphasizing that it is very difficult to make aggregate estimates of carbon sequestration potentials. It is very difficult to estimate the worldwide potential of these activities, but they would be well below 1 Gt of carbon per year, and probably closer to 0.1 Gt.

From a sustainable development point of view, establishing forest plantations or actively managing forests for maximizing carbon sequestration can have other positive and/or negative side-effects. Plantations can affect biodiversity negatively if they replace biologically rich native grassland, wetland habitats, or natural

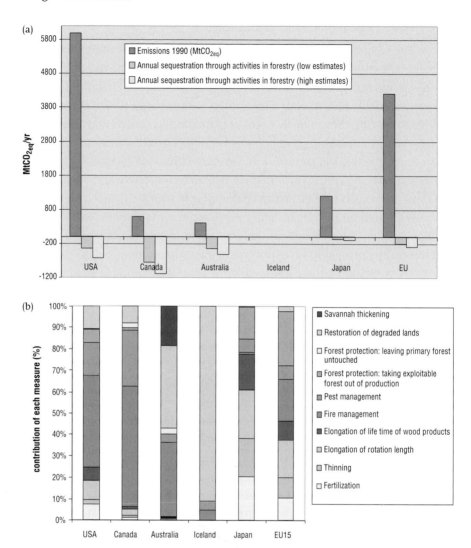

Figure 9.5. Relative size of carbon sequestration opportunities in different countries (a) and the relative importance of each of the 10 forest management techniques to capture them (b). For a colour verion of this graph see the original publication or website. For a colour version of this graph see the original publication or websites http://www.grida.no/climate/ipcc_tar/wg3/fig4-7.htm and http://grida.no/climate/ipcc_tar/wg3/fig4-8.htm. *Source:* Kauppi *et al.* (2001).

forests. In addition, there could be negative social effects, e.g. interference with resource utilization opportunities for local people, displacement of local populations, and reduction of incomes. Utilization of manure, artificial fertilizers and/or pesticides can have negative consequences for local soil and water quality. Forest plantations play three important roles: as (a) physical pools of carbon; (b) substitutes for more energy-intensive materials, and (c) raw materials to generate energy

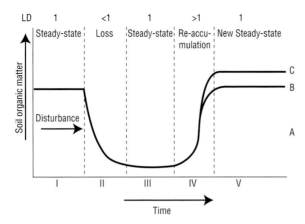

Figure 9.6. Conceptual model of soil organic matter changes following disturbance and changed management practices. The eventual steady state (A, B or C) depends on the new management adopted. *Source: Kauppi et al.* (2001).

(Burschel, Kuerston & Larson 1993; Matthews *et al.* 1996). A positive side-effect of forest plantations is that it reduces the pressure on natural forests, leaving greater areas to provide for biodiversity and other environmental services (Sedjo & Botkin 1997). Forests also help to conserve water resources and prevent flooding.

Agroforestry can both sequester carbon and produce a range of economic, environmental and socioeconomic benefits. For example, trees help to improve soil fertility through control of erosion, maintenance of soil organic matter and physical properties, increased nitrogen availability, extraction of nutrients from deep soil horizons, and promotion of more closed nutrient recycling (Young 1997). Forest cover could also have secondary climate consequences through their feedback in the albedo of the Earth, the hydrological cycle, cloud cover, and the effect of surface roughness on air movements (Garratt 1993).

A third way to expand carbon pools in ecosystems is adding to the carbon in agricultural systems. As natural lands are converted to agriculture, usually carbon is lost as indicated in Figure 9.6 (from equilibrium stage I–III). Management techniques are available that can restore the carbon at various levels, either below (A), equal to (B) or even beyond (C) original levels. Management can imply turning one land-use type into another (e.g. cropland to pasture) or changing techniques for the same type of production, e.g. modified rice paddy management. Reduced tillage reduces the time that organic soil matter decomposes and, hence, contributes to higher soil carbon levels, with often the simultaneous benefit of enhanced productivity. Other options in crop production are enhanced (organic and/or inorganic) fertilization, improved crop varieties, optimized irrigation practices, cover crops, perennial forages in rotation, avoiding bare fallow, and erosion control (Kauppi *et al.* 2001). Agroforestry provides similar options in many circumstances. For pasture, similar options are available: changed plant species, changed grazing intensity, adding nutrients such as phosphorus, fire and erosion control, and irrigation. A very important option from a sustainable development point of view is the restoration of lands degraded by factors such as erosion, over-grazing, and salinization. Such activities only have a chance of success, however, when the basic causes underlying the degradation in the first place are addressed as well. For this category of activities, the global potential has been estimated by various authors, and may be between 0.5–0.8 Gt of carbon per year with a large uncertainty margin.

Substitution of biological products for fossil-fuel or energy-intensive products

Finally, the third main category of options for biological carbon sequestration options is the substitution of biological products such as wood for fossil fuels, or energy intensive products. While the two former categories – conservation and expansion of carbon pools – are limited in time, this third category offers the potential for a continuous contribution to reduced carbon emissions. Wood – and its derivative charcoal – can be used as a renewable source of energy, thereby replacing fossil fuels. It can play this role most effectively if the combustion is as efficient as possible, something that is currently not the case in many applications, e.g. woodstoves for cooking, and brick production, etc. Therefore, an important option is to enhance the efficiency of the combustion process and the production of charcoal. It should be noted, however, that fuelwood combustion often has detrimental effects for local and indoor air quality, and fuelwood supply is increasingly difficult in many regions. Woodburning stoves are therefore often replaced by kerosene stoves if incomes permit. In addition to fuel, wood also can be used as construction material. Options to increase long-term sequestration of carbon in wood products include: (a) the encouragement of production and consumption of wood products; (b) improving their quality and processing efficiency, and (c) enhancing their recycling and reuse (Kauppi *et al.* 2001). Global estimates suggest that the potential of this option is small compared to the other options. It should be realized that much of the carbon in products eventually ends in landfills, which may become carbon storage pools of increasing significance in the future. More important may be the possibility of replacing materials of which the production requires significant amounts of (fossil) energy by wood. This substitution potential may be in the order of a 0.25 Gt carbon per year.

Biological options: their costs, benefits, and linkages with sustainable development

In many circumstances, biological options are economically competitive with others. Costs in tropical countries generally are estimated to be lower than

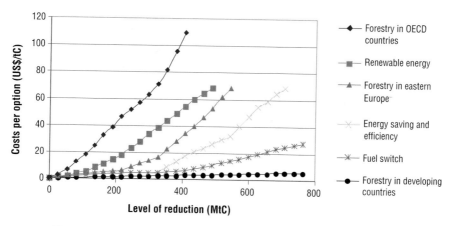

Figure 9.7. Indicative cost curves of forestry and other mitigation options. Note that costs per option are also reported to vary widely at comparable total levels of reduction, mainly because cost studies have not been carried out in the same way. *Source*: Kauppi *et al.* (2001).

in temperate or boreal countries: from US $ 0.1–20 per tonne of carbon in the tropical countries, and from US $ 20–100 elsewhere (Kauppi *et al.* 2001). It should be noted, however, that uncertainties are large. The lower cost estimates are likely to be underestimates, since many cost studies in this area do not take into account, for example, infrastructure costs, appropriate discounting, data collection and interpretation, and opportunity costs. The latter is particularly relevant for large-scale application of biological mitigation, when competition with other land uses becomes increasingly important. Figure 9.7 shows cost curves of forestry options in different regions in comparison with other greenhouse gas mitigation options, suggesting that forestry in tropical countries is an attractive option, but forestry in OECD countries may be more costly than other options. Forestry options in eastern Europe are in-between. Thus, cost estimates may be underestimates, if one addresses purely the carbon sequestration objective, but one may have to take into account that such calculations do not take into consideration that carbon sequestration in practice is likely to be just one of many goals. In other words, land-use change decisions are taken with multiple objectives, carbon sequestration being only one of them. For particular situations, there may be trade-offs between carbon sequestration and other objectives, or no obvious positive or negative interactions, but often there are synergies leading to cost-effective win–win situations. This applies to forestry as well as agricultural options (see Table 9.5 for some examples). In general, by adequately capturing the synergies and avoiding the trade-offs, these options can help, especially in developing countries, to advance development aspirations as well as sequester carbon to mitigate climate change.

Table 9.5. *Examples of synergies and trade-offs between biological mitigation options aimed at enhanced carbon sequestration, and other objectives.*

	Possible trade-offs between carbon sequestration and other objectives	Possible synergies between carbon sequestration and other objectives
Forestry options	• Forest protection versus access to forest for local communities for fuelwood or other forest produce • Carbon uptake enhancement through nutrient and pesticide addition versus soil and groundwater contamination	• Biodiversity protection and carbon sequestration through forest protection • Carbon sequestration in forests or forest plantations, and control of erosion and water resources • Plantations for wood production avoiding loss of natural forests and substitution for fossil-based fuels/materials
Agricultural options	• Intensified management for enhanced soil organic matter versus increased energy use, local water, and soil pollution, e.g. fertilizer, and pesticides • Reduced shifting cultivation versus less available fuelwood from fallow areas	• Enhanced carbon content and crop yields • Reduced tillage increases soil carbon and reduces energy use • Restoration of degraded lands capturing carbon, enhancing productivity, and controlling erosion

9.3   Scope for structural economic changes and behavioural options

The emphasis in Sections 9.1 and 9.2 was on technological options to mitigate greenhouse gas emissions. Many consider such options to be the most important ways to reduce greenhouse gas emissions in a world characterized by continued economic growth and development. However, there are other options directed against the unsustainable consumption patterns in the North and among the élite in the South, as the main reason for increasing pressures on the global environment, e.g. through the enhanced greenhouse effect. These call for changes in lifestyles. The long-term scenarios discussed in Chapter 2 contain elements of both technological options, and structural and lifestyle choices, with some emphasis on the former. The possibilities for, and effectiveness of, economic structural changes and lifestyle adjustments is more difficult to assess than for technological solutions. In most global analyses, the contributions of technological changes are dominant. In addition, in the technocratic societies of today, the feasibility or even desirability of influencing consumption patterns is questioned. A fine line would have to be walked between forcing people into particular types of behaviour (sometimes referred to as 'social engineering') and more cautiously influencing them through information and education programmes. In reality, behavioural, economic and technological changes are intricately interdependent – they cannot be separated. The introduction of information and communication technologies is influencing deeply behavioural patterns worldwide, just as the introduction of agricultural and transportation technologies have influenced them in the past. Consumers choosing 'green electricity' as offered in several European countries, contribute significantly to greenhouse gas mitigation. They may have a slightly higher energy bill, but in general do not change their lifestyles, because they are interested in the services energy provides, not in its source.

Banuri et al. (2001) distinguish between decoupling growth from resource flows, and decoupling wellbeing from production. Decoupling growth from resource flows focuses on reducing the amount of resources needed for a unit of economic output, taking a broader approach than just reducing greenhouse gas emissions from an overall sustainability point of view. Banuri et al. (2001) mention four options, which go beyond merely presenting mitigation technologies, but emphasize choices that combine technological, structural, and behavioural options. First, eco-efficient production systems can be pursued, such as:

1. Eco-efficient innovation, e.g. utilization of biodegradable material, extension of durability, and input minimization.
2. Industrial ecology, from a linear throughput growth to a closed-loop economy.
3. From products to services, shifting the business focus from hardware to service sales.
4. Eco-efficient consumption, reducing waste and pollution.

Second, new approaches to physical planning can lead to 'resource-light' infrastructures. This is particularly relevant for developing countries, in which the infrastructure still has to be developed to a great extent. Striving for compact cities with excellent public transportation systems can avoid the energy-intensive suburbanization of many industrialized countries. Similarly, decentralized electricity in rural areas may have many advantages over centralized systems. Third, appropriate technologies adapted to local circumstances and building on local knowledge can be less resource-intensive than maladapted technologies. Fourth, economic methods such as full-cost pricing, properly taking into account the value of natural resources, can stimulate production with low-intensity resource use.

Even more relevant to the question of to what extent behavioural changes can help in mitigating climate change are the options to decouple wellbeing from production. Research has suggested that there is no direct link between the level of GDP and the quality of life (or satisfaction) beyond certain thresholds (Banuri *et al.* 2001). The GDP may increase without a corresponding increase in human welfare. For example, in the USA the average citizen considers his or her happiness to have decreased since 1957, while consumption has doubled (Toth *et al.* 2001). Expenditures increasingly address protection of what one has (e.g. safety), or compensating for what one may loose (e.g. buying consumer goods for lack of leisure time, and environmental pollution abatement). Banuri *et al.* (2001) distinguish four possible dimensions to exploit possibilities that would decouple wellbeing from production:

1. *Intermediate performance levels*, e.g. avoiding heavy automobiles with powerful engines which cannot be fully exploited because of speed limits anyway.
2. *Regionalization*, e.g. consuming preferentially regionally produced goods to avoid long-distance transportation.
3. *Appropriate lifestyles*, e.g. focusing on non-material satisfaction, vegetarian or beef-poor diets (which would avoid methane emissions and land-use changes related to livestock management), shifts in the modal split in transportation away from private automobiles, and slightly lower heating values in living quarters, etc.
4. *Community resource rights*, e.g. wellbeing derived from access to local resources.

In Section 9.4, we discuss in more detail how these options are inhibited through cultural preferences and habits. 'Lifestyle' can be defined as 'a set of basic attitudes, values, and patterns of behaviour that are common to a social group' (Toth *et al.* 2001). As such, lifestyles are determined by more than only economic rationality as assumed by economic models. Lifestyles can be seen as an expression of individual identity, couched by one's social relationships. In the Western world, individuals consider themselves as more independent of their social environment than in several other cultures, in which the individual considers him or herself more as a part of an interdependent group. Intercultural communication may therefore help to overcome the barrier of consumption-based identity at the expense of others (Toth *et al.* 2001). In principle, there are four combinations of goods: (a) private material goods (e.g. cars); (b) private immaterial goods (e.g. creativity); (c) common material goods (e.g. shared cars or appliances), and (d) common immaterial goods (e.g. a sound environment or art). The emphasis, especially in Western society, is on the first only, e.g. as illustrated by advertising. In many countries, government environmental education programmes aimed at schools as well as the general public try not only to raise awareness of environmental problems such as climate change, but also stimulate more environmentally friendly behaviour. Such programmes may emphasize the other three types of goods (Toth *et al.* 2001). In the real world, however, such options are difficult to implement (see Box 9.3 for an example of a study in The Netherlands).

Box 9.3   Practical options for lifestyle changes in The Netherlands?

In The Netherlands, the possibilities of reducing greenhouse gas emissions through lifestyle changes in households have been studied extensively. One study assessed the potential reduction of greenhouse gas emissions from a number of changes in day-to-day practices:

1. *Food*: eating fewer greenhouse products, less meat, more vegetarian meals, using a shopping by bicycle, using a shopping service, putting the refrigerator/freezer in the cellar, choosing a more efficient refrigerator/freezer, a shift from electric to natural gas cooking, dishwashing by hand, less warm prerinsing, more efficient dishwashers.
2. *Living*: more efficient heating, more efficient warm water provision, lower room temperatures, efficient light bulbs, natural floor covering, lifetime extension furniture (higher quality), art rather than cut flowers for decoration.
3. *Clothing*: replacing synthetic clothing by cotton, life extension shoes (higher quality), less washing, more efficient washing machines, natural drying, more efficient driers, sharing washing machines, lifetime extension washing machines/driers (higher quality).
4. *Miscellaneous*: sharing newspapers/magazines, sharing tool use, sharing camper, no cut flowers as presents, using trains for travelling while on holiday, and not staying in hotels.

Implementing all of these changes would save 27 per cent of the direct and indirect (1990) greenhouse gas emissions. However, analysing the practical feasibility of these options through household surveys revealed that the realistic potential would be closer to 5 per cent, and mainly for those options related to the purchase of energy efficient equipment that would not require real changes in lifestyle. None of the options was considered feasible by all households (Moll *et al.* 2001). This illustrates that the climate problem is not one conducive to changing behaviour: even if people are aware of the serious nature of the risks, there is no clear link between the change in their behaviour and solving the problem. The problem is too large to grasp, and the solutions may only benefit future generations, probably mostly in other regions. Government television advertisements promoting energy efficient behaviour are drowned in commercials advocating energy-intensive lifestyles (fancy cars, electric appliances, etc.). According to an extensive survey, the Dutch consider themselves well-informed, are concerned about environmental problems, feel that the individual can make a difference, display a high self-efficacy, and prioritize ecological concerns over economic gains (Ester & Vinken 2000). However, they do not accept policy choices that limit personal choice.

9.4   Barriers and scope for practical implementation

9.4.1   Introduction: technological change and barriers

In this chapter we have outlined the many technologies and practices known to us today that can reduce global greenhouse gas emissions, in the short to medium term sufficiently to have global emissions peak and start to decline within a few decades, and in the longer term, to continue this decline so as to enable stabilization of carbon dioxide concentrations at 450 ppm or even below. In Chapter 10 we will show that, at least at the aggregate level, costs appear to be modest. The implementation of the Kyoto Protocol, for example, would lead to about 0.2–2 per cent of projected GDP in the OECD countries without emissions trading, and to 0.1–1.1 per cent when full trading in the industrialized countries and economies in transition would be realized (see Chapter 10 for a detailed discussion). These numbers are in the same order of magnitude or lower than the current expenditures on environmental issues in OECD countries. One may expect that for a problem that has been dubbed 'the most serious environmental issue ever', such costs should not be prohibitive and achieve a transition to a low-greenhouse gas economic system. Why, then, is it so difficult to implement the available options? The IPCC has made a systematic assessment of the reasons (Sathaye *et al.* 2001). First, we will recap the conceptual framework, then generically summarize types of barriers, and finally we discuss sector-specific barriers and opportunities in the context of sustainable development. For a more detailed discussion of the barriers, see Sathaye *et al.* (2001).

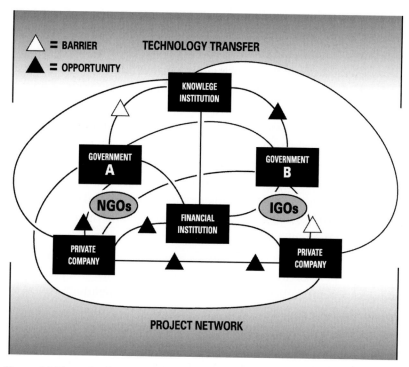

Figure 9.8. The technology innovation system.

In Chapter 8 we presented the often used techno-economic rationality approach to climate mitigation options, in which the key to successful climate change mitigation would be to identify and overcome barriers to the implementation of available mitigation options. This framework is suitable when searching for barriers to technological innovation and change. While in the past, technological innovation and change were seen as a linear process from research and development to wide-scale implementation, more recently one has recognized the complexity of the processes involved. Trindade, Siddiqi & Martinot (2000) see technology development and transfer as part of an 'innovation system' in which different stakeholders (the private sector, governments, academia, civil society) participate (see Figure 9.8). Just as the system is complex in terms of actors, it is also so in terms of the process: rather than a once-through process, there are many interlinkages and feedbacks at various stages. Different actors play different roles in different stages. The process varies also for different types of change and innovation: (a) incremental improvements in particular technologies; (b) radical innovation or radically new technologies; (c) changes in a system of related technologies, or (d) fundamental changes in the underlying techno-economic paradigm with associated changes in production and consumption patterns (Sathaye *et al.* 2001). Some of the processes that play a role in this interconnected system are: (a) the assessment of needs and potential markets; (b) basic research; (c) creation of new ideas; (d) learning from experience; (e) exchange of information and ideas; (f) experimenting and testing; (g) development and demonstration of technologies; and (h) selection of the technology. In this system of partly simultaneous, partly sequential, developments, various feedbacks play a role, e.g. adjustments as the result of experience and evaluation.

Although the dynamic network system may suggest a large potential for rapid change, there are many factors hindering such rapid change. Many technologies are embedded in complex systems of infrastructure and behaviour that have been developed over years. A well-known example is the infrastructure developed around the petrol-fuelled private car, including petrol-distribution systems, design of cities with associated systems of roads and parking areas, as well as the car-dependent social lives to which many people have become accustomed. Other examples are difficulties in introducing fluorescent lightbulbs because existing lamp designs are not adapted to the differently shaped bulbs, and the QWERTY computer keyboard, originally designed to avoid keys sticking in mechanical typewriters. This kind of situation in which inefficient technologies are replaced only with serious difficulties because of existing infrastructure and habits is also called a 'lock-in' effect.

9.4.2   Types of generic barriers

The IPCC distinguishes various types of barriers (see for a more detailed account Sathaye *et al.* 2001):

1. *Unstable or otherwise unfavourable economic conditions.* Such conditions discourage investments in environmentally sound technologies because of the risks involved for the investor. Examples are corruption, unsound borrowing of fiscal policies, and trade barriers. In addition, tied aid can be considered as an impediment to the choice of most efficient and environmentally sound technologies.
2. *Commercial financing institutions.* Often environmentally sound technologies are relatively modest in scale, sometimes have long payback times, and high transaction costs discouraging banks and other funding organizations.
3. *Distorted or incomplete prices.* Most important examples are the usual absence of a market price for externalities such as environmental factors, and tax and subsidy structures encouraging, rather than discouraging, products with high, rather than low, greenhouse gas emissions.

4. *Network externalities.* Often, the successful implementation of environmentally sound technologies requires a whole sequence, or network, of changes in activities posing problems with respect to co-ordination between all involved. This contributes to system inertia, (e.g. in the transport sector) which is dependent on a whole chain of activities and infrastructure, notably petroleum production and distribution systems.

5. *Misplaced incentives.* These play a role if the actor causing the greenhouse gas emission is not responsible for it, e.g. if heating is included as a fixed amount of the rent of a house or apartment, or employees use automobiles leased for them by their employers.

6. *Vested interests.* Climate change mitigation is likely to require a transition in the way we produce and use energy, which is not in the interests of large and powerful companies who would be affected by such a transition. Their influence with governments would favour a conservative approach.

7. *Lack of effective regulatory agencies.* Most countries have good rules and regulations, but there is a lack of institutional capability to implement them, especially in the developing world.

8. *Information.* Information, or rather lack of information, remains an important barrier to the diffusion of environmentally sound technologies. The provision of information about them would be a role of the government.

9. *Current lifestyles, behaviours and consumption patterns.* As discussed in Section 9.4.1, current lifestyles in developed countries are very energy-intensive and hard to change, as they are embedded in the daily lives of people. Relationships between different groups, individual motivations and ambitions, evolving worldviews, scientific, economic, and technological developments and other issues, all play a role, and just one reason – climate change mitigation – is usually not sufficient for changes, even if people acknowledge the problem as being serious.

10. *Uncertainty.* Uncertainty has many faces. Uncertainty not only about the problem to be tackled, but also in the effectiveness and efficiency of available options is an important barrier.

9.4.3   *Sectoral barriers and opportunities to overcome them*

The above barriers hinder penetration of environmentally sound technologies differently in different economic sectors and in different regions (Sathaye *et al.* 2001). In Section 9.5, policies and measures and their pros and cons are discussed in a generic sense. Here, we discuss barriers and opportunities to overcome them at the sectoral level. The scope of this book does not allow for a fully comprehensive discussion of all the options.

In the buildings sector, emissions grow rapidly due to energy consumption for space heating and/or cooling, cooking, appliances, lighting, and water heating. The share of commercial and residential energy consumption is about equal. Lifestyles and customs play important roles. There is a tendency towards larger dwellings and more appliances, while energy efficiency considerations do not play a major role in decision-making of households in these areas. Sometimes the reasons for this are economic: more energy-efficient housing or appliances require higher up-front investments, and customers use short payback times in their considerations. Energy prices sometimes are subsidized specifically, or one could consider them to be subsidized almost everywhere, since environmental externalities rarely are included in the energy price. Sometimes lack of information is a key barrier: one really has to make an effort to have clear information about the energy efficiency of the purchase. In many cases, designers of buildings such as architects, have no incentives to consider energy efficiency explicitly in their design. Something similar applies to tenants paying a fixed price to their landlords, including energy services.

There are many opportunities to overcome these barriers, in which usually the government plays a major role. Important here are targeted energy efficiency standards for both buildings and equipment. This is one of the most popular ways of overcoming barriers to the application of energy-efficient technologies in many countries for both commercial and residential buildings. Voluntary programmes can be agreed with the construction business, the service sector, or with manufacturers of appliances, e.g. for construction of energy-efficient buildings, labelling of energy efficient equipment, and other methods of information exchange. The public sector is an important player in the building sector, and can give the right example in its own building activities. To overcome the barrier formed of many consumers and small- and medium-sized companies who are not interested or informed about energy efficiency opportunities in their buildings, they can be approached by suppliers or intermediaries. The former can be achieved by way of demand-side management, in which energy companies provide rebates for energy efficient equipment to their clients. An example of the latter are energy service companies which offer energy savings contracts guaranteeing a particular amount of savings, paid back through the savings and relieving the clients of the hassle of acquiring the information and acting upon it. Capacity building and strengthening of the information basis is important especially in developing countries. Finally, there are several generic ways to reduce energy consumption in the building sector, e.g. through tax policies, emissions trading, and stimulation of research and development. Such measures are discussed below, in relation to mitigation in other sectors.

The transport sector is the fastest growing, particularly because of the growth in car use, road freight, and air traffic. In this sector, many of the barriers are related to the existing infrastructure, the lifestyles of the rich, and the structure of consumption and production. Comfort, size, safety, and performance generally are more important criteria than energy efficiency. Energy consumption or climate change are not among the most important things car buyers have on their minds when making their choices in the garage or showroom. Even if high-efficiency cars would be cost-effective over their lifetime, consumers are still not willing to make the additional investment. In transport, the lock-in effect is particularly strong for the road-based transport, not only because of physical factors related to the road and fuel distribution systems, but also because of the image of personal freedom, status, identity, and safety associated with the private car. Therefore, a shift to alternative fuels, or even a shift to other modes of transport, is very difficult to make.

Nevertheless, there are opportunities to overcome these barriers. The vast technological potential has been discussed already. Interestingly, the opportuntities often are related to the adverse impacts of transport wider than climate change, notably local air pollution, space requirements, traffic congestion, and safety. Economic instruments (e.g. taxes) appear to be attractive means of capturing such environmental and social externalities, but even if the revenues are used to address the problems, taxes are still seen as merely revenue-raising instruments for the government and are not popular. Road taxes and parking fees would have to be high to change significantly driving and car ownership behaviour. In areas in which traffic problems are most severe, usually at the local (e.g. city) level, a combination of measures to control transport and the associated energy use can be effective. At least in theory, in developing countries the lock-in effect is yet less strong since the infrastructure still has to be largely developed, and there are more opportunities for avoiding the car dependency of the industrialized countries. The fact that there are only a limited number of car manufacturers around the globe suggests that there are opportunities for internationally co-ordinated advances in car efficiency triggered by government interventions.

Also in the *industrial sector*, there are many barriers to the adoption of efficient and environmentally sound technologies. Many of the available options in many large international firms in developed countries may have been implemented already; this is less the case for industries in many developing countries and economies in transition as well as in small- and medium-sized companies in all countries. Nevertheless, advances can be made also in large international companies. As in other sectors, energy efficiency generally does not rank high in corporate decision-making. Often, information is either lacking, or not trusted, or expensive to acquire. Short required payback times in a competitive environment hinder cost-effective adoption of energy-efficient equipment and practices. In developing countries and particularly in small- and medium-sized companies, lack of access to sufficient capital to make the necessary investments is a major hurdle. For the same group, lack of skill in installing, operating, and maintaining the equipment is another frequent barrier.

Opportunities to remove these barriers abound. Policies and measures are likely to be most effective if developed while accounting for specific local circumstances, and often in combination. As in the building and transport sectors, energy efficiency standards are popular methods to stimulate the implementation of environmentally sound technologies in industry. Also, or in combination, specific taxes and subsidies can be used to overcome barriers. In the manufacturing sector, voluntary agreements between private enterprises and the government are increasingly popular in many industrialized countries, although their effectiveness is heavily debated. This effectiveness depends very much on local circumstances and the details of the agreement, e.g. target-setting, supporting policies, reporting requirements, and the nature, size, and structure of the subsector involved. Generic policies (e.g. taxes and emissions trading systems) are discussed in Section 9.5.

In the *energy sector*, access to information may often be somewhat less of a problem, but there are many other barriers hindering the penetration of environmentally sound technologies. Lack of skills to use and maintain them in modern form is among these, as well as problems with capital availability, which is very important for this capital-intensive sector. In addition, a barrier is formed by the prevalent low energy prices, often low because of subsidies or absence of accounting for social and environmental externalities. If companies can recuperate all their operating expenses, there is no incentive to pursue efficiency improvements. In many countries, deregulation and/or privatization of the energy sector has/have profound effects on the structure of the supply. While on the one hand it may lead to more efficient electricity generation, on the other it may bring about a worsened position of non-fossil energy sources because of the low prices of fossil on the world market. Furthermore, segmentation of the energy chain goes at the expense of demand-side management possibilities. Institutional and regulatory complexities may impede the introduction of efficient co-generation equipment.

There are attractive opportunities in the energy sector to leapfrog from outdated technology to the most efficient technologies available especially in developing countries, e.g. from 30 to more than 50 per cent efficiency in power generation. However, in order to achieve this, capital availability has to be secured and skills developed to ensure a proper enabling environment for technology transfer. As for the transport sector, the local environmental problems associated with energy supply operations may provide opportunities to achieve multiple benefits and therefore lower the barriers. In rural areas, opportunities for decentralized energy systems exist, avoiding the need for construction of expensive gridded electricity systems. In all countries, full-cost pricing helps overcome barriers to the implementation of efficient technologies.

The land-use, agriculture, forestry, and waste management sectors may, together, contribute less to global warming than any of the above categories alone (i.e. buildings, transport, industry, and energy), but still provide important opportunities to abate greenhouse gas emissions in a way that also addresses other problems. In agriculture and forestry, barriers often are related to lack of information, skills, and capital in the rural areas, at the farm or forest stand level. Incentives to reduce energy consumption in agriculture are limited, since energy is often not a major production factor in this sector, while the same applies to controlling non-carbon dioxide greenhouse gases, without any rationale other than climate change. Energy and energy-intensive inputs (e.g. fertilizers) sometimes are subsidized. In the waste management sector, usually there are few government incentives or institutional arrangements to control emissions, such as methane from landfills. In this sector, the diversity of actors involved and the complexity of the waste management chain (i.e. generation, collection, transportation, treatment, disposal, or recycling) makes co-ordinated action difficult. Probably because of the relative small contribution of these sectors to energy use, and the relatively large scientific uncertainties regarding the sources of non-carbon dioxide greenhouse gases, little attention has been paid yet to innovative options, and barriers and opportunities for mitigation.

Capacity building, information programmes, and credit or price support schemes can help remove barriers at the farm level in agriculture. For waste management, capacity building to enhance knowledge and institutional capabilities can help overcome barriers as well as methane recovery regulations. Generic instruments (e.g. government regulations, taxes and subsidies) also are part of the portfolio of options in these sectors. The relative lack of research and development in this area could be addressed to reduce uncertainties and generate feasible mitigation options, e.g. in soil and water management, irrigation efficiency, and carbon dioxide sequestration in subtropical and tropical areas. Also in these sectors, mitigation policies can address multiple objectives (see Section 9.4.3), which may provide opportunities for overcoming the barriers.

9.5   Policies, measures, and instruments

9.5.1   *Types of policies, and sectoral examples*

Above, we discussed a technoeconomic rationality approach to the problem of mitigating greenhouse gas emissions: there are technological options and practices, which are further developed over time, but barriers exist to their implementation. According to this approach, policies, measures, and instruments[4] should be applied to overcome these barriers and capture the opportunities. Possible policies could be divided into four main categories:

1. Market instruments.
2. Regulatory instruments.
3. Voluntary agreements.
4. Information campaigns.

There is no simple cookbook that can be used to determine which policy would be the policy-of-choice in a particular situation. Suitability and effectiveness are very dependent on local circumstances (e.g. institutional capabilities, public awareness, and existing legal or regulatory systems etc.), implementation details (e.g. strictness, magnitude of tax/subsidy, and compensatory mechanisms), and local preferences, e.g. based on previous experiences, or the level of influence of stakeholders to which the policies may be addressed. Policies can be used in different combinations to capture the strengths of different types of policies. Various criteria can be used for a systematic evaluation of the possibilities (Bashmakov *et al.* 2001):

1. *Environmental effectiveness.* How well does the policy achieve an environmental goal, e.g. a greenhouse gas emissions reduction target? How reliable is the instrument in achieving that target?, does the instrument's effectiveness erode over time?, and does the instrument create continual incentives to improve products or processes in ways that reduce emissions? Also: what are the wider environmental effects, e.g. local air-quality improvement (usually referred to as the ancillary benefits)?
2. *Cost effectiveness.* Does the policy achieve the environmental goal at the lowest cost, taking transaction, information, and enforcement costs into account? Are there any additional economic benefits depending on if and how the revenues are recycled? What are the wider economic effects, e.g. the potential effects on variables such as inflation, competitiveness, employment, trade, and growth? Are there any secondary effects, e.g. changes in attitudes and awareness, learning, innovation, technical progress, and dissemination and transfer of technology?
3. *Distributional considerations.* How are the costs of achieving the environmental goal distributed across groups within society, including future generations and between regions?

---

[4] In the following, 'policies' is used to include policies, measures, and instruments, terms that may be defined differently, but are also often used interchangeably.

Table 9.6. *Examples of policies, measures and instruments in four categories.*

(a) *Examples of policies and measures in transportation*

Market-based instruments	Regulatory instruments
• Information technologies enabling pricing external costs of transportation, from congestion to environmental pollution • Implementation of more efficient pricing, providing greater incentives for energy efficiency in both equipment and modal structure	• Mandatory energy efficiency regulations • Speed limits
Voluntary agreements	Research and development, information campaigns
• Voluntary energy efficiency regulations • Lifestyle changes influencing model split	• Intensive research and development efforts for light-duty road vehicles road freight, air, rail and marine transport technologies, e.g. to improve hybrid powertrain and fuel cell technologies • Information base and consumer awareness

(b) *Examples of policies and measures in the building sector*

Market-based instruments	Regulatory instruments
• Tax incentives • Subsidies • Energy pricing	• Mandatory standards, building codes • Deregulation of the energy industry • Demand-side management
Voluntary agreements	Research and development, information campaigns
• Voluntary programmes and standards	• Capacity building • Improve marketing approaches and skills • Innovation programmes • Information base and consumer awareness, e.g. energy labelling

(c) *Examples of policies and measures in the industrial sector*

Market-based instruments	Regulatory instruments
• Direct subsidies and tax credits • Phase out distorting taxes • Better marketing approaches • Tradable permits	• Legislation, e.g. mandatory energy efficiency regulations • Environmental permits • Waste-heat regulations • Product bans, e.g. CFCs • Recycling regulations

(*cont.*)

Table 9.6. (*cont.*)

Voluntary agreements	Research and development, information campaigns
• Business-nongovernmental voluntary agreements • Government-business voluntary agreements	• Information programmes • Joint government/industry research and development

(d) *Examples of policies and measures in land-use change and forestry*

Market-based instruments	Regulatory instruments
• Taxes, e.g. energy, and fertilizer • Subsidy reform • Tradable permits • Market payments for capturing and holding carbon	• Land-use regulations • Animal density regulations • Fertilizer regulations

Voluntary agreements	Research and development, information campaigns
• Voluntary programmes	• Research and development and outreach programmes

(e) *Examples of policies and measures in agriculture and waste management*

Market-based instruments	Regulatory instruments
• Taxes, e.g. fertilizer • Expansion of credit schemes	• Government regulations • Policies to reduce methane-producing waste and /or requirements to utilize methane from landfills as an energy source

Voluntary agreements	Research and development, information campaigns
• Voluntary programmes	• Shifts in research priorities • Development of institutional linkages across countries

(f) *Examples of policies and measures in energy supply*

Market-based instruments	Regulatory instruments
• Full-cost pricing • Energy/carbon taxes • Tradable permits • Subsidies	• Efficiency standards • Energy mix requirements • Promotion of leapfrogs in energy supply-and-demand technology

(*cont.*)

Table 9.6. (*cont.*)

Voluntary agreements	Research and development, information campaigns
• Industry-government agreements • Green electricity programmes	• Capacity building • Appropriate mechanisms for transfer of clean and efficient energy technologies • Information systems • Research and development programmes

4. *Administrative and political feasibility.* How flexibly can the policy be adjusted if new knowledge becomes available? Is it understandable and acceptable to the general public? What are the impacts on the competitiveness of different industries?

Weighing the different criteria would be done differently in different countries. In the literature, there is some emphasis on the cost-effectiveness criterion because of the dominance of economic analysis in this area. In Section 9.4, we discussed some sectoral opportunities. It is important to note that in all these sectors, policies in all four categories could be effective. Some examples are given in Table 9.6(a)–(f).

In the following, the different types of policies are discussed in more detail.

9.5.2   National policies

Box 9.5 gives definitions of the possible policies at the domestic scale, including market and non-market policies.

Regulatory instruments

Regulatory instruments (e.g. energy efficiency or emissions standards for products or services) are used widely around the world. They are enforced usually through fines for non-compliance. They would need to be updated regularly to follow technological progress. Announcing new or updated standards well in advance forces companies to plan strategically for innovations. The more the standards prescribe a particular technology, the less flexibility there is to meet environmental goals. Another problem with standards is that they necessarily have to be somewhat generic for broad applications and therefore cannot easily address situation-specific circumstances. Regulatory policies can be accompanied by information programmes (see section below on information and education programmes) to address the information gap barrier to environmentally sound technologies. From an economic perspective, regulatory measures are less efficient than market-based

Box 9.5   Definitions of selected National Greenhouse Gas Abatement Policy instruments

1. An *emissions tax* is a levy imposed by a government on each unit of carbon dioxide equivalent emissions by a source. Since virtually all of the carbon in fossil fuels ultimately is emitted as carbon dioxide, a levy on the carbon content of fossil fuels – a carbon tax – is equivalent to an emissions tax for emissions caused by fossil fuel combustion. An energy tax – a levy on the energy content of fuels – reduces the demand for energy and so reduces carbon dioxide emissions through fossil fuel use.

2. A *tradable permit* (cap-and-trade) system establishes a limit on aggregate emissions by specified sources, requires each source to hold permits equal to its actual emissions, and allows permits to be traded among sources. This is different from a credit system, in which credits are created when a source reduces its emissions below a baseline equal to an estimate of what they would have been in the absence of the emissions reduction action. A source subject to an emissions limitation commitment can use credits to meet its obligation.

3. A *subsidy* is a direct payment from a government to an entity, or a tax reduction to that entity, for implementing a practice a government wishes to encourage. Greenhouse gas emissions can be reduced by lowering existing subsidies that in effect raise emissions (e.g. subsidies to fossil fuel use) or by providing subsidies for practices that reduce emissions or enhance sinks, e.g., for insulation of buildings, or planting trees.

4. A *deposit – refund system* combines a deposit or fee (tax) on a commodity with a refund or rebate (subsidy) for implementation of a specified action.

5. A *voluntary agreement* is an agreement between a government authority and one or more private parties, as well as a unilateral commitment that is recognized by the public authority, to achieve environmental objectives or to improve environmental performance beyond compliance.

6. A *non-tradable permit system* establishes a limit on the greenhouse gas emissions of each regulated source. Each source must keep its actual emissions below its own limit; trading among sources is not permitted.

7. A *technology or performance standard* establishes minimum requirements for products or processes to reduce greenhouse gas emissions associated with the manufacture or use of the products or processes.

8. A *product ban* prohibits the use of a specified product in a particular application (e.g. hydrofluorocarbons (HFCs) in refrigeration systems) that gives rise to greenhouse gas emissions.

9. *Direct government spending and investment* involves government expenditures on research and development measures to lower greenhouse gas emissions or enhance greenhouse gas sinks.

Source: Bashmakov *et al.* (2001).

instruments, because they reduce flexibility to meet environmental goals. Regulatory systems require the build-up of institutional capabilities because they can precede market-based instruments, which also need appropriate institutions and well-educated staff for their implementation and management.

Market-based instruments

Market-based instruments address some of the problems associated with regulatory policies, but have other types of problems. Market-based policies leave the way particular environmental goals are achieved to the market rather than prescribing particular solutions. At the same time, they face various implementation issues that have to be addressed. A generic statement about one type of policy to be preferred over another therefore cannot be made. There are two main types of market-based instruments: (a) taxes and subsidies, and (b) emissions trading. Unlike an emissions trading system or regulation, taxes and subsidies do not guarantee that a particular environmental standard or target will be met. To reach such standards or targets, the tax rate would have to be adjusted periodically. In principle, taxes limit the costs of emissions reduction, because emissions are allowed to rise when costs appear too high. The rate also would have to be increased to account for new emitters. Conversely, an emissions trading system can ensure that a particular target is reached efficiently, but it does not protect against high costs. In case of a *tax system*, those emissions reductions will be undertaken that would be less expensive than the tax which would have to be paid without those reductions. The least expensive reductions would be implemented first. The effectiveness of the tax depends on various implementation choices, e.g. the tax base (e.g. carbon emissions, energy use), the collection point (e.g. energy producers, consumers), level of variation amongst sectors, revenue handling, and combinations with other policies. *Tradable permit systems* face similar implementation choices, e.g. permits could be allocated to end-users directly responsible for emissions (e.g consumers, emitting companies), or to product manufacturers, e.g. halocarbon producers. A key problem with permit systems is the distribution of permits over participants. They can be distributed proportional to current emissions or according to another criterion, be distributed for free or by auction, can apply to a limited amount of emissions (e.g. carbon dioxide from energy) or to a wider set (e.g. including other sources and greenhouse gases); the system can be limited to domestic sources or international trade can be allowed, and the system may be limited to emitters, or also other parties (e.g. brokers) can be allowed to trade. The main problem with a trading system from an economic perspective is that there is no upper bound for the permit price, and thus for the mitigation costs. A hybrid system in which governments would intervene in the market as soon as the price exceeds a particular level may be a compromise solution. *Subsidies* can also help mitigate greenhouse gas emissions, e.g. by stimulating research and development, or supporting the introduction of clean (or 'green') energy.

National voluntary agreements

National voluntary agreements or programmes receive increasing attention. Usually these refer to an agreement between a government and one or more private sector actors to improve environmental performance, or energy efficiency. Voluntary agreements come in various shapes. They may be limited to reporting requirements or research and development, but may also have specific efficiency improvement or emissions targets. It is obvious why voluntary agreements are so popular: they are generally not legally binding, give the private sector much room to manoeuvre, and the government needs to make fewer efforts for monitoring and control (and, hence, lower transaction costs). At the same time, they can stimulate action through a consensus-building process. The disadvantage for governments is that the level of abatement is not guaranteed, which is why sometimes the voluntary agreement is accompanied by the possibility of regulatory measures if it is not implemented satisfactorily. Evaluation of the voluntary agreement (like other policies) is difficult because of the uncertainty about what would have happened in its absence. In addition to agreements between two or more parties, one party can take on voluntary commitments. Margolick and Russell (2001) identified and evaluated voluntary commitments by more than a dozen major companies in North America, Europe, and Japan. These companies adopted different types of commitments such as emissions targets (e.g. Shell: 10 per cent reduction by 2002 from 1990 levels; DuPont: 65 per cent reduction by 2010 of 1990 emissions) or energy efficiency targets, e.g. IBM: 4 per cent of projected energy use in absolute terms by 2002; Toyota: 15 per cent of energy use per unit of production by 2005 as compared to 1990). Key motivations for such voluntary programmes by companies include expectations about long-term economic competitiveness, corporate culture, public profile considerations, environmental co-benefits and political environment.

Information and education programmes

Lack of information about opportunities to increase energy efficiency and reduce greenhouse gas emissions has been identified as the main barrier to implementation of available technologies and practices. In addition, awareness about environmental problems such as climate change is often low. Education to increase awareness in combination with providing guidance on practical mitigation options can be a useful component of national climate change response strategies. Components of such programmes usually include brochures and other publications, information clearing-house web sites, television and radio programmes, or targeted information and education activities such as workshops. Energy-efficiency labelling is an increasingly applied method.

Structural reform policies and energy market liberalization

In addition to specific climate-oriented policies, it is important to note that broader socioeconomic policies have impacts on energy consumption and associated greenhouse gas emissions. In the last decade of the previous century, many developing countries and economies in transition have adopted so-called structural reform policies, usually to develop their economies in a more market-oriented direction. Elements can include privatization of state enterprises, trade liberalization, tax reforms, opening capital markets to attract foreign investments, and exchange rate unification. The effects of these changes are uncertain, since they may work in two directions: on the one hand, the policies may boost economic growth and energy consumption, and in some cases lead to a shift to more carbon-intensive fuels, but on the other, the open markets can lead to rapid penetration of energy efficient technologies, reducing greenhouse gas emissions. In several industrialized, as well as developing and transition countries, steps have been undertaken towards liberalization of the energy market (electricity, natural gas). Although there is a mounting set of research publications, there is no clear guidance as to the effects in terms of greenhouse gas emissions. In countries relying on renewable energy sources (e.g. hydropower in mountainous countries, biofuels in some others) the liberalization is leading to a shift to low-priced fossil fuels. Increased competition leads to increased efficiency, but this may not balance the upward push by increased consumption and the above-mentioned shift to fossil fuels in some countries. Another disadvantage of many energy market liberalization policies is the separation of energy production from consumption-reducing demand-side management opportunities.

9.5.3   *International policies*

Box 9.6 gives some definitions of possible international or internationally co-ordinated policies. Theoretically, the problem of global warming can be addressed most efficiently at the global level. From the perspective of stabilizing global greenhouse gas concentrations, it does not matter where emissions reductions are realized, so long as they could be achieved where it is cheapest to do so. Taking into account the differences in the world today, it would be cheapest in countries that do not yet use the most efficient technologies, i.e. in developing countries and economies in transition. Developing countries do not often have the resources to implement the available advanced technologies. Furthermore, from an ethical perspective, the industrialized countries should take the lead, being the main cause of the problem to date. Because of these considerations, developing countries do not have any commitment to reduce greenhouse gas emissions in the UNFCCC context. Nevertheless, in the 1997 Kyoto Protocol three international mechanisms (i.e. International Emissions Trading, Joint Implementation, and the Clean Development Mechanism) have been included, of which the implementation rules took 4 years to negotiate, at the expense of losing the USA as party to the Protocol. These three mechanisms are examples of a broader set of internationally co-ordinated policies.

Box 9.6   Definitions of selected international greenhouse gas abatement policy instruments

1.   A *tradable quota system* establishes a national emissions limit for each participating country and requires each country to hold a quota equal to its actual emissions. Governments, and possibly legal entities, of participating countries are allowed to trade quotas. Emissions trading under Article 17 of the Kyoto Protocol is a tradable quota system based on the assigned amounts calculated from the emissions reduction and limitation commitments listed in Annex B of the Protocol.

2.   *Joint implementation* allows the government of, or entities from, a country with a greenhouse gas emissions limit to contribute to the implementation of a project to reduce emissions, or enhance sinks, in another country with a national commitment and to receive emissions reduction units equal to part, or all, of the emissions reduction achieved. The emissions reduction units can be used by the investor country or another Annex I party to help meet its national emissions limitation commitment. Article 6 of the Kyoto

Protocol establishes joint implementation among parties with emissions reduction and limitation commitments listed in Annex B of the Protocol.

3. The *Clean Development Mechanism* allows the government of, or entities from, a country with a greenhouse gas emissions limit to contribute to the implementation of a project to reduce emissions, or possibly enhance sinks, in a country with no national commitment and to receive certified emissions reductions equal to part, or all, of the emissions reductions achieved. Article 12 of the Kyoto Protocol establishes the Clean Development Mechanism to contribute to sustainable development of the host country and to help Annex I parties meet their emissions reduction and limitation commitments.

4. A *harmonized tax on emissions, carbon, and/or energy* commits participating countries to impose a tax at a common rate on the same sources.[1] Each country can retain the tax revenue it collects.

5. An *international tax on emissions, carbon, and/or energy* is a tax imposed on specified sources in participating countries by an international agency. The revenue is distributed or used as specified by participant countries or the international agency.

6. *Non-tradable quotas* impose a limit on the national greenhouse gas emissions of each participating country to be attained exclusively through domestic actions.

7. *International product and/or technology standards* establish minimum requirements for the affected products and/or technologies in countries in which they are adopted. The standards reduce greenhouse gas emissions associated with the manufacture or use of the products and/or application of the technology.

8. An *international voluntary agreement* is an agreement between two or more governments and one or more sources to limit greenhouse gas emissions or to implement measures that will have this effect.

9. *Direct international transfers of financial resources and technology* involve transfers of financial resources from a national government to the government or legal entity in another country, directly or via an international agency, with the objective of stimulating greenhouse gas emissions reduction or sink enhancement actions in the recipient country.

Source: Bashmakov *et al.* (2001).

[1] A harmonized tax does not necessarily require countries to impose a tax at the same rate, but to impose different rates across countries would not be cost-effective.

Regulatory policies

Although not legally binding, standardization of measurement procedures and environmental performance (e.g. through the Organization for International Standardization) can stimulate energy efficiency and environmental
performance in companies around the world. Other regulatory policies at the
international level do not seem to be feasible or attractive options and are not
being considered.

Market-based policies

Most of the international policies currently being researched and discussed are market-based. They face many problems of implementation, as illustrated by the strenuous negotiations of the Kyoto mechanisms.

Internationally harmonized taxes

Theoretically, an international tax can help in pursuing environmental
goals such as greenhouse gas emissions reductions in an efficient manner. However, the larger the number and diversity of countries involved, the more difficult an agreement about the implementation will be: (a) who will raise the tax;
(b) how will the revenue be used; (c) how can losers be compensated, and (d) which
advantages does an international tax have over a domestic one? An internationally
harmonized tax may be somewhat easier – but still extremely difficult politically –
to agree, since every country would manage the taxes and revenue individually.

International emissions trading

A very important option for international policies from an economic perspective is emissions trading. After all, it is most efficient to reduce emissions
where it is cheapest. The system applies to entities with emissions caps, currently
only the industrialized countries, including the economies in transition. In the
Kyoto Protocol (Article 17) international emissions trading has been included as
an option for Annex I parties to the Protocol to meet their agreed targets (assigned
amounts) in the first commitment period of 2008–12.

Theoretically, the larger the number of entities participating in the system, the
more advantageous (see also Chapter 10). The Kyoto Protocol states that such trading should be supplemental to domestic action, and it has been fiercely debated
as to what extent a limitation on the usage of international emissions trading
would be appropriate. Proponents of such limitations (e.g. some European countries) argue that not only do countries have an ethical obligation to reduce their
own greenhouse gas emissions, but also to accelerate technological innovation
and allow it to spill over to other countries. Others argue that such limitations

would increase abatement costs and thus make meeting the emissions targets much more difficult. It is a matter of intense debate whether international emissions trading is compatible with the provisions of the World Trade Organization.

The trading system under the Kyoto Protocol can start by 2008. In 2002, the European Commission agreed a Directive on Emissions Trading, establishing a trading regime for selected sources of carbon dioxide in the member states of the European Union. The regime will start by 2005.

Joint implementation

The second mechanism in the Kyoto Protocol is joint implementation, allowing Annex I parties to meet their emissions limitation and reduction obligations jointly through project-based activities. If, according to Article 6 of the Kyoto Protocol, an Annex I country contributes to greenhouse gas emissions reduction (or sink enhancement) in another Annex I country, it can acquire emissions reduction units. These can be used to meet the party's emissions obligations. In Marrakech at COP7, emissions credits obtained from sinks projects under Article 3.3 (Afforestation, Reforestation, and Deforestation) and Article 3.4 were labelled as ReMoval Units. For more information about sinks, see Box 9.2.

Clean Development Mechanism

The third mechanism in the Kyoto Protocol is the Clean Development Mechanism. This project-based mechanism has two main objectives: (a) to help non-Annex I countries to pursue sustainable development, and (b) help Annex I countries to achieve their quantified emissions limitation and reduction obligations in a cheaper way. Annex I parties can use certified emissions reductions acquired by financing projects in non-Annex I countries which reduce greenhouse gas emissions or enhance sinks, to meet their national commitments. An executive board has to ensure that Clean Development Mechanism projects really achieve measurable and long-term benefits for climate change mitigation, to be certified by entities assigned to this task. One of the key problems with the Clean Development Mechanism is the establishment of a credible baseline: what would have happened without the Clean Development Mechanism project? The host country is responsible for determining whether the project contributes to sustainable development objectives.

During COP7, it was agreed that Clean Development Mechanism projects can start immediately. Also, full fungibility between assigned amount units, emissions reduction units, certified emissions reductions and ReMoval units was agreed to meet the Kyoto targets, i.e. they are fully exchangeable between Annex I parties. Assigned amount units can be carried over into the second commitment period (if any). This is not the case for emissions reduction units and certified emissions reductions, which can only be banked up to a limit of 2.5 per cent of the initial assigned units. Removal units cannot be banked.

International voluntary agreements

Thus far, the instrument of voluntary agreement has been applied mainly domestically. However, in a number of industry branches with a limited number of large international firms (e.g. vehicles, electrical or mechanical equiment) international voluntary agreements may be attractive. In fact, in Europe, agreements were established between the European Commission and international car manufacturers. Some multinational companies have initiated voluntary programmes.

9.6   Special issues in developing countries

Three issues are of particular importance for developing countries with reference to policies and measures to encourage mitigation of greenhouse gas emissions – financial resources, transfer of technologies, and capacity building.

9.6.1   Transfer of financial resources

A special example of international policies is the transfer of financial resources and technologies, and capacity building. Agenda 21 calls for additional financial resources for sustainable development in developing countries to support these countries in their pursuance of a sustainable development trajectory. These funds would need to be new and additional to current official development assistance. However, in the 1990s, global official development assistance decreased. The decrease halted in 1998. Also foreign lending through multilateral development banks for climate-relevant sectors has decreased since 1992 (Radka et al. 2001). While for some very poor countries official development assistance still is of crucial importance, in general other forms of financial transfers have become more important, notably foreign direct investment (see Figure 9.9). However, these investments have concentrated on a limited number of developing countries, especially in East and South East Asia and Latin America. Official development assistance also remains important for particular activities not captured by private flows, e.g. supporting some key pillars of development (sound policy environment, human capital investments, institutional build-up, etc.). It is also important for sectors that attract less private investment, e.g. agriculture, forestry, public health, and coastal zone management (IPCC 2001).

The UNFCCC specifies that the Annex II countries should provide financial assistance to developing countries to address climate change and adapt to its adverse effects. During COP7 in 2001, agreement was reached on three new funds. First,

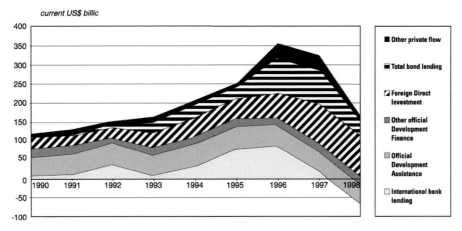

Figure 9.9. Total net resource flows to aid recipient countries.

a.o. since it is believed that the poorest countries are likely to benefit least from the Clean Development Mechanism, a 'least developed countries' fund will support particularly those countries in their response to climate change. Second, a special climate fund has been established to finance activities in the areas of adaptation, technology transfer, energy, transport, industry, agriculture, forestry, and waste management. Special attention is to be given to diversifying the economies of countries dependent on fossil-fuel exports. These funds operate under the UNFCCC. Under the Kyoto Protocol a third, 'adaptation' fund would be financed, among other sources, through a share from the proceeds of Clean Development Mechanism projects.

Technology transfer

In order to meet the ultimate objective of the UNFCCC and stabilize greenhouse gas concentrations, eventually global greenhouse gas emissions will have to be decreased considerably. While according to the UNFCCC, industrialized countries should take the lead, it is inevitable that eventually all countries have to join the effort to reduce emissions. Developing countries should be supported by developed countries, a.o. by technology transfer. For developing countries, there are issues more pressing than climate change, but fortunately controlling greenhouse gases can go hand-in-hand very well with pursuing sustainable development pathways. In general, transfer of environmentally sound technologies contributes automatically to greenhouse gas mitigation. Avoiding unsustainable pathways implies that developing countries move rapidly towards better technologies and institutions. It is important to note that the problem is broader than just the acquisition and adaptation of hardware. It also involves the development of human capacity

Table 9.7. *Principal stakeholders and their decisions or policies in technology transfer.*

Stakeholders	Motivations	Decisions or policies that influence technology transfer
Governments		
• National/federal	Development goals	Tax policies (including investment tax policy)
• Regional/provincial	Environmental goals	
• Local/municipal	Competitive advantage	Import/export policies
		Innovation policies
	Energy security	Education and capacity building policies
		Regulations and institutional development
		Direct credit provision
		Technology
Private-sector business		
• Multinational	Profits Market share	Research and development/ commercialization decisions
• National	Return on investment	
• Local/microenterprise		Marketing decisions
• (including producers, users, distributors, and financiers of technology)		Capital investment decisions
		Skills/capabilities development policies
		Structure for acquiring outside information
		Decision to transfer technology
		Choice of technology transfer pathway
		Lending/credit policies (producers, financiers)
		Technology selection (distributors, users)
Donors		
• Multilateral banks	Development goals	Project selection and design criteria
• GEF	Environmental goals	
• Bilateral aid agencies	Return on investment	Investment decisions
		Technical assistance design and delivery procurement requirements
		Conditional reform requirements
International institutions		
• World Trade Organization	Development goals	Policy and technology focus
• United Nation Sustainable Development	Environmental goals	Selection of participants in forums
	Policy formulation	
• Organization for Economic Co-operation and Development	International dialogue	Choice of modes of information dissemination

Table 9.7. (*cont.*)

Stakeholders	Motivations	Decisions or policies that influence technology transfer
Research/extension		
• Research centres/labs	Basic knowledge	Research agenda
• Universities	Applied research	Technology
• Extension services	Teaching	Research and development/
	Knowledge	commercialization decisions
	transfer	Decision to transfer technology
	Perceived	Choice of pathway to transfer
	credibility	technology
Media/public groups		
• Television, radio, newspaper	Information	Acceptance of advertising
• Schools	distribution	Promotion of selected
• Community groups	Education	technologies
• Nongovernmental	Collective	Educational curricula
organizations	decisions	Lobbying for technology-related
	Collective	policies
	welfare	
Individual consumers		
• Urban/core	Welfare	Purchase decisions
• Rural/periphery	Utility	Decision to learn more about a
	Expense	technology
	minimization	Selection of learning/information
		channels
		Ratings of information credibility
		by source

(knowledge, techniques and management skills) and appropriate institutions and networks (IPCC 2001). On the one hand, transferred technologies should meet local needs, and on the other, their successful adoption depends on a suitable enabling environment. Unfortunately, technology transfer is a very complex process and its success very much dependends on the conditions under which a project is implemented. Usually, a multitude of stakeholders is involved, who can be different in different phases and play very different roles (see Table 9.7). Stakeholders are not limited to governments and private firms, but also include consumers, nongovernmental organizations and other citizens' groups, research and educational organizations, and financial institutions. Technology innovation and transfer can be seen as an intertwined process in which these various stakeholders participate (see Figure 9.8). In most cases, the private sector plays a decisive role, but in many cases also governments can play stimulating and facilitating roles, while in other cases community groups are very important players, determining user needs. In other words, at different stages different pathways can be followed in which different actors take a lead role.

Technology transfer can be seen as a stepwise process. One way of distinguishing between five different stages is the following.

1. *Assessment.* The important first step is the identification of needs, involving those who should use the technology involved.
2. *Agreement.* Based on the assessment of user needs, an agreement about the transfer of the associated technology can be negotiated with the supplier, funder, and others involved.
3. *Implementation.* The acquisition, installation, or implementation of the technology, and usage.
4. *Evaluation and adjustment.* Since in local circumstances the transferred technology may not work in an optimal fashion, adjustments can be needed to suit local circumstances and preferences. To allow this, a careful evaluation is required.
5. *Replication.* Important for the effectiveness of the transfer is that it is not a once-and-for-all affair, but that it is replicated at a broad scale.
6. *Back to 1. for new projects.* The careful evaluation of the above five steps allows for the identification of barriers that can be addressed in follow-up activities, avoiding mistakes made the first time around.

In step 2, the issue of intellectual property rights plays an important role. In the UNFCCC negotiations, the issue is heavily debated between developing and developed countries. The vast majority of IPRs are generated in the industrialized countries. These aim at protecting their business interests, strong intellectual property protection regulation stimulating investments in new technologies. Developing countries, however, need new technologies if they want to develop their economies in a sustainable fashion and too-strict intellectual property protection makes this too expensive. Some arguments for and against tight intellectual property protection regulations are listed in Table 9.8. To some extent, agreement was reached, among other issues, on minimum standards, in the 1994 Agreement on Trade Related Aspects of Intellectual Property in the General Agreement on Tariffs and Trade context. Nevertheless, the controversy has not been resolved because of the disparate and economically very significant interests.

The effectiveness of technology transfer is dependent on many factors. Box 9.7 presents an industry perspective.

Table 9.8. *Technology transfer and intellectual property rights.*

Arguments for strict intellectual property rights protection regime	Arguments for more relaxed intellectual property rights protection regime
• Encourages creative development of new technologies by giving greater confidence for research and development investments • Increase investments because investors can recapture their research and development investments • Can increase the profit of the sellers • Facilitates 'vertical' distribution and transfer of technologies through foreign direct investments in a particular sector • Not all countries need the most advanced, cutting-edge technologies; 'soft' technologies (expertise, skills) may often be more important	• Too-strict intellectual property rights can slow down technology development, relying on old patents • Can decrease the prices for the buyers • Enhances 'horizontal' distribution and transfer of technologies to other sectors and regions • Gives developing countries a better chance through licensed production • Environmentally sound technologies may deserve special treatment because of the overriding importance of sustainable development • Reduce restrictive business practices, e.g. restrictive/prohibitive transfer conditions to maximize rents, licensing refusals, etc.

Source: Adapted from Grubb and Ramakrishna (2000).

Box 9.7 Conditions for success in technology transfer, an industry perspective

The likelihood of maximum benefits to all parties will be greatly facilitated if a number of conditions are met. These include:

1. A stable economic system and an attractive investment opportunity for investing partners.
2. Transparent and equitable legal and financial structure and sound environmental laws.
3. Realistic expectations from the host country and the communities of the benefits that may result from the partnership.
4. A long-term commitment and dedication of resources by all partners.
5. A fair distribution of benefits by all parties.
6. Industry respect for local culture and values.
7. A safe and secure working environment for all employees and contractors.
8. No unnecessary barriers to movement of personnel and materials.

Source: IPIECA (1999).

9.6.2   Capacity-building

But even if financial resources would be (made) available and environmentally sound technologies transferred to developing countries, this does not mean that they will be implemented and maintained, helping the countries following their sustainable development path. A precondition for success is the availability and continuity of human, institutional, and information assessment and monitoring capacity (IPCC 2000). Although the level of local capacity is very different across different developing countries, concerted efforts to improve this local capacity are crucial. In many cases, there is a special role of ODA-related efforts to support this process. However, industry also has a role to train and educate the workforce in developing countries. According to IPIECA (1999), industry should look for opportunities to extend existing local capabilities and strengths that build on long-term economic self-sufficiency and diversification. It is important to realize that new technologies require new skills, new institutions, new infrastructure. This requires a continuous adaptation of people and organizations to new conditions, which by its very nature requires long timescales. Both the donor and the host country or organization have specific responsibilities. The IPCC (2000) distinguishes between three main forms of capacity building in the area of environmentally sound technologies: (a) human capacity; (b) organizational (or institutional) capacity, and (c) information assessment and monitoring capacity. *Human capacity* refers to the capacity at the level of individuals involved in environmentally sound technology transfer, involving technical, business, and regulatory skills. This can involve education and training of employees, forming links with other enterprises, research or other organizations in order to remain up-to-date, and to learn-by-doing, e.g. through twinning with other firms (IPCC 2000). *Organizational* or *institutional capacity* relates to the fact that the effectiveness technology transfer not only depends on the capacity of individuals, but also on the way they interact through networks of institutions. As Figure 9.8 illustrates, a variety of institutional stakeholders are involved, including governments, research institutes, private firms, financial organizations and, often, community groups. In order for the technology transfer process to work effectively at all stages, these institutions should have the ability to communicate effectively and work together towards a common goal. Capacity building efforts therefore also should be aiming at the strengthening of these networks. Examples include: (a) the implementation of participatory approaches with all stakeholders; (b) the encouragement of industry, professional, and consumer organizations; (c) sometimes decentralization of government decision-making; (d) and expansion of opportunities for new actors, e.g. consultants, energy service companies, information brokers, etc.

References

Alcamo, J. and Swart, R. (1998) Future trends of land-use emissions of major greenhouse gases. *Mitigation and Adaptation Strategies to Global Change*, **3**, 343–81.

Anderson, S. O., Carvalho, S. M. M., Devotta, S., Fujimoto, Y., Harnisch, J., Kucnerowicz-Polak, B., Kuijpers, L., McFarland, M., Moomaw, W. and Moreira, J. R. (2001) Options to reduce global warming contributions from substitutes for ozone-depleting substances. In B. Metz, O. Davidson, R. Swart and J. Pan, eds., *Climate Change 2001: Mitigation*. Cambridge: Cambridge University Press.

Banuri, T., Weyant, J., Akumu, G., Najam, A., Pinguelli Rosa, L., Rayner, S., Sachs, W., Sharma, R. and Yohe, G. (2001) Setting the stage: climate change and sustainable development. In B. Metz, O. Davidson, R. Swart and J. Pan, eds., *Climate Change 2001: Mitigation*. Cambridge: Cambridge University Press.

Barker, T., Srivastava, L., Al-Moneef, M., Bernstein, L., Criqui, P., Davis, D., Lennon, S., Li, J., Torres-Martinez, J. and Mori, S. (2001) sector costs and ancillary benefits of mitigation. In B. Metz, O. Davidson, R. Swart and J. Pan, eds., *Climate Change 2001: Mitigation*. Cambridge: Cambridge University Press.

Bashmakov, I., Jepma, C., Bohm, P., Gupta, S., Haites, E., Heller, T., Montero, J-P., Pasco-Font, A., Stavins, R., Turkson, J., Xu, H. and Yamaguchi, M. (2001) Policies, measures and instruments. In B. Metz, O. Davidson, R. Swart and J. Pan, eds., *Climate Change 2001: Mitigation*. Cambridge: Cambridge University Press.

Bolin, B., Sukumar, R., Ciais, P., Cramer, W., Jarvis, P., Khesgi, H., Nobre, C., Semenov, S. and Steffen, W. (2000) Global perspective. In R. T. Watson, I. R. Noble, B. Bolin, N. H. Ravindranath, D. J. Verardo and D. J. Dokken, eds., *IPCC Special Report on Land Use, Land-use Change and Forestry*. Cambridge: Cambridge University Press.

Burschel, P., Kuersten, E. and Larson, B. C. (1993) Die Rolle von Wald und Forstwirtschaft im Kohlenstoffhaushalt. Eine Betrachtung fur die Bundesrepublik Deutschland. (Role for forests and forestry in the carbon cycle: a tryout for Germany). *Forstliche Forschungsberichte Munchen, Schriftenreihe der Forstwissenschaftlichen Fakultat der Universitat Munchen und der Bayerischen Forstlichen Versuchs – und Forschungsanstalt*, **126**.

Ester, P. and Vinken, H. (2000) *Sustainability and the Cultural Factor: Results from the Dutch GOES Mass Public Modukle*, National Research Programme Report No. 410 200 048. Bilthoven: Globus, Institute for Globalization and Sustainable Development, Tilburg.

Garratt, J. R. (1993) Sensitivity of climate simulations to land-surface and atmospheric boundary-layer treatments – a review. *Journal of Climate*, **6**, 419–49.

Grubb, M. and Ramakrishna, K. (2000) International agreements and legal structures. In B. Metz, O. R. Davidson, J.-W. Martens, S. N. M. van Roijen and L. van Wie McGrory, eds., *Methodological and Technological Issues in Technology Transfer: A Special Report of IPCC Working Group III*. Cambridge: Cambridge University Press.

IPCC (2001) B. Metz, O. Davidson, R. Swart and J. Pan, eds., *Climate Change 2001: Mitigation*. Contribution or Working Group III to the *Third Assessment Report* of the IPIECA (1999) *Technology Assessment in Climate Change Mitigation*. Report of the

IPIECA Workshop. Paris: International Petroleum Industry Environmental Conservation Association.

Jepma, C. and Munasinghe, M. (1998) *Climate Change Policy: Facts, Issues and Analyses*. Cambridge: Cambridge University Press.

Kauppi, P., Sedjo, R., Apps, M., Cerri, C., Fujimori, T., Janzen, H., Krankina, O., Makundi, W., Marland, G., Masera, O., Nabuurs, G.-J., Razali, W. and Ravindranath, N. H. (2001) Technological and economic potential of options to enhance, maintain, and manage biological carbon reservoirs and geo-engineering. In *Climate Change 2001: Mitigation*. B. Metz, O. Davidson, R. Swart and J. Pan, eds., Cambridge: Cambridge University Press.

Kram, T., Gielen, D. J., Bos, A. J. M., de Feber, M. A. P. C., Gerlagh, T., Groenendaal, B. J., Moll, H. C., Bouwman, M. E., Daniëls, B. W., Worrell, E., Hekkert. M. P., Joosten, L. A. J., Groeneween, P. and Goverse, T. (2001) *The MATTER Project: Integrated Energy and Materials Systems Engineering for Greenhouse Gas Emission Mitigation*. Dutch National Research Program Report 410 200 055. [location and name of publisher]

Makundi, W. R., Razali, W., Jones, D. and Pinso, C. (1998) Tropical forests in the Kyoto Protocol: prospects for carbon offset projects after Buenos Aires. *Tropical Forestry Update*, 8(4), 5–8. [Yokohama: International Tropical Timber Organization.]

Margolick, M. and Russell, D. (2001) *Corporate Greenhouse Gas Reduction Targets*. Arlington VA: Pew Center on Global Climate Change.

Matthews, R., Nabuurs, G. J., Alexeyev, V., Birdsey, R. A., Fischlin, A., MacLaren, J. P., Marland, G. and Price, D. (1996) WG3 summary: Evaluating the role of forest management and forest products in the carbon cycle. In M. J. Apps, and D. T. Price, eds., *Forest Ecosystems, Forest Management and the Global Carbon Cycle*. NATO Advanced Science Institute Series, NATO-ASI Vol. I, 40, Berlin: North Atlantic Treaty Organization, Proceedings of a workshop held in September 1994 in Banff, Canada, pp. 293–301.

Metz, B., Davidson, O., Swart, R. and Pan, J. (2001) *Climate Change 2001: Mitigation*. Cambridge: Cambridge University Press.

Moll, H. C., Nonhebel, S., Biesiot, W., Kramer, K. J., Wilting, H. C., Groot Marcus, J. P., Brouwer, N. M., Uitdenbogerd, D. E., Blok, K., van den Berg, M., Potting, J., Reinders, A. H. M. E. and Vringer, K. (2001) *Evaluation of Options for Reductions of Greenhouse Gas Emissions by Changes in Household Consumption Patterns*. De Centrum voor Energie en Milieukunde (Center for Energy and Environmental Studies), Groningen: NOP Report No. 410 200 059. Bilthoven.

Moomaw, W. R., Moreira, J. R., Blok, K., Greene, D. L., Gregory, K., Jaszay, T., Kashiwagi, T., Levine, M., McFarland, M., Siva Prasad, N., Price, L., Rogner, H-H., Sims, R., Zhou, F. and Zhou, P. (2001) Technological and economic potential of greenhouse gas emissions reduction. In B. Metz, O. Davidson, R. Swart and J. Pan, eds., *Climate Change 2001: Mitigation*. Cambridge: Cambridge University Press.

Nabuurs, G. J., Dolman, A. V., Verkaik, E., Kuikman, P. J., van Diepen, C. A., Whitmore, A., Daamen, W., Oenema, O., Kabat, P. and Mohren, G. M. J. (2000) Article 3.3 and 3.4. of the Kyoto Protocol – consequences for industrialised countries' commitment, the monitoring needs and possible side effect. *Environmental Science and Policy*, **3**(2/3), 123–34.

Nijkamp, P. (2001) *Globalization, International Transport and the Global Environment.* Amsterdam University and National Research Programme Report No. 410 200 075, Bilthoven: University of Amsterdam.

OECD (1999) *Indicators for the Integration of Environmental Concerns into Transport Policies.* Paris: Organization for Economic Co-operation and Development Environment Directorate.

Penner, J. E., Lister, D. H., Griggs, D. J., Dokken, D. J. and McFarland, M. (1999) *Aviation and the Global Atmosphere.* Cambridge: Cambridge University Press.

Penner, J. E., Lister, D. H., Griggs, D. J., Dokken, D. J. and McFarland, M. (1999) *Aviation and the Global Atmosphere.* A Special Report of the Intergovernmental Panel on Climate Change Working Groups I and II in collaboration with the Scientific Assessment Panel of the Montreal Protocol on substances that deplete the ozone layer. Cambridge: Cambridge University Press.

Radkha, M., Aloisi de Larderel, J., Abeeku Brew-Hammond, J. P. and Xu Huaqing (2001) Trends in technology transfer. In B. Metz, O, Davidson, J. W. Martens, S. N. M. van Rooijen and L. van Wie McGrory, eds., *Methodological and Technological Issues in Technology Transfer.* Intergovernmental Panel on Climate Change. Cambridge and New York: Cambridge University Press.

Ramaswamy, V., Boucher, O., Haigh, J., Hauglustaine, D., Haywood, J., Myhre, G., Nakajima, T., Shi, G. Y., Solomon, S. (2001) Radiative forcing of climate change. In J. Houghton, Y. Ding, D. J. Griggs, M. Nogure, P. J. van der Linden, X. Dai, K. Maskell and C. A. Johnson, eds., *Climate Change 2001: The Scientific Basis.* Cambridge: Cambridge University Press.

Raskin, P., Banuri, T., Gallopin, G., Gutman, P., Hammond, A., Kates, R. and Swart, R. (2001) *Great Transition: The Promise and Lure of the Times Ahead* (in press). Boston MA: Global Scenario Group, Stockholm Environment Institute.

Rosegrant, M., Agcaoili-Sombilla, M. and Perez, N. D. (1995) *Global Food Projections to 2020: Implications for Investment.* Discusson Paper No. 5, Washington: International Food Policy Research Institute.

Sathaye, J., Bouille, D., Biswas, D., Crabbe, P., Geng, Luis, Hall, D., Imura, H., Jaffe, A., Michaelis, L., Peszko, G., Verbruggen, A., Worrell, E. and Yamba, F. (2001) Barriers, opportunities, and market potential of technologies and practices. In *Climate Change 2001: Mitigation.* B. Metz, O. Davidson, R. Swart and J. Pan, eds., Cambridge: Cambridge University Press.

Sedjo, R. A., and Botkin, D. (1997) Using forest plantations to spare natural forests. *Environment*, **30**(10), 14–20 and 30.

Toth, F. L., Mwandosya, M., Carraro, C., Christensen, J., Edmonds, J., Flannery, B., Gay-Garcia, C., Lee, Hoesung, Meyer-Abich, K. M., Nikitina, E., Rahman, A., Richels, R., Ye, R., Villavicencio, A., Wake, Y. and Weyant, J. (2001) Decision-making frameworks. In *Climate Change 2001: Mitigation.* D. B. Metz, O. R. Davidson, R. Swart and J. Pan, eds., Cambridge: Cambridge University Press.

Trindade, S. C., Siddiqi, T. and Martinot, E. (2000) Managing technological change in support of the Climate Change Convention: a framework for decision-making. In B. Metz, O. R. Davidson, J.-W. Martens, S. N. M. van Rooijen and L. van Wie McGrory, eds., *IPCC Special Report on Methodological and Technological Issues in Technology Transfer*. Cambridge: Cambridge University Press.

UNFCCC, 2002

Watson, R. T., Albritton, D. L., Barker, T., Bashmakov, I. A., Canziani, O., Christ, R., Cubasch, U., Davidson, O. R., Gitay, H., Griggs, D., Houghton, J., House, J., Kundzewicz, Z., Lal, M., Leary, N., Magadza, C., McCarthy, J. J., Mitchell, J. F. B., Moreira, J. R., Munasinghe, M., Noble, I., Pachauri, R., Pittock, B., Prather, M., Richels, R. G., Robinson, J. B., Sathaye, J., Schneider, S., Scholes, R., Stocker, T., Sundararaman, N., Swart, R., Taniguchi, T. and Zhou, D. (2001) *Climate Change 2001: Synthesis Report*. Cambrige: Cambridge University Press.

Young, A. (1997) *Agroforestry for Soil Management*, 2nd edn. Oxford: Commonwealth Agricultural Bureaux International.

10

Assessment of mitigation costs and benefits

In Chapter 8, various approaches to the costing of mitigation options were discussed. In this chapter, we discuss how some of these approaches have been applied recently, and what their outcomes are. Section 10.1 sets out the various modelling techniques being used. In Section 10.2, we discuss costs of mitigation in various key greenhouse gas emitting sectors. In Section 10.3, the costs of mitigation at the national to global scale are explored, focusing on the costs of meeting the Kyoto Protocol targets. Section 10.4 addresses a number of key issues in costing analysis that should be taken into account when evaluating results, notably the existence of no-regrets options, the concept of double dividend, carbon leakage, spill-over effects, and avoided damage. Finally, Section 10.5 looks at the costs associated with meeting various stabilization targets for greenhouse gas emissions.

10.1 Costing models of greenhouse gas abatement

Evaluation of abatement strategies usually implies the usage of models. The modelling of climate mitigation strategies is complex, and a number of techniques have been applied including input–output models, macroeconomic models, computable general equilibrium models, and energy-sector-based models. Hybrid models also have been developed to provide more detail on the structure of the economy and the energy sector. The appropriate use of these models depends on the subject evaluated and the availability of data. Usually, different models give different, complementary, answers to different questions.

The costs of climate change policy have been assessed using various analytical approaches, including the following.

1. Input–output models describe the complex inter-relationships within the economic sector in its use of sets of simultaneous linear equations. This model is used when sectoral consequences of mitigation or adaptation options are of interest (Frankhauser & McCoy 1995).

2. Macroeconomic models describe investment and consumption patterns in various sectors, and emphasize short-term dynamics associated with greenhouse gas emission reduction policies. These models are well suited to the consideration of economic effects of greenhouse gas emission reduction policies in the short- to medium-term horizon.

3. Computable general equilibrium models construct the behaviour of economic agents based on microeconomic principles, typically simulating markets for factors of production (e.g. labour, capital, and energy), products, and foreign exchange, with equations that specify supply and demand behaviour.

4. Dynamic energy optimization models minimize the total costs of the energy system, including all end-use sectors, over a 40–50 year horizon and thus compute a partial equilibrium for the energy markets. The costs include investment and operation costs of all sectors. This model is useful in assessing the dynamic aspects of greenhouse gas emissions reductions and potential costs.

5. Integrated energy systems simulation models are bottom-up models that include a detailed representation of energy demand and supply technologies that include end-use, conversion and production technologies. These simulation models are best suited for short- to medium-term studies in which detailed technology information helps explain a major part of energy needs.

6. Partial forecasting models are used to forecast energy supply and demand, either for single time periods or with time development and varying degrees of dynamics and feedback.

In technology-focused, 'bottom-up' models and approaches, the cost of mitigation is derived from the aggregation of technological and fuel costs (e.g. investments, operation and maintenance costs, and fuel procurement) as well as revenues and costs from import and exports, the latter being a more recent trend.

Models can be ranked along two classification axes. First, they range from simple engineering-economic calculations (technology-by-technology) to integrated partial equilibrium models of whole energy systems. Second, they range from the strict calculation of direct technical costs of reduction, to the consideration of observed technology adoption behaviour of markets, and to the welfare losses due to demand reductions and revenue gains and losses due to changes in trade.

This leads to contrasting two generic approaches: (a) the engineering–economics approach, and (b) least-cost equilibrium modelling. In (a), each technology is assessed independently via an accounting of its costs and savings. Once these elements have been estimated, a unit cost can be calculated for each action, and each action can be ranked according to its costs. This approach is very useful for pointing out the potentials for negative cost abatements due to the 'efficiency gap' between the best available technologies and technologies currently in use. However, its most important limitation is that studies neglect, or do not treat in a systematic way, the interdependence of the various actions under examination.

Partial equilibrium least-cost models have been constructed to remedy this defect, by considering all actions simultaneously and selecting the optimal bundle of actions in all sectors and at all time periods. These more integrated studies conclude higher total costs of greenhouse gas mitigation than the strict technology by technology studies. Based on an optimization framework, they give very easily interpretable results that compare an optimal response to an optimal baseline; however, their limitation is that they rarely calibrate the base year of the model to the existing non-optimal situation, and assume implicitly an optimal baseline. Consequently, they provide no information about the negative cost potentials.

Bottom-up approaches have produced a wealth of new results for both Annex I and non-Annex I countries, as well as for groups of countries. Furthermore, they have extended their scope much beyond the classical computations of direct abatement costs by inclusion of demand effects and some trade effects. However, the modelling results show considerable variations from study to study that are explained by a number of factors, some of which reflect the widely differing conditions that prevail in the countries studied (e.g. energy endowment, economic growth, energy intensity, industrial and trade structure), and others that reflect modelling assumptions and assumptions about negative cost potentials.

Differences in costs of mitigation appear to be caused largely by different approaches and assumptions, with the most important being the type of model adopted. Bottom-up engineering models assuming new technological opportunities tend to show benefits from mitigation. Top-down time series econometric models appear to show higher costs than top-down general equilibrium models. The main assumptions leading to reduced costs in these latter models are that:

1. New flexible instruments (e.g. emissions trading and joint implementation) are included.

2. Revenues from taxes or permit sales are assumed to be returned to the economy by reducing burdensome taxes.
3. Ancillary benefits are included in the results.

Finally, long-term technological progress and diffusion are seen largely as exogenous in the top-down models; different assumptions or a more integrated, dynamic, treatment could have major effects on the results.

Short- and medium-term marginal abatement costs, which govern most of the macroeconomic impacts of climate policies, are very sensitive to uncertainty regarding baseline scenarios (rate of growth and energy intensity) and technical costs. Even with significant negative cost options, marginal costs may rise quickly beyond a certain anticipated mitigation level. This risk is far lower in models allowing for carbon trading. Over the long term this risk is reduced as technical change curtails the slope of marginal cost curves.

Mitigation and adaptation usually are modelled separately as a necessary simplification to gain traction on an immense and complex issue. Consequently, the costs of risk reduction action are estimated separately, and therefore each measure is potentially biased. This realization suggests that more attention to the interaction of mitigation and adaptation, and its empirical ramification, is worth while, though uncertainty about the nature and timing of impacts, including surprises, will constrain the extent to which the associated costs can be fully internalized.

The main categories of climate change mitigation policies include: (a) market-oriented policies; (b) technology-oriented policies; (c) voluntary policies, and (d) other policies, e.g. research and development, education, and communication programmes (see Chapter 9). Climate change mitigation policies can include all four of the above policy elements. Most analytical approaches, however, consider only some of the four elements. Economic models, for example, mainly assess market-oriented policies, and in some cases, technology policies – primarily those related to energy supply options – while engineering approaches focus mainly on supply- and demand-side technology policies. Both of these approaches are relatively weak in the representation of voluntary agreements, research and development and education and communication policies.

There are a number of key linkages between mitigation-costing issues and broader development impacts of the policies, including macroeconomic impacts, employment creation, inflation, marginal costs of public funds, capital availability, spillovers, and trade (see Section 10.2 for more details).

Integrated assessment models combine key elements of biophysical and economic systems into one integrated system so that the costs of climate protection can be assessed. They have been of use in evaluating policies and responses, structuring knowledge and prioritizing uncertainties (Parson & Fisher-Vanden 1997).

10.2   Sectoral costs and benefits of both domestic and international mitigation policies and related measures

Policies adopted to mitigate global warming will have implications for specific sectors. A sectoral assessment helps to: (a) put the costs in perspective; (b) identify the potential losers and the extent and location of the losses, and (c) identify the sectors that may benefit. In Chapter 8 we discussed the technological mitigation options in the various sectors, barriers towards their implementation, and policies and instruments to overcome these. Here, we look at the sectoral economic costs and benefits of mitigation.

There is a fundamental problem for mitigation policies. It is well established that, compared to the situation for potential gainers, the potential sectoral losers are easier to identify, and their losses are likely to be more immediate, more concentrated, and more certain. The potential sectoral gainers (apart from the renewables sector and perhaps the natural gas sector) can expect only a small, diffused, and rather uncertain gain, spread over a long period. Indeed, many of those who may gain do not exist, being future generations and industries yet to develop.

It is also well established that the overall effects on gross domestic product (GDP) of mitigation policies and measures, whether positive or negative, conceal large differences between sectors. In general, the energy intensity and the carbon intensity of the economies will decline. The coal industry, and perhaps that for oil, are expected to lose substantial proportions of their traditional output relative to those in the reference scenarios, though the impact of this on them will depend on diversification, and other sectors may increase their outputs but by much smaller proportions.

Carbon dioxide emissions have tended to grow more slowly than GDP in a number of countries over the past 40 years. The reasons for such trends vary, but include:

1.   A shift away from coal and oil and towards nuclear and gas as the source of energy.
2.   Improvements in energy efficiency by industry and households.
3.   A shift from heavy manufacturing towards more service- and information-based economic activity.

These trends will be encouraged and strengthened by mitigation policies.

The direct costs for fossil fuel consumption are accompanied by environmental and public health benefits associated with a reduction in the extraction and burning of the fuels. These benefits come from a reduction in the damages caused by these activities, especially a reduction in the emissions of pollutants associated with combustion, such as sulphur dioxide, nitrous oxides, carbon monoxide and other chemicals, and particulate matter. This will improve local and regional air and water quality, and thereby lessen damage to human, animal, and plant health, and to ecosystems. If all the pollutants associated with greenhouse gas emissions are removed by new technologies or end-of-pipe abatement (e.g., flue gas desulphurization on a power station combined with removal of all other non-greenhouse gas pollutants), then this ancillary benefit will no longer exist. But such abatement is limited at present and it is expensive, especially for small-scale emissions from dwellings and cars.

Researchers make a distinction between project, sector, and economywide analyses. Project level analysis considers a 'stand-alone' investment assumed to have insignificant secondary impacts on markets. Methods used for this level include cost–benefit analysis, cost-effectiveness analysis, and lifecycle analysis. Sector level analysis examines sectoral policies in a 'partial equilibrium' context in which all other variables are assumed to be exogenous. Economywide analysis explores how policies affect all sectors and markets, using various macroeconomic and general equilibrium models. A trade-off exists between the level of detail in the assessment and complexity of the system considered.

A combination of different modelling approaches is required for an effective assessment of climate change mitigation options in economic sectors. For example, detailed project assessment has been combined with a more general analysis of sectoral impacts, and macroeconomic carbon tax studies have been combined with the sectoral modelling of larger technology investment programmes.

10.2.1   Coal

Within this broad picture, certain sectors will be affected substantially by mitigation. Relative to the reference case, the coal industry, producing the most carbon-intensive of products, faces almost inevitable decline in the long term, relative to the baseline projection. Technologies still under development, such as carbon dioxide removal and storage from coalburning plants and in situ gasification, could play future roles in maintaining the output of coal while avoiding carbon dioxide and other emissions. Particularly large effects on the coal sector are expected from policies such as the removal of fossil fuel subsidies, or the restructuring of energy taxes so as to tax the carbon content rather than the energy content of fuels. It is a well-established finding that removal of the subsidies would result in substantial reductions in greenhouse gas emissions, as well as stimulating economic growth. However, the effects in specific countries depend heavily on the type of subsidy removed and the commercial viability of alternative energy sources, including imported coal.

Reduction in coal consumption (and, hence, production) would have certain impacts on economies (IEA 1997, 1999; WCI 1999). These include: (a) reduced economic activity in coal-producing countries due to reduced coal sales; (b) job losses in coalmining, transport, and processing sectors; (c) potential for 'stranding' of coalmining and processing assets; (d) closure of mines (which are expensive to reopen); (e) negative impacts on communities due to mine closures; (f) higher trade deficits from reductions in coal exports, especially for Colombia, Indonesia, and South Africa (Knapp 2000); (g) reduction in national energy security due to increased reliance on imported energy sources, and (h) possible slowdown in economic growth during transition from coal to other energy sources.

Ancillary benefits associated with reduction in coalburning include public health impacts. In the coal industry, mitigation would increase energy efficiency in coal utilization (Li *et al.* 1995). In addition, new, high-efficiency, clean coal technologies (IEA 1998a) could lead to enhanced skill levels and technological capacity in developing nations. Further benefits include increased productivity due to increased market pressures and the extension of the life of coal reserves (IPCC 2001). Enhanced research and development in the coal industry also will result, due to the need to find alternatives and non-emitting applications for coal (IEA 1999). Mitigation is likely to favour coal production in non-Annex B countries due to the migration of energy intensive industries to developing countries.

10.2.2   *Petroleum and oil*

The oil industry also faces a potential relative decline, although this may be moderated by: (a) lack of substitutes for oil in transportation; (b) substitution away from solid fuels towards liquid fuels in electricity generation, and (c) the diversification of the industry into energy supply in general.

Table 10.1 shows a number of model results for the impacts of implementation of the Kyoto Protocol on oil-exporting countries. Each model uses a different measure of impact, and many use different groups of countries in their definition of oil exporters. However, all the studies show that the use of the flexibility mechanisms will reduce the economic cost to oil producers.

Studies show a wide range of estimates for the impact of greenhouse gas mitigation policies on oil production and revenue. Much of these differences are attributable to the assumptions made about: (a) the availability of conventional oil reserves; (b) the degree of mitigation required; (c) the use of emission trading; (d) control of greenhouse gases other than carbon dioxide, and (e) the use of carbon sinks. However, all studies show a net growth in both oil production and revenue to at least 2020, and significantly less impact on the real price of oil than has resulted from market fluctuations over the past 30 years (IPCC 2001).

Table 10.1. *Costs of Kyoto Protocol implementation for oil exporting region/countries*[a].

Model[b]	Without trading[c]	With Annex-I trading	With 'global trading'
G-cubed	− 25% oil revenue	− 13% oil revenue	− 7% oil revenue
GREEN	− 3% real income	'Substantially reduced loss'	N/A
GTEM	0.2% GDP loss	< 0.05% GDP loss	N/A
MS-MRT	1.39% welfare loss	1.15% welfare loss	0.36% welfare loss
OPEC model	− 17% OPEC revenue	− 10% OPEC revenue	− 8% OPEC revenue
CLIMOX	N/A	− 10% some oil-exporters' revenues	N/A

[a] The definition of oil exporting country varies: for G-cubed and the OPEC model it is the OPEC countries, for GREEN it is a group of oil exporting countries, for GTEM it is Mexico and Indonesia, for MS–MRT it is OPEC + Mexico, and for CLIMOX it is West Asian and North African oil exporters.

[b] The models all consider the global economy to 2010 with mitigation according to the Kyoto Protocol targets (usually in the models, applied to carbon dioxide mitigation by 2010 rather than greenhouse gas emissions for 2008–12) achieved by imposing a carbon tax or auctioned emission permits with revenues recycled through lump sum payments to consumers; no co-benefits (e.g. reductions in local air pollution damages) are taken into account in the results.

[c] 'Trading' denotes trading in emission permits between countries.

These studies do not consider some or all of the following policies and measures that could lessen the impact on oil exporters (Pershing 2000):

1. Policies and measures for non-carbon dioxide greenhouse gases or non-energy sources of all greenhouse gases.
2. Offsets from sinks.
3. Industry restructuring, e.g. from energy producer to supplier of energy services.
4. The use of the market power of the Organization of Petroleum Exporting Countries (OPEC).
5. Actions (e.g. of Annex B parties) related to funding, insurance, and the transfer of technology.

In addition, the studies typically do not include the following policies and effects that can reduce the total cost of mitigation.

1. The use of tax revenues to reduce tax burdens or finance other mitigation measures.
2. Environmental co- or ancillary benefits of reductions in fossil fuel use.
3. Induced technical change from mitigation policies.

As a result, the studies may tend to overstate both the costs to oil exporting countries and overall costs.

Ancillary benefits of policies to reduce the growth in demand for crude oil will slow down the rate of depletion of oil reserves, and decrease air and water pollution impacts associated with oil production.

Reductions in fossil fuel output below the baseline will not affect all fossil fuels equally. Fuels have different costs and price sensitivities; they respond differently to mitigation policies. Energy efficiency technology is fuel and combustion device-specific, and reductions in demand can affect imports differently from output. Energy intensive sectors (e.g. heavy chemicals, iron and steel, and mineral products) will face higher costs, accelerated technical or organizational change, or loss of output (again relative to the reference scenario) depending on their energy use and the policies adopted for mitigation.

10.2.3   Gas

Natural gas has the lowest carbon content of the fossil fuels, and is assumed that its use will increase with efforts to decrease carbon dioxide emissions (Ferriter 1997). Modelling studies suggest that mitigation policies may have the least impact on oil, the most impact on coal, with the impact on gas somewhere between. The high variation across studies for the effects of mitigation on gas demand is associated with the importance of its availability in different locations, its specific demand patterns, and the potential for gas to replace coal in power generation. These results are different from recent trends, which show natural gas usage growing faster than the use of either coal or oil. They can be explained as follows. In the transport sector – the largest user of oil – current technology and infrastructure will not allow much switching from oil to non-fossil fuel alternatives in Annex I countries before about 2020. Annex B countries can meet their Kyoto Protocol commitments only by reducing overall energy use, and this will result in a reduction in natural gas demand, unless this is offset by a switch towards natural gas for power generation. The modelling of such a switch remains limited in these models (Brown *et al.* 1998).

Ancillary benefits of reduced demand for natural gas result from decreased rates of depletion of this natural resource, reduced air and water pollution associated with this industry, and lower potential for natural gas explosions (IPCC 2001).

10.2.4   Electricity

Fossil fuels continue to dominate heat and electric power production and account for about one-third of the carbon dioxide emissions from the energy sector worldwide (EIA 2000). Electricity generation accounted for 2100 million tonnes of carbon (Mt carbon) or 37.5 per cent of global carbon emissions in 1990. Baseline scenarios without carbon emission policies anticipate emissions of 3500 and 4000 Mt carbon equivalent for 2010 and 2020, respectively.

Given the extensive use of fossil fuel in the production of electricity, a variety of proposals have been put forth to mitigate greenhouse gas emissions. These include proposed renewable technologies (SDPC 1996; Piscitello & Bogach 1997), carbon taxes, nuclear power, etc. In the electricity sector, mitigation policies either mandate or directly provide incentives for increased use of zero-emission technologies (e.g. nuclear, hydropower, and other renewables) and lower-greenhouse-gas-emitting generation technologies, e.g. as combined-cycle natural gas. Or they drive their increased use indirectly by more flexible approaches that place a tax on, or require a permit for emission of, greenhouse gases. Either way, the result will be a shift in the mix of fuels used to generate electricity towards increased use of the zero- and low-emission generation technologies, and away from the high-emission fossil fuels (Criqui, Kouvaritakis & Schrattenholzer 2000).

Nuclear power would have substantial advantages as a result of greenhouse gas mitigation policies, as it produces negligible amounts of greenhouse gases. In spite of this advantage, nuclear power is not seen as the solution to the global warming problem in many countries. The main issues are: (a) the high costs compared to alternative CCGTs; (b) public concern over operating safety and waste; (c) safety of radioactive waste management and recycling of nuclear fuel; (d) the risks of nuclear fuel transportation, and (e) nuclear weapons proliferation (Hagen 1998). Combined-cycle gas turbines are expected to be the largest provider of new capacity between 2005 and 2020 worldwide, and will be strong candidates to replace coal-fired power stations where additional gas supplies can be made available. Biomass, based mainly on wastes, agricultural and forestry by-products and wind power, are also potentially capable of making major contributions by 2020. Hydropower is an established technology, and further opportunities exist beyond those anticipated to contribute to reducing carbon dioxide equivalent emissions. Finally, while costs of solar power are expected to decline substantially, it is likely to remain an expensive option by 2020 for central power generation, but likely also to make increased contributions in niche markets and off-grid generation. The best mitigation option is likely to be dependent on local circumstances, and a combination of these technologies.

Ancillary benefits expected from greenhouse gas mitigation would include sales and employment growth for the manufacture of new-generation technologies and the fuels used in these technologies. Ancillary benefits associated with increased use of renewable resources include: (a) increased social and economic development in rural areas (i.e enhanced employment in rural areas, decreased rural urban migration and poverty); (b) land restoration (new rural development, prevention of erosion etc.); (c) reduced emissions, and potential for fuel diversity, and (d) elimination of the need for costly disposal of waste, e.g. crop residues, and household refuse, (IPCC 2001).

Ancillary costs associated with mitigation of greenhouse gases in the electricity sector include the loss of employment and revenue for those producing and previous technologies (e.g. coal plants) and fuels (e.g. coal and oil) prior to mitigation (OECD 1998). Some environmental impacts associated with renewable technologies include the increased concern about the ecological impacts of intensive cultivation of biomass energy, loss of land, and other negative impacts of hydropower development, and noise, killing of birds, and visual interference by wind power (Pimental *et al.* 1994; IEA 1997, 1998b; Miyamoto 1997).

10.2.5   *Transport*

Transport energy has been growing steadily around the world with the largest increases in Asia, the Middle East, and North Africa (Schafer 1998). Unless highly efficient vehicles (e.g. fuel cell vehicles) become available rapidly, there are few options available to reduce transport energy use in the short term which do not involve significant economic, social, or political costs (Michaelis & Davidson 1996). No government has yet demonstrated policies that can reduce the overall demand for mobility, and all governments find it politically difficult to contemplate such measures (IEA 1997b).

In 1995, the transport sector contributed 22 per cent of global energy-related carbon dioxide emissions; globally, emissions from this sector are growing at a rapid rate of approximately 2.5 per cent annually. Since 1990, principal growth has been in the developing countries (7.3 per cent per year in the Asia–Pacific region) and is actually declining at a rate of 5 per cent per year for the economies in transition. Hybrid petrol–electric vehicles have been introduced on a commercial basis with fuel economies 50–100 per cent better than those of comparably sized four-passenger vehicles. Nevertheless, large-scale penetration of such novel technologies is still hampered by various kinds of barriers, including prices, which would have to be removed by policy initiatives.

Initiatives to improve fuel economy have led to manufacturers agreeing voluntarily to reduce the fuel consumption of cars by 20 per cent before 2010, with only a small increase in costs. Bose (1998) found that improving public transport to meet 80 per cent of the travel demand, and promoting cleaner fuels and engine technologies in six Indian cities, can reduce significantly both emissions and fuel consumption. Automotive emissions are reduced by 30–80 per cent compared to the baseline case. In the case of freight transportation, the impact of carbon fees to reduce US carbon emissions to 3 per cent below 1990 levels to meet the Kyoto Protocol commitments, was estimated to increase the cost of diesel fuel by

US $0.68 per gallon, but resulted in only 4.9 per cent reduction in US freight travel. The cost of marine fuel is projected to rise by US $0.84 per gallon, but domestic shipping is expected to decline only by 10 per cent (EIA 1998). Greenhouse gas mitigation policies lead to reduced air emissions associated with less fuel use and corresponding reductions in damages caused by these emissions (Ross 1999). Reduced congestion, fewer traffic crashes, less noise and less road damage are some other co-benefits of greenhouse gas mitigation policies. A recent study of air-pollution-related health costs in Switzerland, Austria, and France shows the total of US $49.7 billion, or between 1.1 and 5.8 per cent of GDP (Sommer *et al.* 1999).

Technical efficiency improvements, in the absence of complementary fiscal policies, are subject to a 'rebound effect', in that they reduce the fuel cost of travel. Rebound effects in the USA amount to about 20 per cent of the potential greenhouse gas reductions (Greene 1999). In Europe, where fuel prices are higher, rebound effects may be as large as 40 per cent (Michaelis 1997).

Local concerns, traffic congestion, and air pollution are key drivers for transport policy (Bose 2000). Attempts have been made to improve fuel consumption of passenger cars in order to reduce greenhouse gas emissions. In 1993, the USA aimed to develop a car with triple the current fuel economy (80 miles per gallon). The incremental cost of these vehicles was estimated to be as low as $2500 per car (DeCicco & Mark 1997) to as high as $6000 per car (Duleep 1997). Since these vehicles are designed to meet the emission standards anticipated to be in effect when they are produced, no ancillary local air pollution benefit is expected. Dowlatabadi, Lave and Russell (1996) found that increasing fuel economy to 60 miles per gallon had little beneficial effect on urban ozone concentration, and could decrease the safety of passenger cars unless offsetting measures were taken.

The Australian Bureau of Transport and Communication economics (BTCE 1996) found that: (a) metropolitan road user charges; (b) reduced urban public transport fares; (c) citywide parking charges; (d) labelling of new cars to inform buyers about fuel efficiency, and (e) shifting intercapital freight from road to rail had zero or negative costs to society as a whole.

Improving public transport, promoting cleaner fuels and improved engine technologies (i.e. catalytic converters, unleaded petrol, electric vehicles, and phasing out two-stroke engines) can reduce significantly both emissions and fuel consumption (Bose 1998). Projections to 2010 and beyond reflect the belief that transport growth will continue to outpace efficiency improvements, and that, without significant policy intervention, global transport greenhouse gas emissions will be 50–100 per cent greater in 2020 than 1995 (Bose 1999; Denis & Koopman 1998; Jansen & Denis 1999; van Wee & Annema 1999).

Ancillary benefits of reductions in traffic congestion include: reduced damage from air pollution (WHO 1999); reduced congestion (Barker, Johnstone & O'Shea 1993), fewer traffic crashes (Ross 1999), less noise, and less road damage.

Substantial additional improvements in aircraft energy efficiency are needed to control future emissions, possibly complemented by policies that increase the price of, and therefore affect the amount of, air travel. Raising the price of air travel by taxes is fraught with a number of political hurdles. Many of the bilateral treaties that currently govern the operation of the air transport system contain provisions for exemptions of taxes and charges, other than for the cost of operating and improving the system. Although improvements in aircraft and engine technology and in the efficiency of the air traffic control system will bring environmental benefits, these will not offset fully the effects of increased emissions from the projected growth in aviation (Penner *et al.* 1999).

10.2.6   *Industry and manufacturing*

The effect of greenhouse gas mitigation on manufacturing sectors is likely to be very mixed, depending on the use of carbon based fuels as inputs and the ability of the producer to adapt production techniques and to pass increases in costs to the consumer (IPCC 2001). Industrial emissions account for 43 per cent of carbon released in 1995. Industrial sector carbon emissions grew at a rate of 1.5 per cent per year between 1971 and 1995, slowing to 0.4 per cent per year since 1990. Industries continue to find more energy-efficient processes and reductions of process-related greenhouse gases. This is the only sector that has shown an annual decrease in carbon emissions in OECD economies (−0.8 per cent per year between 1990 and 1995). The carbon dioxide from economies in transition declined most strongly (−6.4 per cent per year between 1990 and 1995, when total industrial production dropped).

Possible mitigation options include: (a) conserve energy (adopt more efficient technologies); (b) shift to products with lower carbon intensities (e.g. electronics and pharmaceuticals); (c) accept extra taxation or emission permits and the possible effects on profits and product sales; (d) shift production abroad; (e) improve material efficiency (including recycling, more efficient product design, and material substitution); (f) switch fuel; (g) remove carbon dioxide and store it, and (h) apply blended cements (IPCC 2001).

Ancillary benefits of proposed mitigation options include: (a) the adoption of energy conservation technologies; (b) the accumulation of scientific and technological knowledge that contributes to the development of new products and processes, and (c) the internationalizing of manufacturing that stimulates technological transfer to developing regions and greater equity in wealth distribution (IPCC 2001). Mitigation policies would increase output and employment in the energy-equipment industries. Under an innovation scenario for greenhouse gas mitigation of 10 per cent relative to 1990, it is anticipated that by 2010, the GDP of the USA is expected to increase by 0.02 per cent (Laitner, Bernow & DeCicco 1998).

The potential for carbon dioxide emission reductions in China based on energy-saving technology options under various tax and subsidy measures show that, for example, by adopting advanced coking oven systems by the iron and steel industry, only 15.9 per cent of existing ovens will be replaced by advanced ones by 2010. With a carbon tax, the replacement share rises to 62 per cent. With a tax, and a subsidy for advanced ovens, this share rises to 100 per cent (Jiang *et al.* 1998). Energy-saving technology (e.g. material and thermal recycling) has a large potential to reduce carbon emissions. For example, Ikeda *et al.* (1995) estimated that the utilization of by-products in the steel and iron industry, and of steel scrap, could reduce carbon dioxide emissions by 2.4 per cent in Japan in 1990.

Some ancillary costs of transferring production to non-Annex B countries include losses in Annex B manufacturing employment, and increases in non-Annex B emissions.

Greenhouse gas mitigation policies would be favourable for various industries in the area of environmentally sound technologies, but could have negative economic effects on inefficient and energy intensive industries.

Use of *hydrofluorocarbons* (HFCs) and, to a lesser extent, perfluorocarbons (PFC) has grown, as these chemicals replaced about 8 per cent of the projected use of chlorofluorocarbons (CFCs) by weight in 1997. Considering energy efficiency simultaneously with ozone layer protection is important, especially in the context of developing countries, in which markets have just begun to develop and are expected to grow at a fast rate. Based on current trends and assuming no new uses outside the ozone-depleting substance substitution area, HFC production is projected to be 370 kt or 170 Mt carbon equivalent per year by 2010, while PFC production is expected to be less than 12 Mt carbon equivalent per year. The largest emissions are likely to be associated with mobile air-conditioning followed by commercial refrigeration and stationary air-conditioning. Hydrofluorocarbon use in foam blowing is currently low, but if HFCs replace a substantial part of the hydrochlorofluorocarbons (HCFCs) used here, their use is projected to reach 30 Mt carbon equivalent per year by 2010, with emissions in the order of 5–10 Mt carbon equivalent per year. The future market for HFCs is very dependent upon the choice of (H)CFC replacement technology. If containment and recycling of HFCs would be considered as an acceptable, feasible, and effective option, HFCs (being potent greenhouse gases that do not, however, deplete stratospheric ozone) could be produced increasingly even in a greenhouse-gas-constrained world. Alternative options (e.g. using other substances, which would have neither an ozone depletion potential nor a greenhouse warming potential, as for several hydrocarbons, or other processes, would have a negative effect on the producers of fluoro- compounds, but offer new opportunities for other industries.

Industries concerned directly with mitigation are likely to benefit from action. These include renewable and nuclear electricity, producers of mitigation equipment (incorporating energy- and carbon-saving technologies), agriculture and forestry producing energy crops, and research services producing energy- and carbon-saving research and development. They may benefit in the long term from the availability of financial and other resources that would otherwise have been taken up in fossil fuel production. They may also benefit from reductions in tax burdens if taxes are used for mitigation and the revenues recycled as reductions in employer, corporate, or other taxes. Those studies that report reductions in GDP do not always provide a range of recycling options, suggesting that policy packages increasing GDP have not been explored. The extent and nature of the benefits will vary with the policies followed. Some mitigation policies can lead to net overall economic benefits, implying that the gains from many sectors will outweigh the losses for coal and other fossil fuels, and energy intensive industries. In contrast, other less well-designed policies can lead to overall losses.

10.2.7 *Building and construction sector*

Buildings require energy for lighting, space heating and cooling, and electricity for equipment. Energy consumption in buildings accounted for 31 per cent of global energy-related carbon dioxide emissions in 1995, and emissions have grown at an annual rate of 1.8 per cent since 1971. Mitigation options include changes to material used, design and heat control, and increase in quality of building.

There are many new cost-effective technologies and measures that have the potential significantly to reduce the growth in greenhouse gas emissions from buildings in both developing and developed countries, by improving energy performance of whole buildings as well as reducing emissions from appliances and equipment within buildings.

Implementation of energy-efficient technologies and measures can reduce carbon dioxide emissions from residential buildings in 2010 by 325 Mt carbon in developed countries and the economies-in-transition region at costs ranging from −250 to −US $150 per tonne of carbon saved; and by 125 Mt carbon in developing countries at from −200 to US $50 per tonne of carbon saved. Similarly, carbon dioxide emissions from commercial buildings in 2010 can be reduced by 185 Mt carbon in developed countries and the economics-in-transition region at costs ranging from −400 to −US $250 per tonne of carbon saved, and by 80 Mt carbon in developing countries at from −400 to US $0 per tonne of carbon saved (Acosta Moreno *et al.* 1996; Brown *et al.* 1998).

Most renewable energy investments (e.g. hydropower and biomass) require inputs from the construction sector. Multisectoral modelling suggests that carbon tax and permit policies will have little impact on construction output and employ-ment (Bertram, Stroonbergen & Terry 1993; Cambridge Econometrics 1998).

10.2.8   *Agriculture and forestry*

This sector contributes only about 4 per cent of global carbon emissions from energy use, but over 20 per cent of anthropogenic greenhouse gas emissions (in terms of million tonnes of carbon equivalent per year), mainly from methane and nitrous oxide (Mosier *et al.* 1998), as well as carbon from land clearing. Uncertainties on the intensity of use of these technologies by farmers are great, since they may have additional costs involved in their uptake. It is as yet uncertain to what extent climate mitigation policies would affect the agricultural sector economically.

If agricultural production per hectare in developing countries could be increased to meet the growing demand as a result of modern farming technologies and improved management systems, there would be less incentive for deforestation to provide more agricultural land. Plantations can affect biodiversity negatively if they replace biologically rich native grassland or wetland habitat (Keenan *et al.* 1997; Lugo 1997).

Ancillary benefits for agriculture from reduced emissions (i.e. volatile organic compounds and nitrogen oxides) include reduced damage from depletion of topospheric ozone, thereby protecting vegetation and human health (EPA 1997). Mitigation strategies to increase the efficiency of energy use or increase the use of non-fossil fuel energy are likely to reduce nitrous oxide emissions. Halting deforestation and forest degradation not only maintains carbon stocks but also preserves biodiversity (Dixon *et al.* 1993; EPA/USIJI 1998). Preserving forests also conserves water resources and prevents flooding (Chomitz & Kumari 1996), and desertification (Kuliev 1996). Agroforestry systems can improve and conserve soil properties (Wang & Feng 1995).

Ancillary costs of forest protection include social effects such as displacement of local populations, reduced income and flow of subsistence products from forests. In the case of agriculture, carbon accumulation in the soil to increase productivity requires higher amounts of nitrogen to be added, thereby increasing the amount of nitrous oxide emissions. Changes in fertilizer use, pesticides, and agricultural machinery, may enhance or offset any gains in soil carbon due to the carbon dioxide released from fossil fuel (Flach, Barnwell & Crosson 1997). Other possible negative effects of practices to retain carbon in the soil include contamination of groundwater with nutrients or pesticides (Cole *et al.* 1993; Insensee & Sadeghi 1996), and possible environmental effects from widespread application of manure (Batjes 1998).

10.2.9   Waste management

There has been increased utilization of methane from landfill and coal beds. Recovery costs are negative for half of landfill methane (EPA 1999). If everyone in the USA increased per capita recycling rates from the national average to the per capita recycling rate achieved in Seattle of 40 per cent (US national average, 29 per cent), the result would be a reduction of 4 per cent of total US greenhouse gas emissions. It appears that greenhouse mitigation policies explicitly including methane abatement offers economic incentives for the waste sector.

In the USA, communities report that ambitious waste reduction, recycling, and composting programmes cost no more than waste disposal, and often significantly less (EPA 1999). Overall average recycling costs are slightly above landfill disposal costs (Ackerman 1997). In view of greenhouse gas emissions, the most favourable management options are those that reduce fossil fuel use in manufacturing, as does recycling. If the whole lifecycle, and not just the disposal of the material, is considered, recycled materials usually are associated with lower greenhouse gas emissions than virgin material. There are many cost-effective methods to reduce greenhouse gases from waste. Source reduction is indisputably the most environmentally sound and cost-effective tool to reduce greenhouse gas emissions from solid waste.'

10.2.10   Sectoral mitigation technological options

The potential for major greenhouse gas emission reductions is estimated for each sector for a range of costs in Table 10.2. In the industrial sector, costs for carbon emission abatement are estimated to range from negative (i.e. no regrets, where reductions can be made at a profit), to around US $300 per tonne of carbon. In the buildings sector, aggressive implementation of energy-efficient technologies and measures can lead to a reduction in carbon dioxide emissions from residential buildings in 2010 by 325 Mt carbon per year in developed and economies-in-transition countries at costs ranging from −250 to −US $150 per tonne of carbon, and by 125 Mt carbon in developing countries at costs of from −250 to US $50 per tonne of carbon. Similarly, carbon dioxide emissions from commercial buildings in 2010 can be reduced by 185 Mt carbon in developed and economies-in-transition countries at costs ranging from −400 to −US $250 per tonne of carbon and by 80 Mt carbon in developing countries at costs ranging from −400 to US $0 per tonne of carbon. In the transport sector, costs range from −200 to US $300, and in the agricultural sector from −100 to US $300 per tonne of carbon. Materials management, including recycling and landfill gas recovery, can also produce savings at negative to modest cost of under US $100 per tonne of carbon. In the energy

Table 10.2. Estimations of greehouse gas emission reductions and cost per tonne of carbon equivalent avoided following the anticipated socioeconomic potential uptake by 2010–20 of selected energy efficiency and supply technologies, either globally or by region and with varying degrees of uncertainty.

Sources: BTM Consult (1999), Chipato (1999), ECMT (1997), Interlaboratory Working Group (1997), IPCC (2000), Kashiwagi et al. (1999), Kroeze and Mosier (1999), Lal and Bruce (1999), Moore (1998), OECD (1999), Reilly et al. (2003), Reimer and Freund (1999), Sulilatu (1998), USDOE/EIA (1998), Wang and Smith (1999), Worrell et al. (1997) and Zhou (1998).

		US per tonne of carbon (−400 −200 0 +200)	2010 Potential[a]	2010 Probability[b]	2020 Potential[a]	2020 Probability[b]
Building / appliances						
Residential sector	OECD/EIT	▬ (around −200)	◆◆◆◆◆◆◆	◇◇◇◇◇	◆◆◆◆◆◆◆	◇◇◇◇◇
	Developing countries	▬ (around −200)	◆◆◆◆◆	◇◇◇	◆◆◆◆◆	◇◇◇◇◇
Commercial sector	OECD/EIT	▬ (around −400)	◆◆◆◆◆	◇◇◇◇	◆◆◆◆◆◆◆	◇◇◇◇◇
	Developing countries	▬ (around −350)	◆◆◆◆	◇◇◇	◆◆◆◆◆	◇◇◇◇◇
Transport						
Automobile efficiency improvements	USA	▬ (around −200)	◆◆◆◆	◇◇◇◇	◆◆◆◆	◇◇◇
	Europe	▬ (around −200)	◆◆◆◆◆◆◆	◇◇	◆◆◆◆◆◆◆◆◆	◇◇
	Japan		◆◆◆◆◆◆◆	◇◇	◆◆◆◆◆◆◆◆◆	◇◇
	Developing Countries	▮ (0 to +200)	◆◆◆◆	◇◇	◆◆◆◆◆◆	◇◇
		▮ (0 to +200)				
Manufacturing						
Carbon dioxide removal — fertilizer, refineries	Global	▬ (around 0)	◆	◇◇◇◇◇	◆	◇◇◇◇

Measure	Region
Material efficiency improvement	Global
Blended cements	Global
Nitrous oxide reduction by chem. indus.	Global
Polyfluoro carbon reduction by Al industry	Global
Hydrofluoro carbon-23 reduction by chemical industry	Global
Energy efficient improvements	Global
Agriculture	
Increased uptake of conservation tillage and cropland management	Developing countries
Soil carbon sequestration	Global
	Global
Nitrogenous, fertilizer management	OECD
	Global
Entric methane reduction	OECD
	USA
	Developing countries
Rice paddy irrigation and fertilizers	Global
Wastes	
Landfill methane capture	OECD

(cont.)

Table 10.2. (*cont.*)

	Region	US$ per tonne of carbon				2010		2020	
		−400	−200	0	+200	Potential[a]	Probability[b]	Potential[a]	Probability[b]
Energy supply									
Nuclear for coal	Global								
	Annex I								
	Non-Annex I								
Nuclear for gas	Annex I								
	Non-Annex I								
Gas for coal	Annex I								
	Non-Annex I								
Carbon dioxide capture from coal	Global								
Carbon dioxide capture from gas	Global								
Biomass for coal	Global								
Biomass for gas	Global								
Wind for coal or gas	Global								
Co-fire coal with 10% biomass	USA								
Solar for coal	Annex I								
	Non-Annex I								
Hydro for coal	Global								
Hydro for gas	Global								

[a] Potential in terms of tonnes of carbon equivalent avoided for the cost range of US$/t carbon given.

◆ = < 20 Mt carbon per year ◆◆ = 20–50 Mt carbon per year ◆◆◆ = 50–100 Mt carbon per year ◆◆◆◆ = 100–200 Mt carbon per year

◆◆◆◆◆ = >200 Mt carbon per year

[b] Probability of realizing this level of potential based on the costs as indicaed from the literature.

◇ = Very unlikely ◇◇ = Unlikely ◇◇◇ = Possible ◇◇◇◇ = Probable ◇◇◇◇◇ = Highly probable

[c] Energy supply total mitigation options assumes that not all the potential will be realized for various reasons including competition between the individual technologies as listed below the totals

supply sector, a number of fuel switching and technological substitutions are possible at costs of from −100 to more than US $200 per tonne of carbon. The realization of this potential will be determined by the market conditions as influenced by human and societal preferences and government interventions.

10.3   National, regional, and global costs and benefits of both domestic and national polices and related measures

10.3.1   Development, equity, and sustainability issues

Climate change mitigation policies implemented at a national level will, in most cases, have implications for short-term economic and social development, local environmental quality, and intragenerational equity. Mitigation cost assessments that follow this line can address these impacts on the basis of a decision-making framework that includes a number of side impacts to the greenhouse gas emissions reduction policy objective. The goal of such an assessment is to inform decision-makers about how different policy objectives can be met efficiently, given priorities of equity and other policy constraints, e.g. natural resources, and environmental objectives.

10.3.2   Domestic policy

Particularly important for determining the gross mitigation costs is the magnitude of emissions reductions required in order to meet a given target; thus, the emissions baseline is a critical factor. The growth rate of carbon dioxide depends on the growth rate in GDP, the rate of decline of energy use per unit of output, and the rate of decline of carbon dioxide emissions per unit of energy use.

In a multimodel comparison project that engaged more than a dozen modelling teams internationally, the gross costs of complying with the Kyoto Protocol were examined, using energy-sector models. Carbon taxes are implemented in order to lower emissions, and the tax revenue is recycled lump sum. The magnitude of the carbon tax provides a rough indication of the amount of market intervention that would be needed, and equates the marginal abatement cost to meet a prescribed emissions target. The size of the tax required to meet a specific target will be determined by the marginal source of supply (including conservation) with and without the target. This, in turn, will depend on such factors as the size of the necessary emissions reductions, assumptions about the cost and availability of carbon-based and carbon-free technologies, the fossil fuel resource base, and short- and long-term price elasticities.

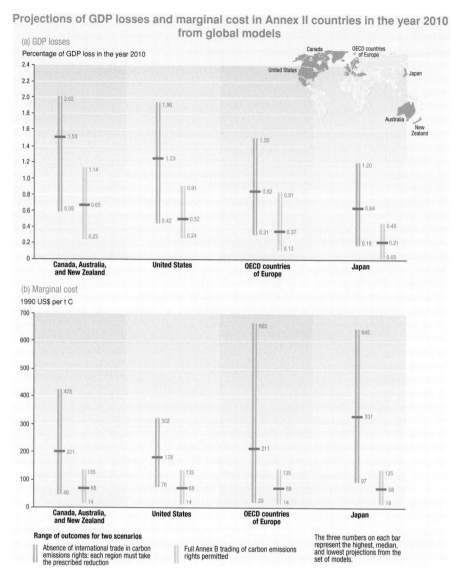

Figure 10.1. Projections of GDP losses and marginal costs in Annex II countries in the year 2010 from global models. For a color version of this graph see the original reference or website http://www.grida.no/climate/ipcc_tar/vol4/english/fig7-2.htm. *Source*: IPCC (2001).

With no international emission-trading, the carbon taxes necessary to meet the Kyoto restrictions in 2010 vary a lot among the models (see Figure 10.1). For the USA, they are calculated to be from 76 to US $322; for OECD Europe from 20 to US $665; for Japan, from 97 to US $645; and, finally, for the rest of the OECD (Canada, Australia, and New Zealand) from 46 to US $425. All numbers are reported at 1990 values. Marginal abatement costs are in the range of from 20 to US $135 per tonne of carbon if international trading is allowed. These models generally do not include no-regrets measures, nor take account of the mitigation potential of carbon dioxide sinks or of greenhouse gases other than carbon dioxide.

The above studies assume that the revenues from carbon taxes (or auctioned emissions permits) are recycled in a lump-sum fashion to the economy. The net social cost resulting from a given marginal cost of emissions constraint can be reduced if the revenues are targeted to finance cuts in the marginal rates of pre-existing distortionary taxes, e.g. income, payroll, and sales taxes. While recycling revenues in a lump-sum fashion confers no efficiency benefit, recycling through marginal rate cuts helps avoid some of the efficiency costs or dead-weight loss of existing taxes. This raises the possibility that revenue-neutral carbon taxes might offer a 'double dividend' by: (a) improving the environment, and (b) reducing the costs of the tax system (see Section 10.4.3 for more details).

10.3.3   Carbon taxes

In all countries in which carbon dioxide taxes have been introduced, some sectors have been exempted by the tax, or the tax is differentiated across sectors. Most studies conclude that tax exemptions raise economic costs relative to a policy involving uniform taxes. However, results differ in the magnitude of the cost of exemptions.

In addition to the total cost, the distribution of the costs is important for the overall evaluation of climate policies. A policy that leads to an efficiency gain may not be welfare-improving overall if some people are in a worse position than before, and vice versa. Notably, if there is a wish to reduce the income differences in the society, the effect on the income distribution should be taken into account in the assessment. The distributional effects of a carbon tax appear to be regressive unless the tax revenues are used either directly or indirectly in favour of the low-income groups.

10.3.4   International emissions trading

It has long been recognized that international trade in emissions quotas can reduce mitigation costs. This will occur when countries with high domestic marginal abatement costs purchase emission quotas from countries with low marginal abatement costs. This is often referred to as 'where flexibility' — allowing reductions to take place where it is cheapest to do so regardless of geographical location. It is important to note that the question of where the reductions take place is independent of who pays for them.

'Where flexibility' can occur on a number of scales. It can be global, regional, or at the country level. In the theoretical case of full global trading, all countries agree to emission caps and participate in the international market as buyers or sellers of emission allowances. The Clean Development Mechanism may allow some of these cost reductions to be captured. When the market is defined at the regional level (e.g. Annex B countries), the trading market is more limited. For example, in 2002, the EU has adopted a Directive for a carbon dioxide emissions trading programme to assist member states to meet its Kyoto Protocol obligations in the most cost-effective manner. Finally, trade may take place domestically with all emission reductions occurring in the country of origin.

Table 10.3 shows the cost reductions from emissions trading for Annex B and full global trading compared to a no-trading case. In each instance, the goal is to meet the emission reduction targets contained in the Kyoto Protocol. All of the models show significant gains as the size of the trading market is expanded. The difference among models is due in part to differences in their baseline, the assumptions about the cost, and availability of low-cost substitutes on both the supply and demand sides of the energy sector, and the treatment of short-term macro-shocks. In general, all calculated gross costs for the non-trading case are below 2 per cent of GDP, and in most cases below 1 per cent. Annex B trading lowers the costs for the OECD region as a whole to less than 0.5 per cent, and regional impacts within this vary from 0.1 to 1.1 per cent. Global trading in general would decrease these costs to well below 0.5 per cent of GDP, with the OECD average below 0.2 per cent.

The issue of the so-called 'hot air' also influences the cost of implementing the Kyoto Protocol. The recent decline in economic activity in Eastern Europe and the former Soviet Union has led to a decrease in their greenhouse gas emissions. Although this trend eventually is expected to reverse, for some countries emissions are still projected to lie below the constraint imposed by the Kyoto Protocol. If this does occur, they will have excess emission quotas that may be sold to countries in search of low-cost options for meeting their own targets. The cost savings from trading are sensitive to the magnitude of excess emissions allowances or 'hot air' used.

Numerous assessments of reduction in projected GDP have been associated with complying with Kyoto-type limits. Most economic analyses have focused on gross costs of carbon-emitting activities, ignoring the cost-saving potential of mitigating non-carbon dioxide gases, using carbon sequestration, and not taking into account environmental ancillary benefits and avoided climate change, nor using revenues to remove distortions. Including such possibilities could lower costs.

A constraint theoretically would lead to a reallocation of resources away from the pattern that is preferred in the absence of a limit, and into potentially

Table 10.3. *Energy Modelling Forum main results: GDP loss in 2010 (in percentage of GDP; 2010 Kyoto targets).* Source: *Hourcade* et al. *(2001).*

Model	No trading				Annex I trading				Global trading			
	US	OECD	Japan	CANZ	US	OECD	Japan	CANZ	US	OECD	Japan	CANZ
ABARE-GTEM	1.96	0.94	0.72	1.96	0.47	0.13	0.05	0.23	0.09	0.03	0.01	0.04
AIM	0.45	0.31	0.25	0.59	0.31	0.17	0.13	0.36	0.20	0.08	0.01	0.35
CETA	1.93				0.67				0.43			
G-CUBED	0.42	1.50	0.57	1.83	0.24	0.61	0.45	0.72	0.06	0.26	0.14	0.32
GRAPE		0.81	0.19			0.81	0.10			0.54	0.05	
MERGE3	1.06	0.99	0.80	2.02	0.51	0.47	0.19	1.14	0.20	0.20	0.01	0.67
MS-MRT	1.88	0.63	1.20	1.83	0.91	0.13	0.22	0.88	0.29	0.03	0.02	0.32
Oxford	1.78	2.08	1.88		1.03	0.73	0.52		0.66	0.47	0.33	
RICE	0.94	0.55	0.78	0.96	0.56	0.28	0.30	0.54	0.19	0.09	0.09	0.19

The results of the Oxford model are not included in the ranges cited in the Technical Summary and Summary for Policy-makers because this model has not been subject to substantive academic review (and hence is inappropriate for IPCC assessment), and relies on data from the early 1980s for a key parametization that determines the model results. This model is entirely unrelated to the CLIMOX model, from the Oxford Institutes of Energy Studies.

costly conservation and fuel substitution.[1] Relative prices will also change. These forced adjustments will lead to reductions in economic performance, which will impact upon GDP. Clearly, the broader the permit trading market, the greater the opportunity for reducing overall mitigation costs. Conversely, limits on the extent to which a country can satisfy its obligations through the purchase of emissions quotas can increase mitigation costs. Several studies have calculated the magnitude of the increase as being substantial, falling in particular on countries with the highest marginal abatement costs. Another parameter likely to limit the savings from carbon trading is the very functioning of trading systems, i.e. transaction costs, management costs, insurance against uncertainty, and strategic behaviour in the use of permits (Ha Duong, Hourcade & Lecocq 1999).

[1] In the Kyoto Protocol negotiations, several European countries have argued for a limit on the usage of the Kyoto mechanisms, not only because of the moral responsibility of the OECD countries to take the lead in greenhouse gas emissions abatement, but also because domestic emissions controls in industrialized countries would lead to technological innovation that would spill over to other countries, a possibility usually not taken into account in traditional economic modelling.

The cost estimates for Annex B countries to implement the Kyoto Protocol vary between studies and regions, and depend strongly upon the assumptions regarding the use of the Kyoto mechanisms and their interactions with domestic measures. The great majority of global studies reporting and comparing these costs use international energy-economic models. The following GDP impacts were observed (IPCC 2001).

1. *Annex II countries.* In the absence of emissions trading between Annex B countries, the majority of global studies show reductions in projected GDP of about 0.2–2 per cent in 2010 for different Annex II regions. With full emissions trading between Annex B countries, the estimated reductions in 2010 are between 0.1 and 1.1 per cent of projected GDP. Models with results reported here assume full use of emissions trading without transaction cost. Results for cases that do not allow Annex B trading assume full domestic trading within each region. Models do not include sinks or non-carbon dioxide greenhouse gases. They do not include the Clean Development Mechanism negative cost options, ancillary benefits, or targeted revenue recycling.

    For all regions, costs are also influenced by the following factors.
    (a) Constraints on the use of Annex B trading, high transaction costs in implementing the mechanisms, and inefficient domestic implementation could raise costs.
    (b) Inclusion in domestic policy and measures of the no-regrets possibilities, use of the Clean Development Mechanism, sinks, and inclusion of non-carbon dioxide greenhouse gases, could lower costs. Costs for individual countries can vary more widely.
    The models show that the Kyoto mechanisms are important in controlling risks of high costs in given countries, and thus can complement domestic policy mechanisms. Similarly, they can minimize risks of inequitable international impacts and help to level marginal costs. The global modelling studies reported above show national marginal costs to meet the Kyoto targets from about US$ 20 up to US $600 per tonne of carbon without trading, and a range from about US $15 up to US $150 per tonne of carbon with Annex B trading. The cost reductions from these mechanisms may depend on the details of implementation, including the compatibility of domestic and international mechanisms, constraints, and transaction costs.

2. *Economies in transition.* For most of these countries, GDP effects range from negligible to a several per cent increase. This reflects opportunities for energy efficiency improvements not available to Annex II countries. Under assumptions of drastic energy efficiency improvement and/or

continuing economic recessions in some countries, the assigned amounts may exceed projected emissions in the first commitment period. In this case, models show increased GDP through revenues from trading assigned amounts. However, for some economies in transition, implementing the Kyoto Protocol will have similar impacts on GDP as for Annex II countries.

3. *Non-Annex I countries.* Emission constraints in Annex I countries have well established, albeit varied, 'spillover' effects on non-Annex I countries.

    (a) Oil-exporting, non-Annex I countries. Analyses report costs differently, including, inter alia, reductions in projected GDP and reductions in projected oil revenues. The study reporting the lowest costs shows reductions of 0.2 per cent of projected GDP with no emissions trading, and less than 0.05 per cent of projected GDP with Annex B emissions trading in 2010. The study reporting the highest costs shows reductions of 25 per cent of projected oil revenues with no emissions trading, and 13 per cent of projected oil revenues with Annex B emissions trading in 2010. These studies do not consider policies and measures other than Annex B emissions trading that could lessen the impact on non-Annex I, oil-exporting countries, and therefore tend to overstate both the costs to these countries and overall costs. The effects on these countries can be reduced further by removal of subsidies for fossil fuels, energy tax restructuring according to carbon content, increased use of natural gas, and diversification of the economies of non-Annex I, oil-exporting countries.

    (b) Other non-Annex I countries may be affected adversely by reductions in demand for their exports to OECD nations and by the price increase of carbon-intensive and other products they continue to import. These countries may benefit from the reduction in fuel prices, increased exports of carbon-intensive products, and the transfer of environmentally sound technologies and expertise. The net balance for a given country depends on which of these factors dominates. The breakdown of winners and losers remains uncertain because of these complexities.

    (c) Carbon leakage. The possible relocation of some carbon-intensive industries to non-Annex I countries, and wider impacts on trade flows in response to changing prices, may lead to leakage of about 5–20 per cent. Exemptions, for example for energy-intensive industries, make the higher model estimates for carbon leakage unlikely, but would raise aggregate costs. The transfer of environmentally sound technologies and expertise, not included in models, may lead to lower leakage and especially in the longer term may more than offset the leakage.

10.4   No regrets, co-benefits, double dividend, spillover effects, leakages, and avoided damages

10.4.1   No regrets options

No-regrets options are by definition actions to reduce greenhouse gas emissions that have negative net costs. This means that some mitigation actions can be realized at negative costs, because these options generate direct or indirect benefits (e.g. those resulting from reductions in market failures, double dividends through revenue recycling, and ancillary benefits) large enough to offset the costs of implementing the options. The no-regrets potential results from existing market or institutional imperfections that prevent cost-effective emission reduction measures from being taken. The key question is whether such imperfections can be removed cost-effectively by policy measures.

The existence of no-regrets potentials is a necessary, but not a sufficient, condition for the potential implementation of these options. The actual implementation also requires the development of a policy strategy that is complex and comprehensive enough to address market and institutional failures and barriers (Cameron, Montgomery & Foster 1999). The no-regrets issue reflects specific assumptions about the working and the efficiency of the economy, especially the existence and stability of a social welfare function, based on a social cost concept:

1. Reduction of existing market or institutional failures and other barriers that impede adoption of cost-effective emission reduction measures can lower private costs compared to current practice. This can also reduce private costs overall.
2. A double dividend related to recycling of the revenue of carbon taxes in such a way that it offsets distortionary taxes.
3. Ancillary benefits and costs (or ancillary impacts), which can be synergies or trade-offs in cases in which the reduction of greenhouse gas emissions has joint impacts on other environmental policies, i.e. relating to local air pollution, urban congestion, or land and natural resource degradation.

10.4.2   Market imperfections

The existence of a no-regrets potential implies that market and institutions do not behave perfectly, because of market imperfections such as lack of information, distorted price signals, lack of competition, and/or institutional failures related to inadequate regulation, inadequate delineation of property rights, distortion-inducing fiscal systems, and limited financial markets. Reducing market imperfections suggests it is possible to identify and implement policies that can correct these market and institutional failures without incurring costs larger than the benefits gained.

10.4.3   Double dividend

The potential for a double dividend arising from climate mitigation policies was studied extensively during the 1990s. In addition to the primary aim of improving the environment (the first dividend), such policies, if conducted through revenue-raising instruments such as carbon taxes or auctioned emission permits, yield a second dividend, which can be set against the gross costs of these policies (IPCC 2001). All domestic greenhouse gas policies have an indirect economic cost from the interactions of the policy instruments with the fiscal system, but in the case of revenue-raising policies this cost is partly offset (or more than offset) if, for example, the revenue is used to reduce existing distortionary taxes (Bohm 1998). Whether these revenue-raising policies can reduce distortions in practice depends on whether revenues can be 'recycled' to tax reduction (Bovenberg, Lans & Mooij 1994).

One can distinguish a weak and a strong form of the double dividend. The weak form asserts that the costs of a given revenue-neutral environmental reform, when revenues are devoted to cuts in marginal rates of prior distortionary taxes, are reduced relative to the costs when revenues are returned in lump-sum fashion to households or firms. The strong form of the double-dividend assertion is that the costs of the revenue-neutral environmental tax reform are zero or negative. While the weak form of the double-dividend claim receives virtually universal support, the strong form of the double dividend assertion is controversial.

Simulation results show that in economies that are especially inefficient or distorted along non-environmental lines, the revenue-recycling effect can, indeed, be strong enough to outweigh the primary cost and tax-interaction effect, so that the strong double dividend may materialize. Thus, in several studies involving European economies, in which tax systems may be highly distorted in terms of the relative taxation of labour, the strong double dividend can be obtained, in any case more frequently than in other recycling options. In contrast, most studies of carbon taxes or permits policies in the USA demonstrate that recycling through lower labour taxation is less efficient than through capital taxation, but they generally do not find a strong double dividend. Another conclusion is that, even in cases of no strong double-dividend effect, one fares considerably better with a revenue-recycling policy in which revenues are used to cut marginal rates of prior taxes, than with a non-revenue recycling policy as, for example, with grandfathered quotas.

10.4.4   Ancillary benefits and costs (ancillary impacts)

Policies aimed at mitigating greenhouse gases can have positive and negative 'ancillary effects' on public health, ecosystems, land use, materials, etc.

'Ancillary benefits' or 'co-benefits' refers to the non-climate benefits of greenhouse gas mitigation policies that are explicitly incorporated into the initial creation of mitigation policies. The term 'co-benefits' reflects that most policies designed to address greenhouse gas mitigation also have other, often at least equally important, rationales involved at the inception of these policies, e.g. related to objectives of development, sustainability, and equity. In contrast, the term 'ancillary benefits' connotes those secondary or side-effects of climate change mitigation policies on problems that arise subsequent to any proposed greenhouse gas mitigation policies.

Positive ancillary effects could result from mitigation policies that reduce health- or environment-damaging emissions of conventional pollutants. Negative ancillary effects may result from policies that increase health or environmental damages, e.g. increased reliance on diesel fuels which have lower greenhouse gas emissions than petrol but can increase health and environmental risks.

For example, transport sector options have an impact on both greenhouse gas emissions and urban air pollution control programmes. Greenhouse gas emission control polices, as with vehicle maintenance programmes, reduce both greenhouse gas emissions and other pollution. Another option (e.g. the introduction of diesel-fuelled trucks as substitutes for petrol-powered ones) decreases greenhouse gas emissions but increases emissions of nitrogenous oxides, i.e. local air pollution. The gross and net costs assessed for these programmes depend on specific baseline and policy case scenarios (IPCC 2001).

There is little agreement on the definition, reach, and size of these ancillary benefits, and on methodologies for integrating them into climate policy. In spite of recent progress in methods development, it remains very challenging to develop quantitative estimates of the ancillary effects, benefits and costs of greenhouse gas mitigation policies. Despite these difficulties, in the short term, ancillary benefits of greenhouse gas policies under some circumstances can be a significant fraction of private (direct) mitigation costs, and in some cases be comparable to the mitigation costs. Ancillary benefits may be of particular importance in developing countries. Moreover, something that is perceived as a greenhouse gas reduction programme from an international perspective may be seen, in a national context, as one in which local pollutants and greenhouse gases are equally important.

The exact magnitude, scale, and scope of these ancillary benefits and costs will vary with local geographical and baseline conditions. In some circumstances, where baseline conditions involve relatively low carbon emissions and population density, benefits may be low. Some models (e.g. computable general equilibrium models) have difficulty in estimating ancillary benefits, as they rarely have the necessary spatial detail (Dessus & O'Connor 1999; Garbaccio, Ho & Jorgenson 2000; Brendemoen & Vennemo 1994; Scheraga & Leary 1993; Boyd, Krutilla & Viscusi 1995). A major component of uncertainty for modelling ancillary benefits for public health is the link between emissions and atmospheric concentrations, particularly in light of the importance of secondary pollutants (see Box 10.1). However, it is recognized that there are significant ancillary benefits in addition to those for public health that have not been quantified or monetized. There is compelling evidence that ancillary benefits may be a significant fraction of, or in some situations even larger than, the mitigation costs, especially where baseline conditions involve relatively high levels of pollution and there are likely to be minor ancillary costs. It appears that there are major gaps in the methods and models for estimating ancillary costs.

Box 10.1   Climate change and air pollution: two sides of the same coin?

A first step towards including issues beyond greenhouse gas emissions in the negotiations about climate change, is explicitly to take into account the linkages with air pollution. Scientific research and political negotiations on these two issues have developed largely in isolation. This is surprising and has been counterproductive, since the two problems are intertwined, with common sources of emissions, linked atmospheric processes, and mutually dependent impacts. The stalling of progress in addressing climate change through the Kyoto Protocol, and the difficulties of meeting the targets of the Gotenburg Protocol and the National Emissions Ceiling Directive in Europe, has meant that changing this situation has started only recently in industrialized countries. Also, the importance of involving developing countries in an international climate change regime without forced commitments has highlighted the connection between urban air pollution problems and greenhouse gas emissions.

The two main links between climate change and air pollution, in terms of atmospheric processes, are: (a) tropospheric ozone, and (b) aerosols. Tropospheric ozone is the third most important greenhouse gas, and its abundance influences the lifetime of other greenhouse gases. Conversely, climate change affects the atmospheric chemistry leading to ozone formation. Ozone is an important air pollutant affecting ecosystem vitality and human health. Particulate matter (e.g. aerosols) has an important effect on radiative forcing as well, depending on the composition: sulphate, nitrate, and organic aerosols have negative radiative forcing, while black carbon has a positive effect. Because of the predominance of the first set of components, the net effect of aerosols is generally a net cooling. Particulate matter also is one of the most important air pollutants. The impacts of human health appear to be dependent on particle size rather than composition. The smaller the particles, the further they penetrate into the respiratory system, and the more severe the health effects.

Impacts of climate change and air pollution on humans and ecosystems interact through changing exposure and changed sensitivity. The exposure of ecosystems to air pollution can vary as a result of phenological factors induced by climate change (e.g. variations in the length of the growing season) and altered weather patterns. Their sensitivity can vary because climate change affects ecosystem vitality, soil processes and ecosystem composition (changing critical loads of air pollutants).

The fact that greenhouse gases and air pollutants largely share the same sources can lead to synergies or trade-offs of abatement measures. Examples of trade-offs that can take place include: (a) increased energy consumption (and associated greenhouse gas emissions) by add-on desulphurization of flue gases from power plants to abate acidification; (b) increased nitrous oxide emissions from some particular options to abate ammonia emissions, or (c) increased emissions or particulates if moving from petrol, to more energy-efficient diesel, engines in transport. A large number of opportunities for synergies, however, is available in moving to a cleaner and more efficient energy system. Meeting the Kyoto targets can reduce costs of meeting air quality goals by up to 70 per cent, dependent on the scenario. The linkages have to be taken into account to maximize synergies and minimize trade-offs, not only in selecting technological abatement options, but also in designing policy instruments. For example, greenhouse gas emissions trading reduces costs of meeting greenhouse gas emissions reductions targets, but can change the spatial distribution of emissions, sometimes making it more difficult to meet air quality standards.

Main *source*: Tuinstra *et al.* (2004).

10.4.5   *'Spillover' effects and leakages*

In a world in which economies are linked by international trade and capital flows, abatement of one economy will have welfare impacts on other abating or non-abating economies. These impacts are called *spillover effects*, and include effects on trade, carbon leakage, transfer and diffusion of environmentally sound technology, and other issues.

Industrial reallocation The 'leakage effect' reflects the extent to which cuts in domestic emissions are offset by shifts in production and therefore increases in emission abroad. Developing nations may not gain, because less capital will be available as the income in developed countries drops and it becomes more costly to import from developed nations the capital goods that promote growth, e.g. machinery and transport equipment.

Technological spillovers Technology policies in one country could affect development in other countries or specific sectors through: (a) research and development, which may increase the knowledge base; (b) increased market access for low carbon dioxide technologies, and (c) domestic regulations on technology performance and standards.

The dominant finding of the effects of emission constraints in Annex B countries on non-Annex B countries in simulation studies prior to the Kyoto Protocol was that Annex B abatement would have a predominantly adverse impact on non-Annex B regions. In simulations of the Kyoto Protocol, the results are more mixed, with some non-Annex B regions experiencing welfare gains and other losses. This is due mainly to a milder target in the Kyoto simulations than in pre-Kyoto ones. It was also found universally that most non-Annex B economies that suffered welfare losses under uniform independent abatement would suffer smaller ones under emissions trading.

A reduction in Annex B emissions will tend to result in an increase in non-Annex B emissions, reducing the environmental effectiveness of Annex B abatement. This is called *carbon leakage*, and can occur in the order of 5–20 per cent through a possible relocation of carbon-intensive industries as a result of reduced Annex B competitiveness in the international market-place, lower producer prices of fossil fuels in the international market, and changes in income due to better terms of trade.

Reduction in the variance of estimates of carbon leakages obtained may result largely from the development of new models based on reasonably similar assumptions and data sources. Such developments do not necessarily reflect more widespread agreement about appropriate behavioural assumptions. One robust result seems to be that carbon leakage is an increasing function of the stringency of the abatement strategy. This means that leakage may be a less serious problem under the Kyoto target than under the more stringent targets considered previously. Furthermore, emission leakage is lower under emissions trading than under domestic abatement. Exemptions for energy-intensive industries found in practice, and other factors, make the higher model estimates for carbon leakage unlikely, but would raise aggregate costs.

Carbon leakage may also be influenced by the assumed degree of competitiveness in the world oil market (Berg, Kverndokk & Rosendahl 1997a). While most studies assume a competitive oil market, studies considering imperfect competition find lower leakage if OPEC is able to exercise a degree of market power over the supply of oil and therefore reduce the fall in the international oil price (Berg, Kverndokk & Rosendahl 1997b). Whether or not OPEC acts as a cartel can have a reasonably significant effect on the loss of wealth to OPEC and other oil producers and on the level of permit prices in Annex B regions.

This is related to induced technical change. The transfer of environmentally sound technologies and knowhow, not included in models, may lead to lower leakage and, especially in the longer term, may more than offset the leakage.

10.4.6   *Avoided damages*

The issues discussed above in this section – no regrets, double dividend, ancillary or co-benefits, spill-over effects and leakages – address short-term economic and other effects of mitigation policies. They do not cover what appears to be the main justification for mitigation policies, i.e. mitigating the expected damages from climate change, which are expected to accrue primarily in the long term. Avoided damages from mitigation activities should be calculated as the difference between the impacts with and without mitigation policies. There has been no systematic analysis of the avoided damages of specific mitigation policies, e.g. policies that stabilize atmospheric concentrations of greenhouse gases at various levels.

Over the years, various attempts have been made to quantify damages from climate change in monetary terms. The IPCC's *Second Assessment Report* discusses the methodology to estimate the potential benefits of emissions control in some detail (Pearce *at al.* 1996). The IPCC's *Third Assessment Report* provides an update of the 1995 estimates of global impacts, expressed as a percentage of GDP (see Table 1.5 in Chapter 1). Such estimates have to be used with caution for several reasons: (a) the choice of the indicators selected is arbitrary and not all indicators can be easily or unambiguously monetized; (b) there are serious gaps in knowledge about future impacts; (c) the effects of adaptation are difficult to quantify, and (d) the estimates are highly dependent on assumptions with reference to future development concerning demography, population, economy, and technology (Aahein 2002). Some estimates have proved to be very controversial, notably the attribution of a monetary value to a 'statistical life'.

Since 1995, market-based estimates of impact damage has decreased, because adaptation has been taken more into account, i.e. the cost of adapting to avoid

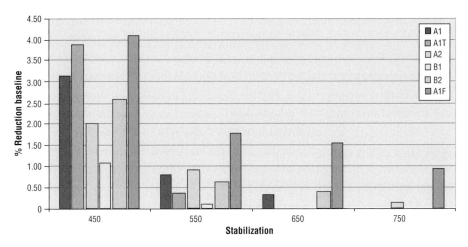

Figure 10.2. Costs of stabilizing atmospheric carbon dioxide concentrations for different stabilization levels and baseline scenarios expressed in global average GDP reduction. For a colour version of this graph see the original publication or website http://www.grida.no/climate/ipcc_tar/vol4/english/fig7-4.htm. *Source:* Hourcade *et al.* (2001).

damage is less than the full potential impact damages so avoided. Because of the large uncertainties in monetary 'aggregate' damage estimates, Smith *et al.* (2001) consider these 'aggregate impacts' as only one of five 'reasons for concern'. Also, it may be questionable whether an estimate of global or even regional damage (or avoided damage), is meaningful, since the primary impacts are felt locally. Because of uncertainties, the long timescales involved, and the incomplete coverage of monetary damage estimates, such estimates are basically indicative and serve merely as a rough guide – from the perspective of more short-term sustainable development priorities.

10.5   Costs of meeting a range of stabilization targets

Cost-effectiveness studies with a century timescale, estimate that the costs of stabilizing carbon dioxide concentrations in the atmosphere increase as the concentration stabilization level declines. Different baselines can have a strong influence on absolute costs. While there is a moderate increase in the costs when passing from a 750 to a 550 ppmv concentration stabilization level, there is a larger increase in costs passing from 550 to 450 ppmv unless the emissions in the baseline scenario are very low (see Figure 10.2). However, these studies do not incorporate carbon sequestration and gases other than carbon dioxide, and did not examine the possible effect of more ambitious targets on induced technological change. In particular, the choice of the reference scenario has a strong influence. Recent studies using the IPCC's *Special Report on Emissions Scenarios (SRES)* reference scenarios as baselines against which to analyse stabilization show clearly that the average reduction in projected GDP in most of the stabilization scenarios reviewed here is under 3 per cent of the baseline value (the maximum reduction across all the stabilization scenarios reached 6.1 per cent in a given year). At the same time, some scenarios showed an increase in GDP compared to the baseline because of apparent positive economic feedbacks of technology development and transfer. The modelled GDP reduction (averaged across storylines and stabilization levels) is lowest in 2020 (1 per cent), reaches a maximum in 2050 (1.5 per cent), and declines by 2100 (1.3 per cent). However, in the scenario groups with the highest baseline emissions, the estimated size of the GDP reduction increases throughout the modelling period. Due to their relatively small scale when compared to absolute GDP levels, GDP reductions in these so-called 'post-*SRES* stabilization scenarios' do not lead to significant declines in GDP growth rates over this century. For example, the annual 1990–2100 GDP growth rate across all the stabilization scenarios was reduced on average by only 0.003 per cent per year, with a maximum reduction reaching 0.06 per cent per year.

The concentration of carbon dioxide in the atmosphere is determined more by cumulative, rather than year-by-year, emissions. That is, a particular concentration target can be reached through a variety of emissions pathways. A number of studies suggest that the choice of emissions pathway can be as important as the target itself in determining overall mitigation costs. The studies fall into two categories: (a) those that assume that the target is known, and (b) those that characterize the issue as one of decision-making under uncertainty.

For studies that assume that the target is known, the issue is one of identifying the least-cost mitigation pathway for achieving the prescribed target. Here the choice of pathway can be seen as a carbon budget problem. This problem so far has been addressed in terms of carbon dioxide only, and very limited treatment has been given to non-carbon greenhouse gases.[2]

A concentration target defines an allowable amount of carbon to be emitted into the atmosphere between now and the date upon which the target is to be achieved. The issue is how best to allocate the carbon budget over time. Most studies that have attempted to identify the least-cost pathway for meeting a particular target conclude that such a pathway tends to depart gradually from the

---

[2] A notable exception is the work of the Massachusetts Institute of Technology, which suggests that mitigation strategies that include non-carbon dioxide gases in addition to carbon dioxide can lead to significant cost savings, e.g. Reilly *et al.* (2003).

model's baseline in the early years, with more rapid reductions later on. There are several reasons why this is so. A gradual near-term transition from the world's present energy system minimizes premature retirement of existing capital stock, provides time for technology development, and avoids premature lock-in to early versions of rapidly developing low-emission technology. On the other hand, more aggressive near-term action would decrease environmental risks associated with rapid climatic changes, stimulate more rapid deployment of existing low-emission technologies, provide strong near-term incentives to future technological changes that may help to avoid lock-in to carbon-intensive technologies, and allow for later tightening of targets should that be deemed desirable in the light of evolving scientific understanding. It should also be noted that the lower the concentration target, the smaller the carbon budget and, hence, the earlier the departure from the baseline. However, even with higher concentration targets, the more gradual transition from the baseline does not negate the need for early action. All stabilization targets require future capital stock to be less carbon-intensive. This has immediate implications for near-term investment decisions. New supply options typically take many years to enter the market-place. An immediate and sustained commitment to research and development is required if low-carbon low-cost substitutes are to be available when needed. In Table 8.5, we summarize the arguments for modest versus stringent early abatement.

The above addresses the issue of mitigation costs. It is also important to examine the environmental impacts of choosing one emission baseline pathway over another. This is because different emission pathways imply not only different emission reduction costs, but also different benefits in terms of avoided environmental impacts.

The assumption that the target is known with certainty is, of course, an oversimplification. The United Nations Framework Convention on Climate Change (UNFCCC), recognizing the dynamic nature of the decision problem, has called for periodic reviews 'in light of the best scientific information on climate change and its impacts'. Such a sequential decision-making process aims to identify short-term hedging strategies in the face of long-term uncertainties. The relevant question is not: what is the best course of action for the next 100 years?, but rather, what is the best course for the near term?, given the long-term uncertainties. Several studies indicate that the desirable amount of hedging depends upon one's assessment of the stakes, the odds, and the cost of mitigation. The risk premium – the amount that society is willing to pay to avoid risk – ultimately is a political decision that differs among countries.

Most models used to assess the costs of meeting a particular mitigation objective tend to oversimplify the process of technical change. Typically, the rate of technical change is assumed to be independent of the level of emissions control. Such change is referred to as 'autonomous'. In recent years, the issue of induced technical change has received increased attention. Some argue that such change might lower substantially, and perhaps even eliminate, the costs of carbon dioxide abatement policies. Others are much less sanguine about the impact of induced technical change.

Recent research suggests that the effect on timing depends on the assumed source of technological change. When the channel for technological change is perceived to be primarily research and development, the induced technological change makes it preferable to concentrate more on abatement in the future. The reason is that technological change lowers the costs of future abatement relative to current abatement, making it more cost-effective to place more emphasis on future abatement. But, when the channel for technological change would, in reality, be mainly learning-by-doing, the presence of induced technological change has an ambiguous impact on the optimal timing of abatement. On the one hand, induced technical change makes future abatement less costly, which suggests the need to emphasize future abatement efforts. On the other hand, there is an added value to current abatement, because such abatement contributes to experience or learning and helps reduce the costs of future abatement. Which of these two effects predominates depends on the particular nature of the technologies and cost functions.

References

Aahein, H. A. (2002) Impacts of climate change in monetary terms? Issues for developing countries. *International Journal Global Environmental Issues*, **2**(3/4), 223–39.

Ackerman, F. (1997) *Why Do We Recycle? Markets, Values, and Public Policy*. Washington: Island Press.

Acosta Moreno, R., Baron, R., Bohm, P., Chandler, W., Cole, V., Davidson, O., Dutt, G., Haites, E., Ishitani, H., Kruger, D., Levine, M. D., Zhong, L., Michaelis, L., Moomaw, W., Moreira, J. R., Mosier, A., Moss, R., Nakicenovic, N., Price, L., Ravindranath, H. H., Rogner, H.-H., Sathay, J., Shukla, P., van Wie McGrory, L. and Williams, T. (1996) Technologies, policies and measures for mitigating climate change. In R. T. Watson, M. C. Zinyowera and R. H. Moss, eds., *IPCC Technical Paper 1*. Geneva: Intergovernmental Panel on Climate Change.

Barker, T., Johnstone, N. and O'Shea, T. (1993) The CEC carbon/energy tax and secondary transport-related benefits. *Energy-Environment-Economy Modelling Discussion Paper No. 5*. Department of Applied Economics. University of Cambridge.

Batjes, N. H. (1998) Mitigation of atmospheric CO_2 concentrations by increased carbon sequestration in the soil. *Biology and Fertility of Soils*, **27**, 230–5.

Berg, E., Kverndokk, S. and Rosendahl, K. (1997a) Gains from cartelisation in the oil market. *Energy Policy*, **25**(13), 1075–91.

Berg, E., Kverndokk, S. and Rosendahl, K. E. (1997b) Market power, international CO_2 taxation and petroleum wealth. *The Energy Journal*, **18**(4), 33–71.

Bernstein, L. and Pan J., eds. (2000) *Sectoral Economic Costs and Benefits of GHG Mitigation*. Proceedings of an IPCC expert meeting, 14–15 February 2000, Technical Support Unit, Working Group III. Geneva: Intergovernmental Panel on Climate Change.

Bertram G., Stroonbergen, A. and Terry, S. (1993) *Energy and Carbon Taxes Reform Options and Impacts*. Wellington: Prepared for the Ministry of the Environment, New Zealand by Simon Terry Associates and Business and Economic Research Ltd.

Bohm, P. (1998) Public investment issues and efficient climate change policy. In *The Welfare State, Public Investment and Growth*. H. Shibata and T. Ihori, eds., Tokyo: Springer-Verlag.

Bose, R. K. (1998) Automotive energy use and emission control: a simulation model to analyse transport strategies for Indian metropolises. *Energy Policy*, **23**(13), 1001–1016.

Bose, R. K. (1999) Engineering-economic studies of energy technologies to reduce carbon emissions in the transport sector. Proceedings of the international workshop on *Technologies to Reduce Greenhouse Gas Emissions: Engineering-Economic Analyses of Conserved Energy and Carbon*. Paris: International Energy Agency.

Bose, R. K. (2000) Mitigating GHG emissions from the transport sector in developing nations: synergy explored in local and global environmental agendas. In L. Bernstein and J. Pan, eds., *Sectoral Economics Costs and Benefits of GHG Mitigation*. Proceedings of IPCC expert meeting, 14–15 February 2000, Technical Support Unit, Working Group III. Geneva: Intergovernmental Panel on Climate Change.

Bovenberg A., Lans, R. and de Mooij, A. (1994) Environmental levies and distortionary taxation. *American Economic Review*, **84**(4), 1085–9.

Boyd, R., Krutilla, K. and Viscusi, W. K. (1995) Energy taxation as a policy instrument to reduce CO_2 emissions: a net benefit analysis. *Journal of Environmental Economics and Management*, **29**(1), 1–25.

Brendemoen, A. and Vennemo, H. (1994) A climate treaty and the Norwegian economy: a CGE assessment. *The Energy Journal*, **15**(1), 77–93.

Brown, M. A., Levine, M. D., Rom, J. P., Rosenfeld, A. H. and Koomey, J. G. (1998) Engineering-economic studies of energy technologies to reduce greenhouse gas emissions: opportunities and challenges. *Annual Review of Energy and Environment*, **23**, 287–385.

BTCE (1996) *Transport and Greenhouse Costs and Options for Reducing Emissions*. Bureau of Transport and Communication Economics Report No. 94. Canberra: Australian Government Publishing Service.

BTM Consult (1999) *International Wind Energy Development – World Market Update 1998, plus forecast for 1999–2003*. Copenhagen: BTM Consult.

Cambridge Econometrics (1998) *Industrial Benefits from Environmental Tax Reform*. A report to the Forum for the Future and Friends of the Earth, Technical Report No. 1. London: Forum for the Future.

Cameron, L. J., Montgomery, W. D. and Foster, H. L. (1999) The economics of strategies to reduce greenhouse gas emissions. *Energy Studies Review*, **9**(1) 63–73.

Chipato, C. (1999) *Ruminant methane in Zimbabwe.* Washington: Global Livestock Group.

Chomitz, K. M. and Kumari, K. (1996) *The Domestic Benefits of Tropical Forests: A Critical Review Emphasising Hydrological Functions.* Policy Research Working Paper No. WPS1601. New York NY: World Bank.

Cole, C. V., Flach, K., Lee, J., Sauerbeck, D. and Stewart, B. (1993) Agricultural sources and sinks of carbon. *Water, Air, and Soil Pollution,* **70,** 111–22.

Criqui, P., Kouvaritakis, N. and Schrattenholzer, L. (2000) The impacts of carbon constraints on power generation and renewable energy technologies. In L. Bernstein and J. Pan, eds., *Sectoral Economic Costs and Benefits of GHG Mitigation.* Proceedings of an IPCC Expert Meeting, 14–15 February 2000, Technical Support Unit, Working Group III. Geneva: Intergovernmental Panel on Climate Change.

DeCicco, J., and Mark, J. (1997) Meeting the energy and climate challenge for transportation. *Energy Policy,* **26,** 395–412.

Denis, C., and Koopman, J. (1998) *EUCARS: A Partial Equilibrium Model of European CAR Emissions (Version 3.0).* II/341/98-EN. European Commission: Directorate General II.

Dessus, S. and O'Connor, D. (1999) Climate Policy Without Tears: CGE-Based Ancillary Benefits Estimates for Chile. OECD Development Centre. Paris: Organization for Economic Co-operation and Development.

Desvousges, W., Naughton M. and Parsons, R. (1992) Benefits transfer: conceptual problems in estimating water quality benefits. *Water Resources Research,* **28,** 675–83.

Dixon, R. K., Winjum, J. K. and Schroeder, P. E. (1993) Conservation and sequestration of carbon: the potential of forest and agroforest management practices. *Global Environmental Change,* 3(2), 159–73.

Dowlatabadi, H., Lave, L. and Russell, A. G. (1996) A free lunch at higher CAFÉ: a review of economic, environment, and social benefits. *Energy Policy,* **24,** 253–64.

Duleep, K. G. (1997) Evolutionary and revolutionary technologies for improving fuel economy. In J. de Cicco, and M. Delucchi, eds., Transport, energy and environment: how far can technology take us? *Sustainable Transportation Energy Strategies.* Washington: American Council for an Energy-Efficient Economy.

ECMT (1997) *CO Emissinos from Transport.* Paris: Organization for Economic Development.

EPA (1997) *The Benefits and Costs of the Clean Air Act, 1970 to 1990.* Washington: US Environmental Protection Agency.

EIA (1998): *Impacts of the Kyoto Protocol on US Energy Markets and Economic Activity.* Report No. SR/OIA/98-03. Washington: US Energy Information Agency.

EIA (2000) *International Energy Outlook 2000.* Washington.

EPA (1999) *The Benefits and Costs of the Clean Air Act, 1990 to 2010.* Report No. EPA-410-R-99-001. Washington: Environmental Protection Agency.

EPA/USIJI (1998) *Activities Implemented Jointly.* Third Report of the Secretariat of the UN Framework Convention on Climate Change. Report No. 236-R-98-004. Washington: Environmental Protection Agency and US Initiative on Joint Implementation, p. 19 (vol. 1) and p. 607 (vol. 2).

ExternE (1995) *Externalities of Energy,* Vol. 3 *Coal and Lignite.* Luxembourg: Commission of the European Communities, DGXII.

ExternE (1997) *Externalities of Fuel Cycles 'ExternE' Project: Results of National Implementation.* Draft Final Report. Brussels: Commission of the European Communities, DGXII.

ExternE (1999) In M. Holland, J. Berry, D. Forster, eds. *Externalities of Energy*, Vol. 7: *Methodology 1998 Update.* Luxembourg: Office for the Official Publications of the European Communities.

Ferriter, J. (1997) The effects of CO_2 reduction policies on energy markets. In Y. Kaya and K. Yokobori, eds., *Environment, Energy and Economy.* New York: United Nations University Press.

Flach, K. W., Barnwell Jr., T. O. and Crosson, P. (1997) Impacts of agriculture on atmospheric carbon dioxide. In E. A. Paul, E. T. Elliot, K. Paustian and C. V. Cole, eds., *Soil Organic Matter in Temperate Agrosystems, Long Term Experiments in North America.* Boca Raton: CRC Press, pp. 3–13.

Frankhauser, S. and McCoy, D. (1995) Modelling the economic consequences of environmental policies. In H. Folmer, L. Gabel, J. Opschoor, eds., *Principles of environmental and resource economics: A guide to Decision Makers and Students.* Aldershot: Edward Elgar. pp. 253–75.

Garbaccio, R. F., Ho. M. S. and Jorgenson, D. W. (2000) The health benefits of controlling carbon emissions in China. Expert Workshop on *Assessing the Ancillary Benefits and Costs of Greenhouse Gas Mitigation Policies*, March 27–9. Washington: Harvard University Press.

Greene, D. L. (1999) *An Assessment of Energy and Environment Issues Related to the Use of Gas-to-Liquid Fuels in Transportation.* Report No. ORNL/TM-1999/258. Oak Ridge: Oak Ridge National Laboratory.

Greenpeace, 1999

Ha-Duong, M., Hourcade, J.-C. and Lecoq, F. (1999) Dynamic consistency problems behind the Kyoto Protocol. *International Journal Environment and Pollution*, **11**(4), 426–46.

Hagan, R. (1998) The future of nuclear power in Asia. *Pacific and Asian Journal of Energy*, **8**(1), 9–22.

Hourcade, J. C., Shukla, P., Cituentes, L., Davis, D., Edmonds, J., Fisher, B., Fortin, E., Golub, A., Hohmeyer, O., Krupnik, A., Kverndokk, S., Loulou, R., Richels, R., Segenovic, H. and Jamali, K. (2001) Global, regional and national costs and ancillary benefits of mitigation. In B. Metz., O. Davidson, R. Swart and J. Pan, eds., *Climate Change 2001: Mitigation.* Intergovernmental Panel on Climate Change. Cambridge and New York: Cambridge University Press.

IEA (1997a) *Renewable Energy Policy in IEA Countries.* Paris: International Energy Agency.

IEA (1997b) *Transport, Energy and Climate Change.* Paris: International Energy Agency.

IEA (1998a) *World Energy Outlook*, 1998 edn. Paris: International Energy Agency.

IEA (1998b) *Biomass: Data, Analysis and Trends.* Paris: International Energy Agency.

IEA (1999) *Coal Information 1998*, 1999 edn. Paris: International Energy Agency.

Ikeda, A., Ishikawa, M., Suga, M., Hujii, Y. and Yoshioka, K. (1995) Application of input-output table for environmental analysis (7) — Simulations on steel-scrap, blast furnace slag and fly-ash utilisation. *Innovation and IO technique — Business Journal of PAPAIOS*, **6**(2), 39–61 (in Japanese).

Insensee, A. R. and Sadeghi, A. M. (1996) Effect of tillage reversal on herbicide leaching to groundwater. *Soil Science*, **161**, 382–9.

Interlaboratory Working Group (1997) *Scenarios of US Carbon Reductions: Potential Impacts of Energy Technologies by 2010 and Beyond.* Reports LBNL-40533 and ORNL-444, respectively. Berkeley: Berkeley National Laboratory/Oak Ridge: Oak Ridge National Laboratory.

IPCC (1996) J. P. Bruce, H. Lee and E. Haites, eds., *Climate Change 1995: Economic and Social Dimensions of Climate Change.* Cambridge: Cambridge University Press.

IPCC (2000) (R. T. Watson, I. R. Noble, B. Bolin, N. H. Ravindranath, D. J. Verardo and D. J. Dokken, eds) *Land Use, Land Use Change, and Forestry: A Special Report of the Intergovernmental Panel for Climate Change.* Cambridge: Cambridge University Press.

IPCC (2001) Metz, B., Davidson, O., Swart, R. and Pan, J., eds., *Climate Change 2001: Mitigation.* Cambridge and New York: Cambridge University Press.

Jansen, H. and Denis, C. (1999) A welfare cost assessment of various policy measures to reduce pollutant emissions from passenger road vehicles. *Transportation Research-D*, **4d**(6) 379–96.

Jiang K., Hu, X., Matsuoka, Y. and Morita, T. (1998) Energy technology changes and CO_2 emission scenarios in China. *Environmental Economics and Policy Studies*, 141–60.

Kashiwagi, T., Saha, B. B., Bonilla, D. and Akisawa, A. (1999) Energy efficiency and structural change for sustainable development and CO_2 mitigation. In *Costing Methodologies*, a contribution to the Intergovernmental Panel on Climate Change expert meeting. Tokyo: Tokyo University of Agriculture and Technology.

Keenan, R., Lamb, D., Woldring, O., Irvine, T. and Jensen, R. (1997) Restoration of plant biodiversity beneath tropical tree plantations in Northern Australia. *Forest Ecology and Management*, **99**, 117–31.

Knapp, R. (2000) Discussion on coal. In L. Bernstein and J. Pan, eds., *Sectoral Economic Costs and Benefits of GHG Mitigation.* Proceedings of an IPCC Expert Meeting, 14–15 February 2000, Technical Support Unit, IPCC Working Group III. Geneva: International Panel on Climate Change.

Kroeze, C. and Mosier, A. (1999) New estimates for emissions of nitrous oxides. In J. van Ham, A. P. M. Baede, L. A. Meyer and R. Ybena, eds., *Second International Symposium on Non- CO_2 Greenhouse Gases.* Dordrecht: Kluwer.

Kuliev, A. (1996) Forests — an important factor in combating desertification. *Problems of desert development*, **4**, 29–31.

Laitner, S., Bernow, S. and DeCicco, J. (1998) Employment and other macroeconomic benefits of an innovation-led climate strategy for the United States. *Energy Policy*, **26**(5), 425–32.

Lal, R. and Bruce, J. P. (1999) The potential of world cropland soils to sequester C and mitigate the greenhouse effect. *Environmental Science and Policy*, **2**(2), 177–86.

Johnson, T. M., Li, J., Jiang, Z. and Taylor, R. P. (1995) Energy demand in china: overview report. *Issues and Options in Greenhouse Gas Emissions Control.* Subreport No. 2. Washington: World Bank.

Lugo, A. (1997) The apparent paradox of re-establishing species richness on degraded lands with tree monocultures. *Forest Ecology and Management*, **99**, 9–19.

Michaelis, L. (1997) *CO2 Emissions from Road Vehicles*. Working Paper 1, Annex 1, UNFCCC Expert Group on the United Nation Framework Convention on Climate Change. Paris: Organization for Economic Co-operation and Development.

Michaelis, L. and Davidson, O. (1996) GHG mitigation in the transport sector. *Energy Policy*, **24**(10–11), 969–84.

Miyamoto, K., ed. (1997) Renewable biological systems for alternative sustainable energy production. *FAO Agricultural Series Bulletin 128*. Rome: Food and Agriculture Organization.

Moore, T. (1998) Electrification and global sustainably. *Electric Power Research Institute Journal*, 43–52.

Mosier, A. R., Kroeze, C., Nevison, C., Oenema, O., Seitzinger S. and van Cleemput, O. (1998) Closing the global N_2O budget: nitrous oxide emissions through the agricultural nitrogen cycle. *Nutrient Cycling in Agrosystems*, **52**, 225–48.

Munasinghe, M. (1990) *Energy Analysis and Policy*. London: ButterworthHeinemann Press.

Munasinghe, M. (1992) *Environmental Economics and Sustainable Development*. Paper presented at the UN Earth Summit, Rio de Janeiro, and reproduced as Environment Paper No. 3. Washington: World Bank.

Munasinghe, M., ed. (1996) *Environmental Impacts of Macroeconomic and Sectoral Policies*. Washington: World Bank.

Munasinghe, M., Meier, P., Hoel, M., Wong, S. and Aaheim, A. (1996) Applicability of techniques of cost-benefit analysis to climate change. In J. P. Bruce, H. Lee and E. H. Haites, eds., *Climate Change 1995: Economic and Social Dimensions*. Geneva: Intergovernmental Panel on Climate Change.

OECD (1998) *Projected Costs of Generating Electricity: Update 1998*. Paris: Organization for Economic Co-operation and Development, International Energy Agency and Nuclear Energy Agency.

OECD (1999) *New Release: Financial Flows to Developing Countries in 1998. Rise in Aid: Sharp Fall in Private Flows*. Paris: Organization for Economic Co-operation and Development.

Palmquist, R. B. (1991) Hedonic methods. In J. Braden and C. Klostad, eds. *Measuring the Demand for Environmental Quality*. New York: North Holland Publication Company.

Parson, E. A. and Fisher-Vanden, K. (1997) Integrated Assessment Models of Global Climate Change. *Annual Review of Energy and the Environment*, **22**, 589–628.

Pearce, D. W., Cline, W. R., Achanta, A. N., Fankhauser, S., Pachauri, R. K., Tol R. S. J. and Vellinga, P. (1996) The social costs of climate change: greenhouse damage and the benefits of control. In J. P. Bruce, Hoesung Lee and E. F. Haites, eds., *Climate Change 1995: Economic and Social Dimensions of Climate Change*. Cambridge: Cambridge University Press.

Penner, J. E., Lister, D. H., Giggs, D. J., Dokken, D. J. and McFarland, M., eds. (1999) *Aviation and the Global Atmosphere*. Geneva: Intergovernmental Panel on Climate Change.

Pershing, J. (2000) Fossil fuel implications on climate change mitigation responses. In L. Bernstein and J. Pan, eds., *Sectoral Economic Costs and Benefits of GHG Mitigation.* Proceedings of an IPCC Expert Meeting, 14–15 February 2000, Technical Support Unit, Working Group III. Geneva: Intergovernmental Panel on Climate Change.

Pimental, D., Rodrigues, G., Wane, T., Abrams, R., Goldberg, K., Staecker, H., Ma, E., Brueckner, L., Trovato, L., Chow, C., Govindarajulu, U. and Boerke, S. (1994) Renewable energy: economic and environmental issues. *Bioscience,* **44**(8).

Piscitello, E. S. and Bogach, V. S. (1997) *Financial Incentives for Renewable Energy.* Proceedings of an International Workshop, 17–21 February 1997. World Bank Discussion Paper, No. 391. Amsterdam.

Reilly, J., Tubiello, F., McCarl, B., Abler, D., Darwin, R., Fuglie, K., Hollinger, S., Izaurralde, C., Jagtap, S., Jones, J., Learns, L., Ojima, D., Paul, E., Paustian, K., Riha, S., Rosenberg, N. and Rosenzweig, C. (2003) US agriculture and climate change: new results. *Climatic Change,* **57**, 43–69.

Reimer, P. and Freund, P. (1999) *Technologies for Reducing Methane Emissions.* Paris: International Energy Agency.

Ross, A. (1999) *Road Accidents: A Global Problem Requiring Urgent Action.* Roads and Highways Topic Note No. RH-2. Washington: World Bank.

Schafer, A. (1998) The global demand for motorised mobility. *Transportation Research A,* **32**(6), 455–77.

Scheraga, J. D. and Leary, N. A. (1993) *Costs and Side Benefits of Using Energy Taxes to Mitigate Global Climate Change.* Proceedings 86th Annual conference. Washington: National Tax Association, pp. 133–8.

SDPC (1996) *National Strategy for New and Renewable Energy Development Plan for 1996 to 2010 in China.* Beijing: State Development Planning Commission.

Smith, J. B., Schellnhuber, H.-J., Mirza, M. Q., Fankhauser, S., Leemans, R., Erda, L., Ogallo, L., Pittock, B., Richels, R., Rosenzweig, C., Safriel, U., Tol, R., S. J., Weyant, J. and Yohe, G. (2001) Vulnerability to climate change and reasons for concern: a synthesis. In J. M. McCarthy, O. F. Canziani, N. A. Leary, D. J. Dokken and K. S. White, eds., *Climate Change 2001: Impacts, Adaptation and Vulnerability.* Cambridge: Cambridge University Press.

Sulilatu, W. F. (1998) *Co-combustion of Biofuels.* Bioenergy Agreement Report T13. Paris: International Energy Agency.

Sommer, H., Seethaler, R., Chanel, O., Herry, M., Masson, S. and Vergnaud, J. C. (1999) *Health Costs Due to Road Traffic-related Air Pollution: An Impact Assessment Project of Austria, France and Switzerland.* Economic Evaluation, Technical Report on Economy. Rome: World Health Organization.

Tuinstra, W., Eerens, H. C., van Minnen, J., Petroula, D., Swart, R. J., Brink, C. J., Kalognomou, E. A., Moussiopoulos, N., Amann, M., Cofala, J., Klimont, Z., Dentener F. and Raes, F. (2004). *Air Quality and Climate Change Policies in Europe.* EEA Technical Report. Copenhagen: European Environment Agency.

USDOE/EIA (1998) *Impacts of the Kyoto Protocol on US Energy Markets and Economic Activity.* Report No. SR/OIAF/98–03. Washington: US Department of Energy.

US EPA (1999) *Cutting the Waste Stream in Half: Community Record-Setters Show How.* Report No. EPA 530-R-99-013. Washington: Environmental Protection Agency.

Van Wee, B. and Annema, J. A. (1999) *Transport, Energy Savings and CO_2 Emission Reductions: Technical-economic Potential in European Studies Compared.* Paris: International Energy Agency.

Wang, X. and Feng, Z. (1995) Atmospheric carbon sequestration through agroforestry in China. *Energy,* **20**(2), 117–21.

Wang, X. and Smith, K. R. (1999) *Near-term Health Benefits of Greenhouse Gas Reductions: A Proposed Assessment Method and Application in Two Energy Sectors in China.* Report No. WHO/SDE/PHE/99.1 Geneva: World Health Organization.

WCI (1999) *Coal – Power for Progress.* London: World Coal Institute.

WHO (1999) *Transport, Environment, and Health.* Third Ministerial Conference on Environment and Health. London, 16–19 June. New York NY: World Health Organizations.

Worrell, E., Levine, M., Price, L., Martin, N., van den Broek, R. and Blok, K. (1997) *Potentials and Policy Implications of Energy and Material Efficiency Improvement.* New York: United Nations Divisions for Sustainable Development.

Zhou, P. P. (1998) *Energy Efficiency for Climate Change – Opportunities and Prerequisites for Southern Africa*, Report No. 20. Copenhagen: International Network for Sustainable Energy.

11

Climate change and sustainable development: a synthesis

11.1 Summary of main findings

11.1.1 Climate change

In the previous chapters we summarized the contemporary scientific understanding of the two interlinked issues of climate change and sustainable development and its relevance to policy. Based mainly on the assessment reports of the Intergovernmental Panel on Climate Change (IPCC), we discussed changes in the climate system and associated impacts on vulnerable natural and human systems, as well as their implications for sustainable development. Options to respond to climate change through adaptation and mitigation, and methodologies to evaluate those options, were described in detail.

During the last 5–10 years, our understanding of the climate problem and the options to address it, have advanced rapidly. Many new insights have been acquired, and it is difficult to select the most important ones, not least because of the fact that for different people, different things will be important. Nevertheless, in Box 10.1 we list the twelve most striking new findings that stand out in our personal view, not only because they are important scientifically, but also because they have a large potential impact on the development of climatic and other policies.

It is intriguing to note that the main reasons for the lack of progress in the international efforts to address climate change appear neither to include a lack of technological and other options, nor prohibitive economic costs at national and international levels. Policy-makers who claim that there are insufficient options to address climate change, or that the costs of the options would endanger their economies, generally do not appear to base these views on sound science or

Box 11.1   Selected key findings with respect to climate change and climate change response options

1.   The evidence has become stronger that the global climate is changing, and that human activities are at least partly responsible.

2.   The understanding has increased that non-linear, large-scale and possibly irreversible changes in climate system components cannot be excluded, although the risk that these will occur in this century is limited.

3.   Evidence has increased that many physical and biological systems have been affected already by climatic changes at the regional level.

4.   Poor people in developing countries are the most vulnerable to climate change impacts because they have the lowest capacity to adapt.

5.   Inertia is an important factor in both natural and socioeconomic systems, emphasizing the need for a long-term perspective and possibly safety margins in the design of climate response strategies.

6.   Technology has developed very fast, and many technological and biological options are known today that can enable us to meet short-term emissions reduction targets (e.g. those of the Kyoto Protocol), as well as the requirements for stabilizing greenhouse gas concentrations at low levels in the long term.

7.   In addition to, and related to, the many technological and biological options, measures related to changes in economic structure, institutional arrangements, and behaviour can provide important opportunities to adapt to, or mitigate, climate change.

8.   There are many cultural, political, institutional, economic, and technological barriers to the adoption of the various options, but also many types of domestic and international policy instruments that can help to overcome the barriers.

9.   At the macroeconomic level, for most countries, costs associated with mitigation will be limited to a few tenths of a percentage point of projected annual growth rates, dependent on the way policies and measures are combined and implemented. However, for particular sectors (e.g. energy-intensive industries) and countries (e.g. oil-exporting countries), costs can be higher.

10.   The characteristics of the general socioeconomic pathway that countries will follow are as important for greenhouse gas emissions as the specific climate policies that can be implemented.

> 11.   While mitigation and adaptation options can usefully be evaluated individually, they can be assessed also in a more holistic manner, acknowledging their dependency on complex and interlinked human and natural systems.
>
> 12.   Sustainable development, including improving equity, can be mutually reinforcing with climate change adaptation and mitigation at national, regional, and local levels.

economics.[1] So why is progress slow?, how can negative developments be turned into positive ones?, and how can the pace of positive change be accelerated?

Climate change will affect several fundamental sectors that will shape the sustainability of future human development, notably energy and food. The changes in the energy and food supply systems, which would be needed to address the problem, are intertwined and have important economic, technological, environmental, social, and institutional dimensions. This complexity makes simple technological solutions relatively ineffective, e.g. like the ones developed successfully to address environmental problems such as stratospheric ozone depletion and acidification. Necessary changes would have to be brought about in an integrated fashion, covering various sectors of modern societies, and taking into account their deeply normative nature. One way to stimulate this process would be to broaden the debate about climate change response from a narrow focus on individual technological solutions (e.g. in the energy system), to a more fundamental discussion on alternative development paths that the whole world could take. In some way, the 1992 and 2002 World Conferences on Environment and Development have done this, but the promises of these international negotiation forums have not yet been followed through at the regional, national, or local levels – notwithstanding the rhetoric of the world leaders present at these sessions. For this reason, the central theme of this book has been the opportunities provided by the search for solutions to the climate change problem that also – or possibly even more so – address wider development objectives.

11.1.2   Sustainable development linkages

We discussed in the previous chapters the various ways in which sustainable development can be approached. While there is no agreement about one appropriate definition of the concept, there is broad agreement that it includes

---

[1] Exceptions may be countries dependent upon revenues from fossil fuel exports.

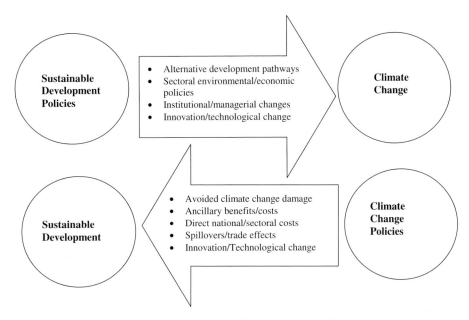

Figure 11.1. Linkages between sustainable development, climate change, and policies.

three broad domains: (a) social; (b) economic, and (c) environmental. We have presented a generic transdisciplinary approach to use scientific knowledge in making development more sustainable in these three dimensions, called *sustainomics*. This framework recognizes that, although the precise definition of sustainable development may be an elusive (and less relevant) goal, an incremental approach based on making existing development efforts more sustainable (including climate change linkages) is more practical and promising—because many unsustainable activities are relatively easy to identify.

Figure 11.1 depicts the mutual interactions between climate change policies and broader development issues. Climate change policies can affect development opportunities in various ways, e.g. by: (a) reducing climate change impacts; (b) providing ancillary benefits; (c) imposing costs associated with mitigation and adaptation; (d) causing positive or negative spillovers to other countries, and (e) inducing particular technological developments. Economic, social, and environmental development policies (which do not explicitly target climate change), can seriously affect climate change and the adaptive and mitigative capacity to deal with it, by: (a) the pursuance of particular development priorities; (b) specific sectoral environmental, social, or economic policies; (c) institutional changes, and (d) stimulating particular technological innovation and change.

While this way of framing the two problems highlights the very important linkages between them, it may also exaggerate their separation. From this perspective, climate policy-makers may continue to focus on specific climate policies, but use a wider set of arguments to 'sell' these to other government departments, the public, and other societal stakeholders. Also, from this perspective, economic- and social-policy-makers may increasingly acknowledge that climate change should be added to their spectrum of policy evaluation criteria, but may still consider it as a separate, add-on criterion of relatively low priority. Therefore, we suggest that most progress can be made if climate change and sustainable development can be addressed in a far more integrated and holistic fashion, while acknowledging that climate change is an important problem – the solutions to which can coincide with those for other social, economic, and environmental problems. By focusing attention on the vast potential of technological and social innovation, the risks posed by climate change can be turned into opportunities for more sustainable development. At the same time, such an integrated approach allows for the identification and subsequent avoidance of trade-offs. Only a decade ago, climate change and sustainable development were seen mostly as issues that could be dealt with through rather independent, parallel, strategies. More recently, it was acknowledged that both issues are closely related, and that strategies in one area may be improved by taking the linkages with the other issue into account. Now, we hope we have presented compelling evidence for adopting a fully integrated approach that is a prerequisite for effective action. It will not be possible to move towards sustainable development without an adequate response to climate change, while policies for mitigation of, or adaptation to, climate will not be effective without broader efforts to make development more sustainable (MDMS).

Figure 11.2 clarifies this critical interdependence, taking the three dimensions of sustainable development into account. We select the following six key examples out of a longer list of possible integrated strategies that address sustainable development and climate change objectives simultaneously.

1. *Explore alternative sustainable development pathways.* During the 1992 United Nations Conference on Environment and Development in Rio de Janeiro, and the subsequent 2002 World Summit on Sustainable Development in Johannesburg, countries have reaffirmed their commitment to sustainable development as a central objective of their development. This calls for reflective debate about future development pathways, involving stakeholders from government, the private sector, and civil society. Priorities in societal objectives independent of climate change have a huge impact on climate change.

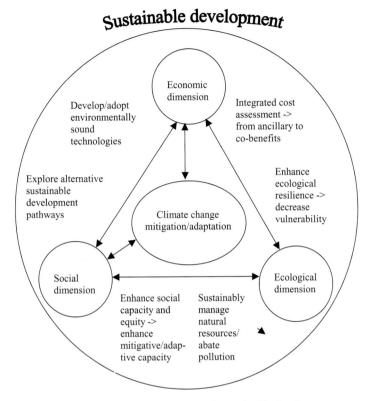

Figure 11.2. Integrating climate change and sustainable development.

2. *Develop and adopt environmentally sound technologies.* Technological innovation and change are vital to resolving various environmental problems, including climate change. (Re-)directing technological development in an environmentally sound, and thus clean and efficient manner, serves multiple purposes.

3. *Fully integrate environmental and social externalities in cost assessment in a balanced manner.* This requires that we give equal attention to all criteria when evaluating costs and benefits of policies, and not regard climate change as the only or major criterion for developing policies in important sectors (e.g. energy and transport) while ignoring other reasons.

4. *Protect biodiversity and enhance ecological resilience.* Biological diversity can be protected in ways that account for possible future climatic changes. Enhancing the resilience of ecosystems can reduce their vulnerability to climate change, and other stresses, at the same time.

5. *Sustainably manage natural resources and abate pollution.* All societies depend to varying degrees on natural resources. Many of the latter are being threatened by multiple anthropogenic stresses. By managing them in a sustainable manner, the vulnerability of these resources and the social groups depending on them is decreased. Abating local and regional air pollution can coincide with climate change mitigation, while abatement of pollution in general, which often has negative health consequences, decreases the vulnerability of the population to climate change and other stresses.

6. *Enhance social capacity and equity.* Both internationally and within nations, the enhancement of social capital and equity is essential for the sustainable development of the economy. This includes empowerment of social groups and inclusion in decision-making, as well as strengthening of appropriate institutions and governance systems. Such actions will at the same time enhance the capacity to mitigate and adapt to climate change, and reduce vulnerability to its impacts.

11.2   Developing a science of sustainable development and applying the knowledge base to climate change

11.2.1   Future directions in thinking

We believe that decision-making for the economic, social, and environmental dimensions of sustainable development should be based on a sound assessment of the evolving science base, including the natural and social sciences, engineering, and the humanities. All three IPCC Working Groups have formulated lists of research needs. Nevertheless, is the current scientific agenda adequate for dealing with the complex linkages between components of the socioecological system discussed above? Climate change was framed in the 1980s by naturalists as a global environmental problem, largely divorced from its social context (Cohen *et al.* 1998). Although gradually the social dimension has received increasing attention, both in science and the policy debate, it was considered an 'add-on' rather than as being of help in defining the issue. We do not advocate that the ongoing debate about climate policies should be abandoned. However, we do argue that a promising, if not indispensable, way forward is to address climate change response options in a more holistic manner, within the broader framework of efforts to MDMS.

This approach also has important consequences for science. In Box 11.2, we describe the emergence of the so-called sustainability science. In the 1980s, the World Climate Research Programme was established to respond to the then emerging concern about climate variability and change. Shortly afterwards, the importance of the geosphere and biosphere was acknowledged as key components of the changing Earth systems, and the International Geosphere – Biosphere Programme (IGBP) and DIVERSITAS research programmes were established. Only in the 1990s was it acknowledged that humanity is a factor in bringing about global change, and the International Human Dimensions Programme (IHDP) was developed as a companion to the earlier three programmes. In 1995, the IGBP Global Analysis, Integration and Modelling (GAIM) taskforce was established to integrate the knowledge generated in the various IGBP core projects. In 1997, in a landmark address to the American Association of the Advancement of Science, Jane Lubchenko referred to the increasing human impacts on the ecological systems of the Earth and called for a new 'social contract' for science, requiring scientists to devote their energies and talents to the most pressing problems of the day (Lubchenco 1998). In the last few years of the twentieth century and at the beginning of the twenty-first, a series of international conferences and initiatives (e.g. Budapest, and Amsterdam, and the US BSD; (ICSU 2002; IGBP/IHDP/WCRP/Biodiversitas 2001; NAS 1999; UNESCO 1999) explored these ideas further, all calling for a more integrated approach between the natural and the social sciences, and for linking the scientific enterprise better with current sustainability problems.

11.2.2   *Some ideas on MDMS*

Contemporary approaches (e.g. sustainability science and ecological economics) have been developed mainly in a Western context, with a major focus on environment and natural resources. From the developing country perspective, development is the key objective, where people-oriented issues (e.g. poverty and equity) play the major role. Therefore, what we should aim for is a comprehensive 'science of sustainable development', that can maintain the balance between the perspectives of the South and North. Addressing the sustainability science questions would help to build the knowledge base for this purpose. Going beyond such a knowledge base, the 'Sustainomics' approach has been proposed as a balanced, transdisciplinary, holistic framework for making development more sustainable, and will take us closer to the science of sustainable development. We propose further to pursue such an approach through learning by doing, and by developing and applying practical tools. Examples of such tools discussed in this book include: (a) sustainable development assessment, which evaluates the social, environmental, and economic dimensions of development; (b) the action impact matrix for assessing policies and projects, and (c) multicriteria analysis for the taking into account of social, environmental, and economic indicators of development.

Box 11.2   Towards a science of sustainable development

1980: World Climate Research Programme.

1986: International Geosphere–Biosphere Programme.

1987: 'Sustainable development' proposed by the Bruntland
Commission.

1990: DIVERSITAS.

1992: United Nations Conference on Environment and
Development, Rio de Janeiro – Agenda 21 and Rio
Declaration ('sustainomics' framework outlined
to make development more sustainable; and
sustainable development triangle elaborated
(Munasinghe 1992).

1995: Global Analysis, Integration and Modelling, taskforce.

1996: International Human Dimensions Programme (IHDP) [1].

1997: Call for 'New Social Contract for Science' (Lubchenco,
1998).

1999: Budapest 'Science for the Twenty-first Century'.

2001: Amsterdam Global Change Conference (IGBP/IHDP/
WCRP/Biodiversitas (2001)).

2001: Proposal for Sustainability Science (Kates *et al.* 2001).

2001: Sustainomics framework elaborated (WSSD PrepCom
Sessions).

2002: WSSD – Earth Summit, Johannesburg.

2003: IPCC endorses sustainable development as major
cross-cutting theme in *Fourth Assessment Report*.

Development of
the knowledge
base

Knowledge
integration and
policy linkage

[1] The IHDP on Global Environment Change was launched initially in 1990 by the ISSC
as the Human Dimensions Programme (HDP). The International Council for Science
joined the ISSC as co-sponsor of the HDP in February 1996. At this time, the name of
the HDP was changed to IHDP.

The knowledge and the tools themselves are important. At the same time, the
processes by which they are developed, and the context in which they are applied,
also are crucial to their effectiveness. One important reason for climate policy-
making being proved to be a controversial and arduous task, is that different
stakeholder groups and countries have different ways of framing the problem, and
have varied preferences for choosing a practical decision-making framework. As a
result, there is no consensus either on preferred strategies or ways to implement
them.

Making development more sustainable is a promising practical approach, which could become a key common goal that people agree upon. It will require a participatory approach, in which strategies are developed that start with what is known about the various problems, as well as the interests of diverse stakeholders. We hope that this volume will contribute to the knowledge base that will help the global community in their efforts simultaneously to address the interlinked issues of climate change and sustainable development.

11.2.3   *Addressing climate change issues*

The United Nations Framework Convention on Climate Change (UNFCCC) seeks to avoid 'dangerous' anthropogenic intervention with the climate system. Both greenhouse gas emissions and signs of climate change are increasing. Thus, there is a growing gap between the current situation and the desirable future target of a stable global climate. The Kyoto Protocol – the only outcome of lengthy negotiations so far – still is not in force, and rejected by the USA, the largest emitter. The Protocol envisages greenhouse gas emissions reductions of about 5–10 per cent by 2012, which is felt widely to be far less than the level required to achieve UNFCCC objectives. While this agreement is only small step forward, it may help to signal a change in policy and favourably influence long-term trends in technology and economic policy. Nevertheless, Annex I countries (the emissions per capita of which are significantly higher than those from non-Annex I nations), need to show far more effective leadership in beginning this process.

We feel that addressing climate change in the context of sustainable development will increase greatly the effectiveness of any recommendations that might emerge. Climate change research could benefit from the greater enthusiasm and funding from governments, donor agencies, and private sources, for example, for economic development and growth. The communities of researchers and policy analysts have key roles to play in this regard, and need to be more active in providing better short-term insights and guidance to decision-makers (e.g. shaping the post-Kyoto agenda) while addressing the longer-term challenges, e.g. determining what might constitute desirable and practical future levels of stable greenhouse gas concentrations. A passive attitude of simply providing the facts may be too conservative. Researchers (especially those from the social sciences) need to engage more proactively in explaining outcomes and their implications for public policy (including thorny issues such as equity and values).

As explained earlier, one fundamental problem is that different stakeholders will start from a variety of premises and assumptions, thereby hindering convergence to practical solutions. Decision-makers need to assess both impact risks and the costs and benefits of remedial policies against the background of an uncertain future. The approach outlined in this volume provides a practical starting framework towards the ultimate goal of identifying relevant values, defining options, and clarifying the roles of stakeholders and actors in both the consensus building process and implementation of agreed actions. To limit the scope to more manageable proportions, the focus should be on the nexus between climate change and sustainable development. Identifying and implementing climate change responses that make ongoing development more sustainable will help to 'mainstream' climate policy within the national agenda.

Most climate research has been done in terms of large, abstract models, at high levels of aggregation, e.g. global and regional modelling. However, meaningful assessment of impacts and effective remediation will require greater focus on practical fieldwork at the local level, where vulnerabilities and adaptation can be captured realistically, the behaviour of agents be incorporated appropriately, and stakeholder participation be enhanced. Furthermore, the analyses will need to integrate and span the range from the global long term to the local short term, e.g. linking the financing of carbon markets globally with sustainable livelihood programmes locally. Large gaps need to be filled in our understanding of the dynamics of evolution and adaptation of social systems and institutions, as well as ecosystems. Another key issue is the vulnerability of disadvantaged groups and their livelihoods to multiple stresses and specific actions of other stakeholders, over a range of time and geographic scales. Durable strategies are required that reduce vulnerability to climatic risks, and increase the capacity to both mitigate and adapt to climate change. These ideas highlight promising avenues of research that should be explored, as a key to facilitating effective climate negotiations.

References

Cohen, S., Demeritt, Robinson, J. and Rothman, D. (1998) Climate change and sustainable development: towards dialogue. *Global Environmental Change*, 8(4): 341–71.

ICSU (2002) *Report of the Scientific and Technological Community to the World summit on Sustainable Development*. Science for Sustainable Development, Report No. 1.

IGBP/IHDP/WCRP/Biodiversitas (2001) International Geosphere–Biosphere Programme/International Human Dimensions Programme/World Climate Research Programme/Biodiversitas international biodiversity research programme). (The Amsterdam Declaration.)

Kates R. W., Clark, W. C., Corell, R., Hall, J. M., Jaeger, C., Lowe, I., McCarthy, J. J., Schellnhuber, H.-J., Bolin, B., Dickson, N. M., Facheux, S., Gallopin, G. C., Gruebler, A., Huntley, B., Jäger, J., Jodha, N. S., Kasperson, R. E., Mabogunje, A., Matson, P., Mooney, H., Moore III, B., O'Riordan, T. and Svedin., U. (2001) Sustainability science. *Science*, **27** (292), 641–2 (extended version published by the Belfer Center for Science and International Affairs, John F. Kennedy School of Government. Boston: Harvard University.

Lubchenco, J. 1998. Entering the century of the environment: a new social contract for science. *Science*, **279**, 491–7.

NAS (1999) *Our Common Journey: A Transition Toward Sustainability*. Report of the Board on Sustainable Development. Washington: National Academy Press.

UNESCO (1999) *Declaration on Science and the Use of Scientific Knowledge*. World Conference on Science. New York: United Nations Educational, Scientific and Cultural Organization.

Index